THOMAS T. STRUHSAKER is a research zoologist at the New York Zoological Society and adjunct associate professor at the Rockefeller University. He has spent many years doing fieldwork in Africa, has also had field experience in South America and India, and has published numerous articles on primates.

THE RED COLOBUS MONKEY

A photograph taken in Colombia, South America, September 1974, of the author and a young woolly monkey (*Lagothrix lagotricha*).

WILDLIFE BEHAVIOR AND ECOLOGY
George B. Schaller, Editor

THE RED COLOBUS MONKEY

Thomas T. Struhsaker

THE UNIVERSITY OF CHICAGO PRESS
Chicago and London

Thomas T. Struhsaker is a research zoologist with the New York Zoological Society
and adjunct associate professor at the Rockefeller University.

The University of Chicago Press, Chicago 60637
The University of Chicago Press, Ltd., London

Printed in the United States of America

Library of Congress Cataloging in Publication Data

Struhsaker, Thomas T
   The red colobus monkey.

   (Wildlife behavior and ecology)
   Bibliography: p.
   Includes index.
   1. Red colobus monkey—Behavior.   2. Red colobus
monkey. I. Title. [DNLM: 1. Apes. 2. Behavior,
Animal. 3. Ecology. QL737.P9 S927r]
QL737.P93S78   599'.82   74-21339
ISBN 0-226-77769-3

To my father and mother

Contents

Preface

The scientific study of primate behavior and ecology under field conditions had its origins in the studies of Carpenter on howlers and spider monkeys in Panama and on gibbons in Siam (1934, 1935, and 1940) and in the studies of Japanese macaques by Imanishi and Itani in the 1950s. Following these early studies there was a break of several years. Renewed interest in primate field studies was stimulated through the efforts of Washburn and DeVore (1961), Altmann (1962), and Jay (1962). These efforts were followed by a great many studies, including those of Schaller, Hall, Kummer, Struhsaker, Gartlan, Ripley, Simonds, and Sugiyama. Speculation as to the evolution of primate social behavior and organization and their relation to ecology and implications for understanding the evolution of human behavior soon followed this renewed interest in primate field studies. However, with the exception of Carpenter's early studies on howlers, spiders, and gibbons, these pioneering studies all dealt with open-country monkey species. Unfortunately, this bias toward species living in savanna or other kinds of open habitats led to premature and some rather erroneous theorizing (DeVore 1963 and Crook and Gartlan 1966), such as the hypothesis correlating large group size with open-country habitats and small group size with forest habitats. Because the majority of primate species live in the rain forest, it soon became apparent that detailed studies on rain-forest species were needed for the development of more realistic hypotheses on the subject of the evolution of primate social behavior. This realization led to several studies of rain-forest monkeys in Africa (Struhsaker 1969, Gautier and Gautier-Hion 1969, and Gartlan and Struhsaker 1972), Ceylon (Rudran 1973), and in Central and South America (Thorington 1967, Mason 1968, Hladik and Hladik 1969, Klein and Klein in prep., Neville 1972). At present this interest in rain-forest species continues, and one finds that the contemporary studies are concentrating more and more on quantitative ecology. Although the early investigators represented a variety of disciplines, including psychology, anthropology, and zoology, with an even more diverse array of motives, it appears that the more recent studies are being done by investigators with stronger training in animal behavior and ecology and with an interest in more general ecological problems.

In spite of the impressive increase in primate field studies, there is still a paucity of published information on the African representatives of the

colobinae and the virtual absence of information for any colobinae living in a
tropical rain forest. The majority of field studies on the colobinae have been
done in open-country habitats or dry woodlands of India and Ceylon
(Sugiyama 1966, 1967, Ripley 1967, 1970, Vogel 1971, and Rudran 1973),
although one study has been done in the montane forest of Ceylon (Rudran
1973). Even more remarkable is the fact that so few field studies dealing with
most aspects of behavior and ecology have been done on any mammal of the
tropical rain forest. A notable exception to this is the detailed study of the
coati in Panama by Kaufmann (1962). Extending the literature survey to all
vertebrates further demonstrates the paucity of detailed and comprehensive
studies of rain-forest species.

    This book provides a detailed account of the behavior and ecology of one
group of *Colobus badius tephrosceles* and compares the results from this
group with the less extensive data from other groups and other subspecies. It
is intended to contribute not only to a better understanding of the African
colobinae, but also to the development of more realistic theories on the
evolution of primate social systems and their relation to ecology. This is not
to imply that the book presents a synthesis on the evolution of primate
behavior. Rather, it presents a detailed study, which, when compared with
future studies of similar depth, will enhance such a synthesis.

CONSERVATION

Of even greater importance is the hope that this book will stimulate more
interest in the study of tropical rain forests and their animals, not only
because the rain forest is a fascinating biome, an understanding of which will
contribute greatly to theoretical science, but also because it is an extremely
fragile habitat which is in severe danger of being totally and irreversibly
destroyed. In most tropical countries having rain forests, the official government attitude is that the forests are there solely for the exploitation of timber.
Consequently, their policies are of rapid and immediate exploitation, notably
lacking in a long-term, ecological perspective. Exceptions do exist, in which
long-range planning is attempted. Some forests are left intact as important
watershed areas, but these are few in number. In some countries, such as
Uganda, exploitation of forests operates on a cycle: a given plot of forest is
selectively timbered once every 80 years or so. But even after 80 years the
forest trees will not have attained their full size, which for many of the larger
species may take as long as 200 to 300 years. Consequently, the rain forest is a
habitat which should not be dispensed with as rapidly and casually as it is in
most tropical countries. For example, it is estimated that more than 90% of
the rain forest in Equatorial Guinea (ex Rio Muni) has been felled. The Ivory
Coast is rapidly approaching this situation, with the Tai Reserve the last
stronghold of rain forest in that country. French timber companies are
rapidly leaving the Ivory Coast and shifting their operations to Cameroun,

where the rape will continue unless strong conservation action is taken immediately.

The problem of rain-forest conservation in the tropics is an extremely complex one, involving sociology, politics, economics, ecology, tourism, and aesthetics. Pressure for forestry exploitation comes from the increasing consumer demands of Western countries which import tropical woods, primarily for decorative purposes, and from local increases in human populations. Most tropical countries have phenomenal growth rates in their human populations, approaching 2-3% per annum with a doubling time of 17-20 years. This rapid growth combined with an agriculturally-based economy, which is typical of most tropical countries, results in ever increasing demands for agricultural land. The human population explosion also results in increasing demands for wood on the local scene. Clearly, the greatest threats to the tropical rain forests are increased consumer demands from the Western world, the human population explosion in the tropics, and the consequent human encroachment on the forests.

I need not elaborate on the value of rain-forest conservation to the theoretical field biologist of the tropics, for without sufficiently large examples of undisturbed and mature rain forest his work is impossible. I will, however, summarize some economic rationales for conserving rain forests, which I have elaborated on earlier (Struhsaker 1972). Areas of rain forest receiving complete protection against all forms of human exploitation are invaluable for the scientific management of neighboring areas insofar as they provide ecological standards with which to compare the effects of man's exploitation on the soil, vegetation, animals, and rainfall. Such areas also provide a reservoir of plants and animals for the recolonization of regenerating forests. H. T. Odum (1970), in pointing out the importance of rain forests to the stabilization of the gases of the atmosphere, states: "An acre of tropical forest land is apparently 30 to 100 times more important metabolically to gaseous exchange than an acre of tropical sea." This is the more remarkable when one bears in mind that marine algae have been considered the major source of the earth's oxygen supply. Other practical uses of undisturbed rain forest include scientific research of an applied nature, such as the study of diseases and the basic biology of commercially valuable trees and animals. The role of tourism as a potential source of income from undisturbed rain forests should not be underestimated. Tourism has already achieved prominent standing in the economies of many tropical countries, and there is no a priori reason why the rain forests could not become a major attraction. (In Puerto Rico, Sunday visitors to the Luquillo Forest recreational areas are said to number up to 4,000 a day; Odum 1970.) It is evident that even on a short-term basis the financial returns from tourism are potentially higher than those from timber exploitation. On a long-term basis the difference is even greater.

It is hoped that this book will stimulate interest not only in rain-forest research but also, more importantly, in rain-forest conservation.

THE FIELD STUDY: BACKGROUND

My own introduction to the tropical rain forest came in 1963 during a brief visit to the area west of Lake Kivu in the eastern Congo. My strongest impressions of the forest at that time were of towering trees, fleeing monkeys, and the incredible array of sounds. It was all very exciting but seemed bewildering and far beyond the comprehension of science, and it was certainly not the place for someone who wanted long and continuous hours of observation on primate social behavior. Slightly more than a year later and after I had completed a detailed field study of vervet monkeys in the semiarid savanna of Amboseli, Kenya, I felt a desire to learn something about rain-forest monkeys.

It was not until late 1966 that I revisited the rain forests of Africa. This visit lasted for 19 months and made me realize all too clearly why we knew so little of the magnificent-looking creatures living there. Seventeen of these 19 months were spent in Cameroun, West Africa. Another month was spent in Rio Muni (now Equatorial Guinea), one of Cameroun's southerly neighbors, and the final month in Ghana. The problems afforded by the forest vegetation and heavy rainfall (10,305 mm in one year) were magnified by the widespread hunting practices of the indigenous people. Almost everywhere I searched in Cameroun the monkeys were being intensely hunted. Even areas that were 2 or 3 days' walk from motorable roads were under the pressure of the homemade muzzle-loader. As a consequence, all observations of anything but fleeing monkeys had to be made from concealment. I soon learned how to stalk, but even the finest stalker has difficulty eluding the acute perception of the cercopithecine monkeys for any length of time. Most of my observations in this study were of short duration and, in terms of understanding social behavior in rain-forest monkeys, were inadequate and certainly far short of my naïve expectations.

Perhaps the most valuable outcome of this Cameroun experience was an even deeper interest in rain-forest biology and an awareness of the perilous state of this unique biome. At the end of this study in May 1968 I felt rather uncomfortable about leaving the rain forest. The subsequent year and a half, most of which was spent in New York City, made me even more uncomfortable. It was with good fortune, however, that I was able to break up this 18-month stretch with a few months in Trinidad and Panama, where I taught a field course in animal behavior along with Professor Peter Marler. It was also about this time that Peter Marler began telling me of his experiences with the red colobus monkeys in western Uganda. During the course of his studies on the vocalizations of monkeys in the Budongo Forest of Uganda in 1964–65, Marler had made a couple of short visits to the Kibale Forest, which is also in western Uganda. Here he had discovered not only that there was an abundance of red colobus monkeys, but also that they were quite observable and were behaviorally very different from the black and white colobus which he had been studying. We both agreed that the red colobus sounded like an ideal species on which to focus for basic and comprehensive information on the behavior and ecology of the African rain-forest monkey—something which had not been accomplished at that time. Furthermore, such a study offered an excellent chance to compare the red colobus with the black and white colobus and thereby to gain better understanding of the interrelation of ecology and phylogeny with social behavior and social organization. It was with Peter Marler's encouragement that I then set about organizing for a field study of the red colobus monkey and my long overdue return to the rain forest.

ACKNOWLEDGMENTS

Innumerable people assisted me with this project, and I am grateful to all of them for their help. Upon the sound advice of Professor François Bourlière I contacted Dr. Gérard Morel in Senegal, who proved to be a wealth of useful information, particularly on red colobus localities. Messrs. Diengo Derma of IFAN, Dakar, and J. C. Bille kindly identified plant specimens, which were collected in Senegal. Dr. Fauck, director of ORSTOM in Dakar, kindly provided accommodation during my stay in Dakar.

   In the Ivory Coast the logistic support and hospitality generously provided by Dr. Pierre Hunkeler (then director of C.S.R.S. at Adiopodoumé) and his wife Claudine were invaluable. M. Adjami, Directeur des Eaux et Forêts at Guiglo, and the Sous-Préfet of Tai also assisted with transport and accommodation. The staff of the ORSTOM Herbarium at Adiopodoumé identified all plant material I collected in the Tai Reserve.

   Mr. C. J. Chorlton, manager of the Ndian oil-palm estate in Cameroun, gave me hospitality and assistance that were most welcome, both before and after my trip into the Korup Reserve. I am also grateful for the accommodation provided by Dr. B. O. L. Duke, Director of the Helminthiasis Research Unit, Kumba, and for the cooperation of the West Cameroun Forestry Department, Kumba. The staff of the herbarium at the University of Yaoundé kindly identified plant specimens collected in the Korup. Of greatest importance to the success of the Korup survey was the ever ready Ferdinand Namata, who acted as my most able field assistant.

   The fieldwork in Uganda was assisted during the initial stages by Mr. Xavier Kururagire and sporadically at later stages by Mr. Pius Kayenga. All plant material collected in Uganda was identified by Dr. Alan Hamilton, previously with the Department of Botany, Makerere University of Kampala, and Mr. A. B. Katende, assistant curator of the Makerere University Herbarium. Dr. Hamilton also assisted with most of the tree enumeration in my major study area. I am extremely grateful to these two for their important contribution to this study. Dr. M. H. Woodford kindly collected 2 specimens of red colobus for me. The Department of Zoology, Makerere University of Kampala acted as my local sponsor during the project, for which I am grateful. The Regional Forest Officer and the District Forest Officer of the Fort Portal Office of the Uganda Forestry Department and the staff at the Kanyawara Forestry Station have assisted me on innumerable occasions and in a multitude of ways. I extend my thanks to all of them. Mr. A. M. Stuart Smith, Senior Conservator of Research, Uganda Forestry Department, provided encouragement throughout this study. I am particularly grateful for his efforts which led to the establishment of a Forestry Department research plot that encompassed my study area in the Kibale Forest and thus conserved it against timber exploitation. This study would not have been possible without official government approval, and I extend my gratitude to the President's Office of Uganda, the National Research Council of Uganda, and Mr. M. L. S. B. Rukuba, the Chief Conservator of Forests, Uganda Forestry Department, for permission to study in Uganda and in the Kibale Forest.

Much of the basic research strategy, data analysis, and conclusions in this study were influenced by valuable discussions with several people, including: Drs. Peter Marler, Tim Clutton-Brock, Steven Green, J. F. Oates and P. Waser. Some of the data analysis was done in New York, where I was ably assisted by Mr. Robert Deutsch. A program which computes intermonthly overlap in ranging patterns was written and prepared for me by Messrs. J. Kavulu and J. Tumushabe and Miss E. Mulindwa of the Computer Center, Makerere University of Kampala. Valuable comments on various sections of the first draft of this book were received from several able critics. The section dealing with vocalizations was commented on by Drs. P. Marler, S. Green, and P. Waser and by Mrs. Mary Sue Waser; the sections on correlates of ranging behavior and species' censuses by the Wasers, S. Green, and J. F. Oates; that on food habits by S. Green and P. Waser; and those sections dealing with interspecific relations, time budgets, and heights of activity by Mr. R. Rudran. Preparation of most of the graphic illustrations and all of the final typing of the manuscript were done in the field by Mrs. Ranjinee Rudran, whose competence and cheerful patience made it all much easier. I am grateful to both Peter Marler and Lysa Leland for their photographic contributions. I extend my warm thanks to all these people.

Last, but certainly not least, I wish to thank the financial sponsors of this study: the New York Zoological Society, The Rockefeller University, the National Science Foundation of the U.S.A. (grant No. GB 15147X, November 1969 to August 1973) and the National Institutes of Mental Health of the U.S.A. (grant No. 1 RO1 MH 23008-01, October 1972 to September 1973).

THE RED COLOBUS MONKEY

1 Red Colobus: General Description and Survey Techniques

The red colobus monkey belongs to the family Cercopithecidae of old world monkeys, which is subdivided into 2 subfamilies. The cercopithecinae is comprised of the baboons, macaques, and guenons, and the colobinae of the langurs, proboscis monkeys, and colobus monkeys. The colobinae have large, sacculated, and rumen-like stomachs, which allow efficient digestion of cellulose (Drawert, et al. 1962; Kuhn 1964; Bauchop and Martucci 1968; Oxnard 1969). Colobus monkeys are found only in Africa, whereas langurs are found in India, Ceylon, and Asia. There are 3 major groups of colobus monkeys: the olive colobus, the black and white colobus, and the red colobus. The olive colobus (*Procolobus verus*) is monotypic and occurs from Sierra Leone eastward to Togoland. There are many forms of black and white colobus extending from the west coast of Africa eastward to the Kenya coast. Dandelot (1968) recognizes 4 species and at least 16 different forms of black and white colobus, all of which he places in the subgenus *Colobus* (a division of the genus *Colobus*). Rahm (1970) accepts Dandelot's division of the genus *Colobus* into 4 species, but recognizes 21 different subspecies. The taxonomy of the red colobus is in the most confused state of all the colobus monkeys. Verheyen (1962) places red colobus monkeys in the genus *Colobus,* whereas Kuhn (1972) prefers the genus *Procolobus*. Dandelot (1968) subdivides the genus of *Colobus* into 3 subgenera, placing all of the red colobus in the subgenus *Piliocolobus*. Within this subgenus Dandelot recognizes 6 species of red colobus and a total of 14 different forms. Rahm (1970) recognizes the same 14 forms, but places all of them as subspecies of *Colobus badius* (fig. 1). I have adopted the latter system for reasons given in chapter 4.

Red colobus are found in equatorial Africa from Senegal to Zanzibar. Their distribution is, however, not continuous, and several of the subspecies are isolated by great distances (fig. 1). In addition, a number of the red colobus subspecies have extremely limited distributions and, in combination with their low numbers, the ease with which they are hunted, and human encroachment on their habitats, are on the verge of extinction. Examples are *Colobus badius preussi* of Cameroun and *C. b. rufomitratus* of the Tana River, Kenya.

The majority of red colobus subspecies live in rain forest. *C. b. temminckii* of Senegal and Gambia is exceptional in that it often lives in savanna

woodland. Some other populations, such as *C. b. rufomitratus,* live in relic
forest patches or narrow riparian forest strips. Like all members of this
subfamily, the red colobus is considered to be primarily a leaf-eating monkey.

Compared with other rain-forest monkeys in Africa, red colobus are relatively large. Unfortunately, there are few data on weights and linear measurements for red colobus. Kingdon (1971) reports weights ranging from 9 to 12.5
kg for adult males and from 7 to 9 kg for adult females. Data from 5 red
colobus which I obtained in the Kibale Forest of western Uganda are consistent with these findings, although the adult female weighed slightly less than
indicated by Kingdon (see table 1). The amount and shade of red in the
pelage varies considerably between the different subspecies. *C. b. badius* has
more red than any of the other subspecies I studied, and its red color is really
more of a deep chestnut. The subspecies I studied most, *C. b. tephrosceles*
(see pls. 1-5), has very little red. In this subspecies only the cap is rusty red.
The back and tail are very dark and essentially black. The limbs are gray with
varying amounts of tan or brown on the lateral surface of the arms. Some
individuals lack this brown. The face is dark and is outlined with gray
whiskers. This gray extends to the sides of the body and the entire ventral
surface. Adult males not only are considerably larger in body weight than
adult females, but also have more robust temporal musculature (giving their
head a more massive appearance), larger canines, and longer cheek whiskers.

Although generally considered to be a species adapted to life in the trees,
the red colobus is, in my opinion, one of the clumsiest climbers in the forest.
Its general mode of locomotion can best be described as suicidal. It usually
employs a galloping gait while moving from one place to another on the same
tree branch. The slow crossed-alternating walk in which fore- and hindlimbs
of opposite sides move in unison is less frequently used. But in moving from
one branch to another or from tree to tree the typical mode of locomotion is a
great leap with all 4 limbs spread-eagled (see pl. 5). Some of these leaps are
most impressive and often cover a vertical drop of 5-10 m. The impression I
have is that in making these leaps the red colobus aims for a general part of
the tree and not for a specific spot, in marked contrast to the precision
climbing of *Cercopithecus* monkeys. When moving to either a higher or lower
branch that is within close reach (about one body length), the red colobus of
Uganda often employs a kind of brachiation, in which the monkey grips a
branch with one or both hands and then swings onto the other branch, with
the body moving in an arc below the point where the hands grip the branch.

Most red colobus live in social groups numbering about 50 and include
several adult males. There are more adult females than adult males in these
groups. In contrast, the red colobus of Senegal-Gambia and of the Tana
River, Kenya, live in groups of about half this size. Intergroup relations are
generally aggressive but, at least for those groups living in the Kibale Forest in
Uganda, not territorial.

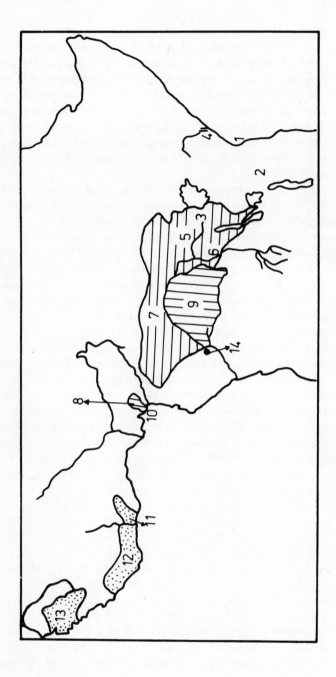

Fig. 1. General distribution map of red colobus: (1) *Colobus badius kirkii*; (2) *C. b. gordonorum*; (3) *C. b. tephrosceles*; (4) *C. b. rufomitratus*; (5) *C. b. ellioti*; (6) *C. b. foai*; (7) *C. b. oustaleti*; (8) *C. b. preussi*; (9) *C. b. tholloni*; (10) *C. b. pennantii*; (11) *C. b. waldroni*; (12) *C. b. badius*; (13) *C. b. temminckii*; (14) *C. b. bouvieri*. Modified from Kingdon (1971).

THE SURVEY

Prior to selecting an area for intensive study, it was necessary to survey a number of potential study sites in order to locate the one affording the best conditions. I was looking not only for a place having adequate numbers of red colobus, but also one in which the habitat was of a relatively undisturbed and mature state. In addition, I wanted an area where the monkeys were under minimal hunting pressure or other forms of human disturbance. It was, of course, necessary that the site be accessible throughout the year. I also wanted a forest which had a number of other primate species so that I might make comparative studies.

The survey was begun in December 1969 with a reconnaissance of the Casamance Province of Senegal. The Casamance lies south of the Gambia River between Senegal's borders with Gambia in the north and Portuguese Guinea in the south. The most southerly area surveyed in Senegal where red colobus (*C. b. temminckii*) were seen was near the village of Niadio (12° 30' N., 16° W.). However, most of the work was done from my camp in the Forêt Classée des Narangs (13° 8' N., 16° 30' W.). Following the survey of Senegal I traveled the full length of Gambia. The most easterly point at which I observed red colobus was near the village of Dembakunda (13° 15' N., 14° 16' W.). Red colobus were studied at a total of 13 different localities. Observations of these monkeys were made between 18 December and 13 January 1970 on 26 different days, plus one day on which only tape recording was done. Fifteen of these days were spent at the Narangs camp. A total of 93 hours of observation, plus 11 hours of tape recording, were logged. The rainy season in this area is generally between July and September, and at the time of my survey there was no rain whatsoever. The red colobus of Casamance were neither hunted nor otherwise harassed by the people; in fact, they were often found living in small relic patches of forest on the edge of villages. The habitat is described more fully in chapter 5. In general, the savanna wood-land and relic forest patches provided excellent observation conditions.

After surveying Senegal and Gambia I moved onto the Ivory Coast, where I surveyed the red colobus (*C. b. badius*) of the Tai Reserve, which is located in the western part of the country between the rivers Cavally and Sassandra. This is an immense rain forest, comprised of about 70,000 ha. Due to lack of transport I restricted my survey to the general vicinity of Troya and Sakré villages, which are located about 22 and 25 kms south of Tai on the Tai-Tabou Road. The section of mature lowland rain forest which I surveyed is within the area delimited by 5° 40' and 5° 43' N. and 7° 15' and 7° 24' W. Red colobus were observed between 4 February and 7 March 1970 on a total of 19 days plus 3 other days on which only tape recording was made. Seventy-four hours of observation and 29.5 hours of tape recording were logged. Most of the observations were made in the Troya-Sakré area, but on 2 of the 22 days observations were made near the Atlantic coast on the Béréby-Tabou Road and near the village of Itragi. Conditions were relatively dry at the time of the survey, as months having 100 mm or more of rainfall are usually restricted to April through July and September through December.

Preuss's red colobus (*C. b. preussi*) is now restricted in its distribution to the Korup Reserve of Cameroun. This is also a huge lowland rain forest and comprises about 100,000 ha. It is located near the Atlantic coast at Cameroun's border with Nigeria (5° to 5° 25' N. and 8° 40' to 9° 05' E.). Only the southern part of this reserve was surveyed between the villages of Ekundukundu One and Erat and the Ndian oil-palm plantation. Observations were made between 20 March and 8 April 1970 on 13 days plus one day of tape recording only. Seventy-three hours of observation and 14 hours of tape recording were made at 4 different localities in the Korup. This is an area of heavy rainfall, with all months usually having more than 100 mm of precipitation. The period of my survey was not exceptional.

The last phase of the survey was made in Uganda, May through July 1970. It was readily apparent during my first few days in the Kibale Forest Reserve of western Uganda that this was the best of all the areas I had surveyed for red colobus monkeys. It had an abundance of red colobus (*C. b. tephrosceles*) plus 8 other anthropoids. Many parts of the reserve had relatively mature forest; it was of adequate size (56,000 ha.); and, most important, the monkeys were not hunted for food by the local people. Other forests surveyed in Uganda included Itwara, Semliki (Bwamba), Impenetrable (Bwindi), Maramagambo, Kasyoha-Kitomi, the Sango Bay forests (Malabigambo, Namalala, and Tero East), the riparian *Cynometra* forest along the Mpanga River just above the Mpanga Falls, and the small relic-forest patches of Miranga and Kasenda. Aside from the Kibale, red colobus were seen only in the latter 3 areas. The Miranga and Kasenda forests are extremely small in area and are being rapidly destroyed and replaced by cultivation. Both are located only a few km west of the Kibale Forest and were surely contiguous with it in the recent past. The strip of the Mpanga River forest just above the falls which contains red colobus is probably less than 8 km long.

The rare and endangered Tana River red colobus (*C. b. rufomitratus*), found only along the lower reaches of the Tana River in Kenya, was studied very briefly on 2 occasions; November 1971 and February 1973. In the first visit, observations were made near the village of Mnazini on 2 days only (2 hours of observation and 11.6 hours of tape recording). Only the results of the first trip are reported here. The second visit was concerned primarily with the initiation of a detailed study of the red colobus by Mr. Clive Marsh.

The Intensive Study

The first few months of study in the Kibale Forest Reserve were of a general nature and involved reconnaissance of the entire reserve and familiarization with the vegetation and animals. In August 1970 I selected the section of forest near the Kanyawara Forestry Station as the place for my detailed study. August and September were spent putting in a basic gridwork of trails, habituating specific groups of red colobus, and mastering individual recognition of some of the animals in these groups. In October 1970 I began the monthly systematic sampling of the main study group, which was called the CW group after one of the adult males whose cheek whiskers reminded me of a cat's whiskers. These monthly samples were usually conducted during the first week of each month and consisted of 5 continuous days of observation on the CW group. Initially it was intended that each of these 5 days should consist of at least 11.5 hours of observation, i.e., from sunrise to sunset. However, this was achieved for only 53 of the total 83 days of systematic sampling between November 1970 and March 1972, inclusive. In the remaining 30 days fewer than 11.5 hours per day were spent with the CW group. Most of the data collected in the October 1970 sample have not been included in the analysis of systematic monthly samples, because the various sampling methods were not yet stabilized.

A total of 1,593.7 hours were spent observing *C. b. tephrosceles* (table 2). Of these, 1,112.5 hours were with the CW group, of which 873.6 were made during the monthly systematic samples. In addition to the hours of observation listed in table 2, 19 hours were spent tape recording and 23 hours photographing *b. tephrosceles*. In mid-March 1972 I left the Kibale Forest for 5 months of data analysis in New York, but returned in late August 1972 and continued with my studies of red colobus until the present (December 1974).

In general I was interested in collecting normative, quantitative, and systematic data on ranging patterns, food habits, and social behavior during the monthly samples. I attempted to establish and maintain a state of neutrality with the monkeys, and at no time did I feed them or otherwise interact with them intentionally. Immediately prior to each monthly sample of the CW group I made a phenological analysis of 85 individually marked trees representing 11 different species within the home range of the CW group. This was intended to give a measure of food availability for each month.

I found that the adult males habituated to me much faster than did the adult females and immature monkeys. Most of the males were quite tame after one to 2 months, but the females required from 5 to 6 months before they were equally tame. During this habituation period I found that, by moving quietly and slowly and remaining as inconspicuous as possible, I did not obviously disturb the animals by my presence. Two other factors made observation difficult. These were the dense foliage and the relatively great heights in the trees at which the monkeys spent much of their time. As a consequence, it was impracticable to conduct longitudinal samples of their behavior, i.e., continuous observations of one particular animal's behavior over a relatively long period of time. Therefore, most of the sampling was opportunistic (scoring certain kinds of behavior whenever seen) or latitudinal, (scoring the behaviors of many animals during a short period of time at regular time intervals). The problems of observation are best appreciated when one considers the fact that I was able to make complete and accurate counts of the CW group (circa 20 monkeys) about once in every 60 hours of observation. The difficulties of habituation, individual recognition, and general observation made it impracticable to concentrate on more than one group, but whenever possible I observed other groups as a check on the general applicability of my results from the CW group.

TECHNIQUES

Sampling methods and data analysis are considered in the context of the various kinds of data examined. I would like to offer here a few remarks on field techniques which may prove useful to others in their studies.

The ease of recognizing individuals and the criteria used differ from species to species. Among the red colobus, adult males were most easily distinguished as individuals. This was partly because they habituated faster and, consequently, one could observe them for longer periods and thus pick out identifying marks. In addition, their appearance was more individualistic than that of the females. Invariably, I could pick out individual males on the basis of their facial and whisker characteristics. I did not need "natural" deformities such as ear notches or tail kinks to identify the males, because their faces all looked different. As an initial check on this means of identification I also kept notes on "natural" deformities. The most common natural mark was one or more stiff fingers. The digit number, combined with a notation of left versus right hand, provided a check against the identification based on facial appearance. Ear notches and tail kinks were notably absent from adult males, although some had tails that were ¼ to ⅓ shorter than normal. In contrast, I found it quite difficult to identify adult females. For them I had to rely on tail kinks, whisker shape and extent of whisker development, and general tone of pelage coloration. Stiff fingers were not seen among the females, although some had pink tips on one or more digits.

The extensive trail system which evolved in my major study area proved invaluable. It was absolutely imperative for the detailed mapping of group movements and for rapid and quiet following of the group by the observer. All trails were cut along compass bearings either parallel to one another or at right angles, so that a grid system of squares was created. In most parts of the study area, trail intersections occurred at 100 m intervals, and in many places trails intersected at every 50 m. All trails were marked at 50 m intervals by a small tin plate placed on a tree and a red ring painted around the tree. Each plate indicated the appropriate trail number and the distance in meters from the zero point of that trail. This information was punched into the plate with a hammer and nail, and then the indentations were painted over with red. Red was found to be the color most readily seen in the forest. These red marks at 50 m intervals were of great help in rapidly locating one's exact location. In nailing the plate to a tree it was found best not to drive the nail all the way in, but rather to leave it sticking out by one or two inches. This allowed the plate to dangle from the nail and prevented elephants from forcing the plate off the tree, which they were fond of doing, and also allowed for the tree to increase in girth without forcing the plate off.

Everyone certainly has his own preference for a type of binocular. I have found Leitz Trinovid 10 x 40 to be excellent for rain-forest work. They are extremely lightweight and can be used reasonably well with one hand, for example, when one is holding the microphone of a tape recorder with the other. The 10-power is critical for good observations of animals at the top of 40 m trees. The central focus allows rapid focusing but, because of the additional "sealing" of this central focus, does not allow ready passage of air into the optical chambers. This means one does not have the usual problems of condensation and fungus development inside the binocular. I have used the same pair of trinovids since 1966 (mostly in rain forest) and have never stored them in silica-gel, never had them cleaned or adjusted, and never had problems with fungus or condensation.

I take all of my notes on paper manifolds. Each manifold consists of an original sheet of good quality bond paper, a sheet of inexpensive single-copy carbon, and a copy sheet. The reason for taking notes in duplicate is obvious to anyone experienced in fieldwork; it constitutes a cheap form of insurance. Once every 6 to 8 weeks I send the note copies back to the United States, and I always store them separately from the originals. These manifolds can be made to custom order by any stationery supplier. At the end of each day's observation I edit and catalogue the notes. The back of each facing page is always blank, and additional notes or comments can be added here. Each note entry is catalogued according to one or more subjects. In the catalogue and under the appropriate subject I enter the date, location, and page number in the notes of the entry, and, usually, a brief description or note of what the entry specifically relates to. Cataloguing permits rapid reference to specific events, which is often necessary during the course of the field study; it also goes a long way toward the eventual data analysis. Notes can be stored in ring files (loose-leaf notebooks), and in the field the blank sheets and notes of the day are best carried in an aluminum loose sheet holder, which is available in a variety of sizes from most forestry suppliers.

2 Social Structure

An understanding of the social group size and composition of a species is imperative for an understanding of all its social behavior. Furthermore, data on group size and composition and intergroup relations is essential for estimating population and biomass densities, as will be discussed in the chapter on ecology.

Methods'

One would like to know the exact size and composition of every group of monkeys on which any observations are made. However, the observation conditions afforded by forest habitats, especially by rain forest, preclude this ideal. In practice, one counts the group under observation whenever the opportunity avails itself, and such opportunities were unfortunately rare in this study.

The savanna woodland of Senegal and Gambia provided the best conditions for counting red colobus groups, because (1) the relatively open habitat usually permitted clear observations at distances greater than 50 m from the monkeys, (2) the red colobus were not hunted by man, and (3) in areas where they lived near villages they were already habituated to man. The worst conditions existed in the rain forest of the Ivory Coast and Cameroun, where the observation conditions were poor and the monkeys were hunted by man for food. Uganda and Kenya provided situations intermediate to these extremes. The vegetation often made observation difficult, but the monkeys were not hunted by man. In spite of this, it required a relatively long period of time to obtain a complete and reliable count even of a habituated group. For example, my main study group of red colobus (the CW group) in the Kibale Forest, Uganda, was counted with reliability about once in every 60 hours of observation. Most group counts, especially of large groups, were incomplete, but they were valuable for estimating and giving the order of magnitude in group size. For example, if 40 monkeys were counted, but the amount of noise and shaking foliage indicated that at least half again as many were present, the group size was estimated as 60.

In the Kibale Forest there was an additional difficulty in group counts: the extensive overlap in home ranges between groups of red colobus, which often resulted in 2 or more groups being virtually contiguous with one another.

11

Only with long study was I able to distinguish the integrity of these groups and, consequently, make reliable counts of some of them.

The criteria for determining age and sex of red colobus are as follows:

1. *Adult males:* large body size (weight); heavier and broader appearing heads; scrotum pendulant and conspicuous in *b. badius,* but because of more anterior position not so conspicuous in *b. temminckii, b. preussi, b. tephrosceles,* and *b. rufomitratus;* nipples not apparent. The perineal organ, as described in detail by Kuhn (1972), was conspicuous only in adult males of *b. badius.* It was not noted in *b. temminckii, b. preussi,* or *b. rufomitratus.* Among *b. tephrosceles* it was relatively small and knob-like in appearance.

2. *Adult females:* prominent and large clitoris in *b. preussi, b. tephrosceles* (one measured 3 cm long, pl. 6), and *b. rufomitratus,* but not prominent in *b. temminckii* and *b. badius;* sexual swelling very large in *b. badius* and *b. preussi,* but small in *b. temminckii* and *b. tephrosceles* and not noted for *b. rufomitratus.* Small sexual swellings were noted only 3 times for *b. temminckii,* but this could be an artifact of the small sample and possible seasonality in reproduction for this subspecies (see chapter 3). The same applies to *b. rufomitratus.* Nipples were prominent in adult females of all subspecies.

3. *Subadults:* slightly smaller than adults; head not so broad as adult male; nipples just visible among females; in *b. tephrosceles* the subadult females have an obvious clitoris and can be distinguished from males, who lack any structure resembling a clitoris but who have a bare and pink-colored pad on the perineum, which is triangular in shape with the apex of the triangle directed anteriorly and ventrally.

4. *Juveniles:* ½ to ⅘ the size of adult females; difficult to determine the sex in the field, but in larger juveniles the clitoris and perineal triangle become apparent in *b. tephrosceles.* Kuhn (1972) also describes the difficulty of determining the sex of young badius on the basis of a superficial examination of the external anatomy.

5. *Young juveniles:* ⅓ to ½ the size of adult females; coloration is virtually that of adults; sex cannot be determined under field conditions; infrequently carried by adult female.

6. *Old infants:* ¼ to ⅓ the size of adult females; coloration is virtually that of adults; sex cannot be determined under field conditions; often carried by adult female.

7. *Infants:* about ¼ the size of adult females; coloration intermediate between neonatal color and adult color, sometimes virtually the color of adults; pelage tends to be of silky texture rather than coarse like that of adults; sex cannot be determined under field conditions.

8. *Young infants:* about ¼ the size of adult females; pelage silky texture; color distinct from that of adults for all subspecies studied:
   b. temminckii: general color light beige or light gray, but darker on

back; tail with a bit of orange coloration, but lighter than that of adult; very little orange on the distal ends of the arms and legs and much less extensive than in adults; sides of the body gray and not orange as in adults; palms and soles pink; perineum bright pink with a knob-like structure;

b. badius: back is gray and not black as in adults; tail rusty but lacks the black tip of that of adults; ventral surface is whitish, not chestnut as in adults; sides and limbs are orange, not rich chestnut as in adults; pelage silky; muzzle pink, but face dusky around the eyes, not like the completely dark face of adults; perineum bright pink and appears swollen;

b. preussi: black on back not smoky as in adults; ventral surface lighter gray than in adults; no sandy color anywhere, but gray on limbs, sides, and tail, unlike adults with sandy limbs, sides, and tail; pelage silky; perineum pink with clitoris or pseudoclitoris (see Kuhn 1972 for description of this structure);

b. tephrosceles (pl. 7): general color black with gray ventral surface; no red or brown coloration anywhere, unlike adults with red cap and rufous color on proximal part of limbs; soles and palms pink; silky pelage; perineum dark with clitoris or pseudoclitoris present but not conspicuous. (Pl. 8 of a young infant male resembles Kuhn's 1972 description for *b. badius.*)

b. rufomitratus: completely dark without any orange, red, or brown as in adults (based on one sighting only).

Group Size and Composition

Red colobus typically live in heterosexual social groups ranging in size from 12 to about 80, with at least 2 fully adult males per group (tables 3–7). The social groups of *b. temminckii* living in the savanna woodland of Senegal and Gambia are smaller than those of the subspecies which live in rain forest; 25 vs. 45. In all groups of all subspecies the adult females outnumber the adult males.

The mean of estimated group sizes is probably an underestimate of actual group size in most cases. For *b. badius, b. preussi,* and *b. tephrosceles* this means that the typical group is somewhat larger than indicated by tables 4, 5, and 6 and may number about 50. In spite of the imperfections of these group counts, it is evident that rain forest groups of red colobus are nearly twice as large as those of the savanna woodlands, riparian forests, and small remnant patches of forest. These observations suggest that with lower densities of food, more pronounced seasonality in food availability, and, presumably, more marginal ecological conditions red colobus live in smaller groups. This hypothesis is further supported by the fact that, even within the subspecies of *b. temminckii,* the smallest groups are found in the drier parts of their range and in very small patches of forest. The 2 small groups (21 and 12) in table 3 were both observed in eastern Gambia in small relic patches of forest. In addition, the social groups of *b. rufomitratus* living in the small and remnant riparian forests along the lower Tana River of Kenya are also small. Although no accurate and complete counts were made of social groups of this sub-species, of 3 groups clearly observed all seemed to be less than 25 in number.

Adult sex ratios in these heterosexual groups range from about 1.5 to 3.0 females per male. Some of this discrepancy may be accounted for by the possibility of differential growth rates between the sexes — the females attaining physical maturity earlier than the males. Reliable data on this factor are not available. The existence of solitary males has been established for *b. tephrosceles,* which partially explains the differential sex ratio in social groups.

Dynamics of Group Membership for *C. b. tephrosceles*

The CW group of red colobus, which lived in compartment 30 of the Kibale Forest, Uganda, was the main study group. Observations were first begun on them in August 1970, but reliable identification of the 3 adult males in this group was not achieved until late September 1970. The counts made on this group between August and the end of October 1970 were, therefore, erratic and are not considered in this analysis.

In the period from early November 1970 through 5 March 1972, the CW group ranged in size from 19 to 22 (table 7). Adult membership was extremely stable. The 3 adult males (CW, LB, and ND) who could be readily identified on 29 September 1970 remained together in this group until the end of the study. In addition, 4 adult females became recognizable, and they too remained with the group until the end of the study (5 March 1972). Their names and the dates on which they first were recognized are as follows: GCW, 1 March 1971; BT, 17 April 1971; KT, 9 May 1971; and PGCW, 30 September 1971. Male SAM was first distinguished in the group in April 1971, when he was an old juvenile. He too remained throughout the study, becoming an adult male during that time. (Adult males, CW, LB, ND, and SAM and adult females GCW, BT, and KT were still together in the CW group at the time of writing, June 1973. Adult female PGCW disappeared between 5 October and 5 November 1972.)

Most of the changes in the size and composition of the CW group can be attributed to births, apparent deaths, maturation, and probable immigration and emigration of young monkeys (table 7). The difficulty of recognizing individual red colobus of the juvenile class or at even earlier ages does not permit direct verification of the suggestion that they are the mobile elements in the population. However, the indirect evidence provided in table 7 suggests that juveniles joined and left the CW group relatively frequently. It seems more than coincidental that some members of the juvenile and old juvenile classes were also the animals who seemed least ready to habituate to the observer. In fact, some of them were still shy of me at the end of the study. A possible explanation is that the shy juveniles were recent immigrants to the group.

Data from the ST and BN groups lend support to the conclusions on the group size, composition, and stability for *b. tephrosceles*. Both groups lived in compartment 30 of the Kibale Forest and had extensive overlap in their home ranges with the CW group. The ST group, never completely counted, was a large group with more than 66 animals (maximum count); I estimated that it included as many as 75 to 80. There were at least 10 fully adult males and 21 adult females in the ST group. Some animals in this group were individually recognizable on 12 October 1970. The group was last observed in the period covered by this report on 13 March 1972. Because the ST group was not observed at regular intervals or for long periods, data on group dynamics are less precise than for the CW group. During these 519 days (16.7 months) I was able to recognize 13 individuals in the ST group: 7 adult males, one subadult male, 4 adult females, and one young juvenile. Casual observations — not verified because of insufficient time — indicate that at least 9 of the 13 recognizable monkeys remained in the ST group for most, if not all, of this period. Three adult males and one adult female were not seen in the group after February 1971. Of the remaining 9, all were seen in the group for a period of at least 390 days, and all were seen until 19 January 1972. (At least 5 of these 9 were still together in the ST group during recent studies: young juvenile MTK — matured to juvenile class — and adult female TK at least until January 1973; and adult males ST, RF, and HIF until June 1973.)

The BN group was less well known than either CW or ST groups. The maximum count of this group was 33, and I estimated that there were between 35 and 45 monkeys. At least 7 adult females and exactly 4 adult males were present in the BN group, and one of them (male BN) was first recognizable on 24 November 1970. Male TTT was first distinguishable on 23 February 1971, and male OM on 10 August 1971. All 3 of these males were together in June 1973.

Solitary *C. b. tephrosceles*

Solitary badius were encountered on 11 occasions between 6 May 1970 and 17 March 1972. Two of these encounters were probably with the same subadult male. Another observation was of 2 "solitaries" together; a subadult and an adult male. Thus a total of 11 different solitary monkeys were observed: 4 adult males, 4 subadult males, 2 old juvenile males, and one whose age and sex were not noted. A subadult and an adult male were once seen together with a group of 20-30 mangabeys (*Cercocebus albigena*). They were intermingled with the mangabeys and moved in progression along the same arboreal route with them. No other badius were nearby. On 6 occasions, solitary badius were seen with groups of *Colobus guereza,* and on one of these a group of *Cercopithecus ascanius* was also present. Solitary red colobus were seen completely alone on only 4 of the 11 occasions. Seven of the observations were made in compartment 30, two in compartment 13, and one in the Dura River bridge area of the Kibale Forest. One observation was made in the Miranga Forest, which is a small relic forest about 10 km southwest of Kanyawara, Kibale Forest.

One of the old juvenile male badius who was associated with a group of guereza was called to my attention by Dr. J. F. Oates. This badius was obviously solitary and had, according to Oates, been with his main study group of guereza for 2 consecutive and complete days. In apparent contrast to this attachment to another species group, was the observation of an adult male badius in the Miranga Forest. He was first seen with a group of guereza and ascanius. Four hours and 10 minutes later he was again seen, but in the absence of any other monkeys. Similarly, a subadult male first seen completely alone on 3 March 1972 in compartment 30 was next seen on 13 March about 750 m from the previous sighting and with at least 2 guereza of approximately adult size. These observations indicate that, although solitary badius often form associations with groups of other species, particularly with guereza, these associations are of relatively short duration. Why they form these associations is not known. Perhaps it is advantageous for them to associate with other species from the standpoint of predator detection and/or location of food. Interestingly enough, solitary guereza also tend to associate with social groups of badius, even though social groups of the 2 species rarely associate (see section on interspecific relations in chapter 5).

Information on the spatial relationships of these solitary badius to social groups of badius is available for one case only. The subadult male of 3 March 1972 was observed between 1225 and 1236 hours when he was about 165 m from the CW group and again at 1858 when he was about 185 m from the same group. Because of the extensive overlap in home range of the CW, ST, and BN groups and partial overlap with at least 3 or 4 other groups of badius, it cannot be concluded that this solitary subadult male derived from the CW group or that he was moving in relation to them as a peripheral animal.

The manner in which red colobus become solitary is not known. However, the fact that more than half the solitaries seen were immature males suggests that they may be forced out of the group by adult males.

One of the solitary adult males behaved in a most peculiar manner, worthy of description. On 10 August 1971 I was called to observe this solitary male, who was extremely well habituated to humans. When I found him, he was feeding about 10 m above a group of 4 people along the road just east of the Kanyawara Forestry Station and about 375 m east of the point at which the road first enters the forest. He seemed completely unconcerned by the close proximity of the people and their even closer scrutiny. Miss Linda Nash, a visitor at Kanyawara, was the first to observe him. She said that he walked along the road for at least 175 m, passing within one meter of her, and then climbed into a tree. Apparently he showed no response to her. I observed him for 25 minutes until nightfall. He remained in the same tree the entire time and only occasionally gave chist calls. Aside from his extreme tolerance of humans, he appeared to be physically normal and healthy.

Intergroup Relations of *C. b. tephrosceles*

In compartment 30 of the Kibale Forest home-range overlap is extensive among red colobus social groups. The CW group overlapped most with the ST and BN groups and relatively little with 4 or 5 other groups. Groups were frequently within 50 m of one another (proximal), and sometimes the CW, ST, and BN groups were all virtually contiguous with one another. During these times of proximity, and even when the groups were separated by distances exceeding 50 m, there were frequent vocal exchanges between adult males of different groups (see chapter 4). Although the vocalizations employed in these exchanges (chists, wheets, and barks) do not have the amplitude or range of the loud calls of other monkey species in the Kibale Forest, such as guereza or mitis, they could be detected at distances of at least 300 m and were effective in intergroup communication. The interactions between groups that were proximal to one another were quite variable. Often there was no apparent interaction, and one might say they were tolerant of one another. On other occasions the adult and subadult males exchanged vocalizations, gave branch-shake and leaping-about displays (see chapter 3), and then chased and counterchased one another. The most intense of these intergroup aggressive encounters resulted in the spatial supplantation of one group by the other. Although these conflicts were vigorously aggressive, no physical contact was seen between members of different groups. With very few exceptions the intergroup aggressive encounters were exclusively the activity of adult, subadult, and old juvenile males. At the other extreme there was the infrequent case in which one group would move completely through another group without any overt interaction.

There was no verification of a nonaggressive social interaction between members of different groups.

Range overlap.   The location of the ST group was plotted on 54 days between 15 October 1970 and 13 March 1972. None of these were complete days (≥ 11½ hours of observation) and, in general, the sampling of their ranging patterns was opportunistic. Consequently, estimation of their home range is minimal and biased in favor of areas where I spent most of my time. In this sample they were found to occupy 160 quadrats (each 0.25 ha.; see chapter 5 and tables 46 and 48), 17 of which were not also used by the CW group. Thus, at least 89.4% of the quadrats used by the ST group were also used by the CW group.

During 27 days between 26 November 1970 and 13 March 1972, the BN group was found to occupy 66 quadrats, only one of which was not also used by the CW group (tables 46–49). A first estimate of the range overlap between these 2 groups is therefore 98.5%. The sampling deficiencies mentioned for the ST group are also applicable to the BN group.

The 4 or 5 other badius groups who used the home range of the CW group did so very infrequently (verified on only 1 or 2 days for each), so that an analysis of percentage overlap is of limited value. It seems that an area of about 50 ha. was used intensively by the CW, ST, and BN groups to the virtual exclusion of other badius groups.

Proximity and aggressive encounters. The CW group was proximal (within 50 m) to at least one other group during 49 (59%) of the 83 days of systematic sampling between November 1970 and March 1972 inclusive. These close spatial relationships occurred throughout the CW home range.

Intergroup aggressive encounters were observed less frequently and occurred during 30 (36%) of the 83 sample days of the CW group. A total of 45 intergroup conflicts were tallied during these 30 days. Each conflict was operationally distinguished from other conflicts by a time interval of at least 20 minutes between the end of one and the beginning of the next. Thirty-two of the conflicts occurred on 20 of the 53 complete sample days ($\geqslant 11\frac{1}{2}$ hours of observation), which gives a rate of 0.6 intergroup aggressive encounters per day. The remaining 13 conflicts occurred during 10 incomplete sample days ($< 11\frac{1}{2}$ hours of observation).

All intergroup conflicts were accompanied by loud vocalizations (chists, wheets, wah!s, barks, shrieks, screams, and sqwacks) and by chasing. Counterchasing and branch-shake and leaping-about displays often accompanied these encounters. Only once in the total 45 conflicts did adult females take an active role. Adult and subadult males were the primary participants, and most of their aggression was restricted to similar males of the opposing group (pl. 9). Sometimes only one adult male was able to spatially supplant an entire group. Usually, however, 2 or 3 adult males supplanted the other group and, in so doing, the males of each group demonstrated a striking spatial cohesion and unity in their aggression against the other group.

Of the 45 conflicts which the CW group had with other groups, 19 (42.2%) were with groups whose identity I could not determine (usually because of visibility problems), 12 (26.7%) were with the BN group, 12 (26.7%) with the ST group, and one each with the DL and HT groups (table 8).

The relations between the CW and the ST groups indicate that the latter was usually dominant to the former. The 2 encounters in which the CW group supplanted the ST group occurred on the same day and these supplantations were of a very temporary nature because as soon as the CW group moved away, the ST group moved back into the area from which they had been supplanted. Furthermore, in the one aggressive encounter in which there was no supplantation between these 2 groups, the ST group was clearly the aggressor (table 8). In contrast, the CW group appeared to be slightly dominant over the BN group (table 8). Outside the systematic samples the ST group was once seen to supplant the BN group.

The bi-directional nature of these dominance interactions suggests that parameters other than the identity of the groups affect the outcome. An obvious factor to consider is that of where the encounter occurs. To what extent does the location of the encounter affect its outcome? The extensive overlap in home ranges and the frequent tolerance of intergroup proximity demonstrate that the CW, ST, and BN groups are not territorial among themselves. The location of intergroup conflicts also supports this (figs. 2 and 3). That certain groups are dominant to the others in certain areas is not supported by the data. The ST group supplanted the CW group throughout the home range of the CW group, even in the areas most heavily used by the CW group (fig. 2 and tables 46 and 47). For example, the quadrat (-2 -13) was the second most heavily used by the CW group and yet they were supplanted from it by the ST group. (Specific quadrats are designated first by row and then by column number; see tables 46 and 47.) Furthermore, although the CW group temporarily supplanted the ST group from quadrats (-2 -13) and (-3 -13), 2½ months after this the ST group moved through the CW group in quadrats (-3 -13) and (-3 -12) without any overt interaction. Even more impressive are 2 examples between the CW and BN groups. In quadrats (-10 -16) and (-11 -12) the CW group was seen to supplant the BN group, although on other days the BN group supplanted them from the same quadrats (fig. 3). Clearly, the outcome of an encounter between 2 specific groups of the CW, ST, and BN groups is not rigidly dependent on the location of the encounter. Furthermore, particular areas are not exclusively dominated by one group. The ST group supplanted the CW group from quadrat (-17 -22), but on another day the CW group supplanted the BN group from the same place (figs. 2 and 3).

Conclusions.   On the basis of the preceding data I draw the following conclusions regarding the intergroup relations of *b. tephrosceles* in compartment 30 of the Kibale Forest. It appears that the CW, ST, and BN groups use an area of approximately 50 ha. to the near exclusion of other badius groups and that this exclusion is maintained by aggression. Among these 3 groups, however, relations appear to be independent of spatial parameters. Rather, it appears that the outcome of intergroup encounters is based on intergroup dominance relations, with the largest group (ST) being dominant over the others. These dominance relations are not unidirectional, however, and it appears that the closer 2 groups are in size, the more ambiguous are their dominance relations. Intergroup conflicts are almost exclusively the activity of adult and subadult males, and it is during such conflicts that the cohesion of the males is most apparent. This almost certainly relates to the relatively long-term stability of their membership in the group and the restriction of a stereotyped display (the type II present) to adult and subadult males within the group and the attraction which adult males hold for young males in the group (see chapter 3). The outcome of any given encounter between 2 specific groups may depend on which specific males of the opposing groups meet and/or how many males of each group actively participate. In larger groups, more males are likely to participate than in smaller groups, which would account for the fact

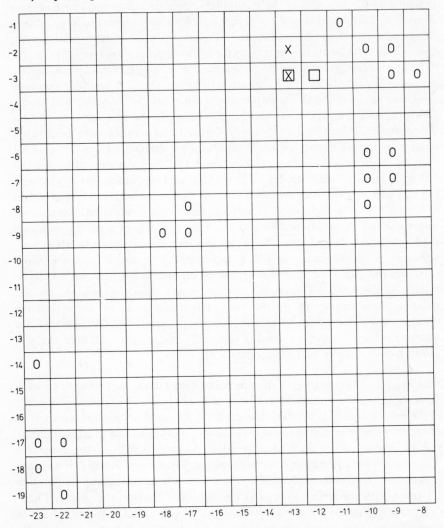

Fig. 2. Location of intergroup conflicts between the CW and ST groups. The row and column numbers permit specific designation of 0.25 ha. quadrats in the study area and are the same as those in fig. 12 and tables 46–49. An X indicates where the CW group supplanted the ST group; an O where the ST group supplanted the CW group; and a square where the 2 groups intermingled but had no conflict.

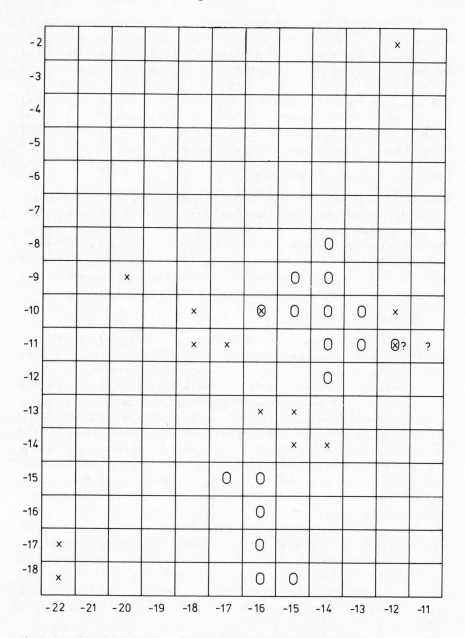

Fig. 3. Location of intergroup conflicts between the CW and BN groups. Row and column numbers as in fig. 2. An X indicates where the CW group supplanted the BN group; an O where the BN group supplanted the CW group; and X and O together indicate where on one day the CW group supplanted the BN group and on another day the BN group supplanted the CW group; and ? indicates that the outcome of the encounter was ambiguous.

that the larger group dominated the majority of encounters. Between groups of similar size the outcome may be more heavily dependent on which particular males happen to meet. This would explain why one group dominates at one time and another group at another time. In cases in which the outcome was unclear, additional males may have joined the encounter after its initiation, thus reversing the direction of the supplantation and chase. Examples of this have been observed between the CW and the BN groups, where one male attacked another group, was chased back to his own group, was there joined by other males, who then reversed the chase again.

Intergroup Relations of Other Subspecies

Only for the *C. b. temminckii* of Senegal were intergroup aggressive conflicts clearly observed. Three cases occurred in the Forêt Classée des Narangs, and all involved much vocalizing (screams, etc.), chasing, and counterchasing. Monkeys fell to the ground during these encounters, and it was common to see 2 or 3 of them clustered together and staring, slapping and screaming toward other clusters of 2 or 3 badius. In addition to these aggressive encounters vocal exchanges between groups were apparent on several occasions.

Comparison with Other Studies

Group size and composition.   Other studies of *C. b. tephrosceles* in Uganda and Tanzania give group sizes ranging from 30 to 82 (Marler 1970, Nishida 1972, and Clutton-Brock 1972) and are consistent with my findings. The most detailed information comes from Clutton-Brock. His main study group in the Gombe numbered 82, and counts of 4 other groups in the same area gave a central tendency of about 55. The 2 groups he studied in the Kibale Forest numbered 64 and 58, also consistent with my results. The ratios of adult males to adult females in the 3 groups studied most intensely by Clutton-Brock were 1:2.55, 1:3.20, and 1:1.62, with several adult males in each group. The ratios were similar to those in my study. The large, multimale groups of red colobus are contrary to the general trend among leaf-eating monkeys of forest habitats. (Eisenberg et al. 1972).

Colobus guereza clearly live in smaller groups than red colobus and generally have only one fully adult male in each group (Ullrich 1961, Schenkel and Schenkel-Hulliger 1967, Marler 1969, Kingston 1971, Clutton-Brock 1972, and Oates 1974). Heterosexual groups range from 3 to 15 in number and average between 8 and 10.4, depending on the location. The ratio of adult males to adult females in these groups is typically about 1:3, but ranges from 1:0.6 to 1:6.

Presbytis entellus, the common langur of India and Ceylon, demonstrates extreme variability in group size and composition. Heterosexual groups range in size from 6 to 120 (Jay 1965, Ripley 1967, Yoshiba 1968, and Vogel 1971). The number of adult males in each of these groups is usually only one, but in some areas there are typically more than 2 per group. The ratio of adult males to adult females in these groups is also variable, ranging from 1:1.5 to 1:5. In all studies of entellus, groups comprised only of males (adults and immatures) have been reported as common. This constitutes a major difference from the 2 African colobinae (badius and guereza), where all-male groups are exceedingly rare.

Both *Presbytis johnii* (Tanaka 1965 and Poirier 1968*a*) and *Presbytis senex* (Eisenberg et al. 1972) tend to live in small groups with only one adult male per group. These groups average about 8 to 15 in number with a ratio of adult males to adult females ranging from 1:1.2 to 1:8. In these respects they resemble guereza more closely than badius.

Presbytis cristatus lives in larger groups than the latter 2 *Presbytis* species, with means of 17.5 and 32 reported by Furuya (1961) and Bernstein (1968), respectively. The 2 groups studied by Furuya had more than one adult male in each group, in contrast to Bernstein's study groups, which typically had only one adult male per group.

Kawabe and Mano (1972) report that the proboscis monkey (*Nasalis larvatus*) lives in multimale social groups averaging about 19 in number. Sex ratios are not given, but the groups contained more adult females than males.

Group dynamics.   Clutton-Brock (1972) found that his main study group of 82 red colobus in the Gombe Park remained relatively constant throughout an 11-month period. All of the recognizable 10 adult males remained with the group. One adult male joined the group and a young male left the group for several months, but later rejoined it. Apparently, no adult females were identifiable.

Only one long-term study has been made of the black and white colobus, and the results indicate that most of the groups have stable adult membership over a period of at least 14 months (Oates 1974).

Group dynamics is another phenomenon in which *Presbytis entellus* demonstrates extremes. Jay (1965) reports very stable groups, whereas Sugiyama (1967) and Mohnot (1971) describe frequent changes in the adult male membership of the group. This male replacement is often followed by the new male killing the young infants in the group. The replacements are the consequence of aggression either by solitary males or by members of all-male groups against the single adult male of the heterosexual group. Sugiyama (1967) reports changes of this kind in 4 out of 9 groups observed for more than one year, and estimates that such changes occur in all groups about once every 3 years. Eisenberg et al. (1972) present evidence indicating that male replacement among *P. senex* is also accompanied by infant killing. They suggest this phenomenon is a response to high population densities and effectively curtails population growth.

Considering other cercopithecidae outside of the colobinae, it appears that immigration and emigration of adults to and from social groups is relatively frequent and common, particularly among adult males, e.g., *Cercopithecus aethiops* (Struhsaker 1967*a* and Gartlan and Brian 1968) and *Papio* spp. (DeVore 1962, Rowell 1966, and Altmann and Altmann 1970). It would seem that the extreme stability in adult membership and the apparent mobility of some immature red colobus is exceptional to the general pattern among the cercopithecidae.

Solitary monkeys. No information is available on solitaries in the other studies of red colobus. Solitary male guereza have been observed on several occasions (Marler 1969, Oates 1974, and see section on interspecific relations in chapter 5). Similarly, solitary males seem to be typical of the other colobinae studied; *Presbytis entellus* (Jay 1965, Sugiyama 1967, and Vogel 1971), *P. johnii* (Tanaka 1965), and *P. senex* (Rudran, pers. comm.). Combined with the data on solitary badius, these observations lend support to an earlier suggestion that solitary monkeys are more typical of forest habitats than of nonforest habitats. This is presumably in response to lower predation pressure and greater ease of concealment in the forest than in a nonforest environment (Struhsaker 1969). If this interpretation is correct, the question remains why solitary red colobus so often associate with other species. Assuming that predation pressures are relatively low in the rain forest, one concludes that the solitary badius do not join groups of other species for purposes of increased predator detection. It is possible that the solitaries gain some advantages in the location of food from such associations. This would explain why they tend to associate most with guereza, whose food habits are more like those of badius than any other primate in the Kibale Forest.

Intergroup relations. Clutton-Brock (1972) observed only 6 intergroup encounters during his 9-month study of one badius group in the Gombe Park. He concluded that "the nature of inter-troop encounters appeared to vary depending on which neighboring troop was involved." These interactions varied from immediate flight by one group from the other, to aggression between males of the groups, to tolerance at close quarters. The qualitative nature of Clutton-Brock's observations in the Kibale Forest is similar to that of his Gombe Park study, although he found that intergroup encounters were much more frequent in the Kibale Forest. These observations are consistent with mine. The lower incidence of intergroup encounters at Gombe may be related to the lower population and group density of red colobus there.

None of the other Colobinae studied seem to have intergroup relations like those described for red colobus. Black and white colobus groups are usually described as territorial (Ullrich 1961, Schenkel and Schenkel-Hulliger 1967, and Marler 1969). However, Oates (1974), in the longest study of guereza, found that although the social groups had preferred areas in their home ranges and would chase other groups from these areas, no area was used exclusively by one group, exclusive use being an essential feature of territoriality. *Presbytis entellus* is extremely variable, with some populations exhibiting territoriality (Sugiyama 1967), some having territories but much overlap in home ranges (Ripley 1967), and others showing extreme intergroup tolerance in the absence of any aggression (Jay 1965). Social groups of *P. johnii* (Poirier 1968c), *P. cristatus* (Bernstein 1968) and *P. senex* (Rudran 1973b) are all described as territorial.

Clearly, the intergroup relations of red colobus monkeys are unique among the colobinae studied to date. In this respect, red colobus appear more like rhesus monkeys. Altmann (1962) called attention to intergroup dominance relations between groups of rhesus (*Macaca mulatta*) in the artificial colony on Cayo Santiago. He stated: "Almost invariably, the larger group (Group 2) displaced the other group, even from the smaller group's center of activities." Similar observations have been made on Cayo Santiago by Boelkins and Wilson (1972). Hausfater (1972) found for the same population that larger groups often dominated smaller ones, but that some of the smaller groups dominated larger ones. He also found that the dominance relationships between some groups were not deterministic or unidirectional, but that on some occasions one group would dominate the other and on other occasions be dominated by it. Similar relations have been described for rhesus monkeys living in urban areas of North India. Southwick et al. (1965) found a pronounced intergroup dominance hierarchy, with one particular group being dominant to all others in all locations. However, the intergroup dominance status between 2 of the subordinate groups depended on the location of the encounter.

3 Social Behavior and Intragroup Social Structure

This chapter centers on *badius tephrosceles* and, in particular, the CW group. Information for the other subspecies is usually incidental.

The frequency data are of limited value as measures of absolute occurrence because observation conditions in the rain forest are far from ideal and the observer probably misses a great many social interactions. Comparison of the frequency of occurrence of a particular kind of social encounter in a rain-forest species with that of a species living in semiarid conditions is of dubious validity. However, comparison of rates of behavior between different species living in the same rain-forest habitat is much more meaningful, especially if the species compared are equally observable.

Methods

The majority of social behaviors included here were of relatively short duration, lasting only a few seconds or less. Detailed qualitative descriptions were made in the field of every social interaction observed. With very few exceptions these descriptions were continued throughout the entire study for all social behaviors. This was done in order to avoid a premature classification of behaviors, which might bias my observations and "force" the data into unwarranted categories. For example, if I had prematurely classified all behaviors in which one monkey oriented his posterior to another monkey simply as *presents,* I would have artificially and unjustly lumped 3 distinct behavior patterns into one. Only after the first 12 months of study at Kanyawara and only with those behavior patterns which, after that period, proved to be extremely stereotyped did I begin to tabulate "kinds" of behavior patterns rather than continue making detailed descriptions in my field notes. In the final analysis I attempted to categorize all of the descriptions of communicative gestures on the basis of their motor patterns. Ideally I wanted this to be a catalogue of natural units of behavior, much like that described by Altmann (1962). I have not, however, divided my list of social behavior patterns as finely as did Altmann. The behavior patterns in my catalogue are distinguished from one another by the uniqueness of one or more elements in the motor pattern or by the sequential arrangement of these elements. If a sequence of motor patterns is always or nearly always given together, such a sequence is regarded as one behavior pattern.

A social interaction was operationally distinguished from other similar interactions by a clear temporal and physical break in the chain of events or by a change in participants. Such distinctions usually posed few problems. For example, sexual mounts were separated from other sexual mounts between the same partners by a dismount. Aggressive encounters were more complex than this, but usually there was a clear beginning and end. Aggression occurred so infrequently that most bouts were usually separated by at least one hour. Grooming bouts proved exceptional. Because grooming between a particular pair of monkeys often continued for several minutes, it was not possible to quickly determine when the bout began and terminated or whether the grooming roles were doubly reversed between the same two monkeys unless one kept them under continuous observation. The systematic sampling of feeding and general activity at 30-minute intervals usually precluded prolonged observation of these grooming bouts (see chapter 5). Consequently, for quantitative purposes I distinguished grooming bouts on the basis of the following criteria: (1) a change in partners, (2) a change in the direction of the grooming action, or (3) a time interval of at least one hour between grooming bouts involving the same pair of monkeys with the same direction of grooming action. For example, if monkey A groomed monkey B, this pair would not be scored again until one hour had elapsed or until B groomed A. Essentially, then, all social interactions were tallied in an opportunistic manner whenever they occurred, with the exceptional qualifications for grooming bouts.

Resumé of Prominent Aspects of Intragroup Social Structure

Within the group there is a pronounced dominance hierarchy, which is expressed through priority of access to space, food, and grooming position. The adult males are dominant to other members of the group. Low-ranking monkeys give the present type I to higher-ranking individuals. The most dominant adult male does by far the most copulation in the group. The stylized present type II is given only among the adult males, with the higher-ranking male presenting to lower males. Adult males are groomed most and groom the least, whereas adult females groomed others more than they were groomed. There is a strong cohesion among the adult males of the group as evidenced by their united effort during intergroup conflicts and by their long-term membership in the group. The type II present may not only function to reinforce the dominance relations between the adult males, but may also enhance this male-male bond. Immature monkeys of the young juvenile and juvenile class frequently harass adult males by approaching them and then jabbing at them or by pulling on their tails or limbs. This is often ignored by the adult males. Conversely, old juvenile and young subadult males are persistently harassed by the adult males of the group. It is possible that young males leave the group at the time of this harassment or, as occurred with one male in the CW group, become the dominant male.

AGONISTIC BEHAVIOR

Behavior patterns and social interactions which either involved physical harm or a high likelihood of such harm can be broadly separated into low-intensity and high-intensity aggression. Such a classification, although artificial, does give some indication of the energy expended by the participants and of the probability that physical harm will be involved.

Low-Intensity Aggression in *C. b. tephrosceles*

Grimace.   This gesture was infrequently seen, and during the period of May 1970 through March 1972 only 10 examples were clearly observed. In all 10 cases the unclenched teeth were exposed, with the mouth open, the lips retracted, and the mouth corners drawn upward. The mouth was opened and closed rapidly, with the teeth being exposed only briefly. Only one grimace was given in 5 cases, and more than one grimace was given in rapid succession in the other 5. There were no vocalizations in 9 cases; in one, an adult female chisted (see chapter 4). Adult males grimaced in 2 cases, a subadult male once, adult females 6 times, and an approximate adult once. In 6 of the examples the grimace was directed toward me. Once an adult female grimaced at a young juvenile suckling from her. In another case a subadult male grimaced as he presented in the type I manner to an adult male. In two cases the grimace was given immediately prior to a grooming bout in which the grimacer was the groomer in one case and the groomee in the other. In one of these latter 2 cases an adult female grimaced toward an approximate adult, and in the other an adult female grimaced toward another adult female. It would appear that the grimace is a gesture of submission and nonaggressive intent.

The grimace differs in form from the gape (see below) in that the teeth are always exposed and much closer together (nearly clenched) in the former. In the gape the head and forequarters are extended slightly forward, which is not so with the grimace. Furthermore, the mouth corners are drawn upward in the grimace, in contrast to the puckered shape of the mouth during the gape.

Touching.   Only 3 examples of this pattern were observed, which probably represent a gross underestimation of its actual frequency of occurrence. All examples involved adult and old subadult males and occurred immediately after high-intensity aggressive encounters. In one case an adult male branch-shook, then approached another adult male and sat beside him; at this, the other male placed one hand on the displaying male's side for about 5 seconds. Following a multipartite aggressive encounter one adult male groomed another adult male while the groomee placed one hand on the forearm of the groomer. The third example also followed a multipartite aggressive encounter. In this case the subadult male (SAM) of the CW group chased adult male ND, at the end of which ND presented type I to SAM and at the same time reached back with one hand and patted SAM on the leg. Touching seems to be a gesture of nonaggressive intent or pacification.

Dangles beneath.   Only 8 examples of this behavior pattern were observed — 6 from the CW group and 2 from the ST group. In this pattern the displayer hangs by 2 or 4 appendages directly beneath the recipient. The displayer squealed during 4 of the cases, screamed in one, and gave no vocalizations in 3 cases. The recipient gave wheets in one case and chists in another. This display was always given toward a larger animal, with adult males being the recipient in 6 cases, an adult female once, and a monkey of undetermined age and sex in another. All but one of these displays were given by immature monkeys: subadults 4 times, juveniles twice, a young juvenile once, and an adult female once. It seemed to be an appeasement gesture employed either when the immature monkey was harassing an adult male or when it was being harassed by an adult male (see below).

Rapid glance.   In this pattern the sender looks toward and then away from the recipient in rapid succession. It is a jerky motion, and when several are given in succession the head is essentially flagged. This gesture is sometimes incorporated into the type I present and is usually given in supplantations by the supplantee (see below). I consider it to be a submissive gesture.

Supplantation.   A supplantation is a social interaction in which one monkey approaches another, whereupon the approached monkey moves away and the supplanter thereby gains access to the space, food, or grooming position occupied by the supplantee. The supplanter approaches with a normal walk and employs no special motor pattern, although sometimes it approaches with a rapid walk. The supplantee moves away quickly and usually makes rapid glances toward the supplanter as it does so. In some cases the supplantee actually leapt away from the supplanter. Infrequently, the supplantee gave the wah! call as it leaped aside. The supplanter rarely vocalized after the supplantation, although adult male supplanters sometimes chisted.

Ten supplantations were observed in at least 3 groups other than the CW group. Adult males were the supplanters 7 times, an adult female once, and an approximate adult twice. The supplantees in these ten cases were: an approximate adult 4 times, adult females 3 times (2 females were simultaneously supplanted by one adult male), an adult male once, a juvenile twice, and a young juvenile once.

Twenty-six supplantations were observed in the CW group (table 9). In the first 9 months, adult male CW was clearly the prime supplanter and, by definition, the dominant animal in the group. In the second half of the study, however, his supplantations were greatly reduced in number and shared about equally with the subadult male (SAM). This apparent change was correlated with the attainment of sexual maturity and rapid increase in body size of SAM (also see section on heterosexual behavior). SAM was first seen to mount an adult female on 2 August 1971. Thereafter, he was clearly in direct competition with male CW for the number one dominance position. In this second half he was never supplanted by CW and even succeeded in supplanting CW once. The results demonstrate that adult males infrequently supplant one another and usually supplant adult females and immatures (compare the results of table 9 with the weighted mean group composition in table 7). However, analysis of the 2 periods summarized in table 9 indicate that in the second half, at a time when SAM was undergoing rapid physical and sexual development and challenging male CW's role as the dominant animal, there were more supplantations among the adult males than would be expected from their proportional representation in the group. Although the data are few, they suggest that, at a time of contest and relative instability in the hierarchy, supplantations among the adult males increase in frequency. This further suggests not only that supplantations are related to the acquisition of some prized item but also that they may function to reinforce dominance relations. In fact, of the 8 supplantations which involved only adult and/or subadult males, 6 were supplantations over space, one was over an estrous adult female, and one was over a grooming position. This contrasts with the remaining encounters, in which the supplantee was neither an adult nor subadult male. Thirteen of these were over space, 11 over food, and 4 over grooming position.

Present type I (fig. 4).   In this gesture all 4 limbs of the presenter are flexed, but the forelimbs are flexed more and thus the posterior is elevated. The tail is elevated and diverted to one side of the body, exposing the perineum. The posterior is directed toward the recipient, and the presenter looks over one shoulder toward the recipient. Usually there is no physical contact between the sender and recipient, but in one exceptional case the sender reached back with one hand and patted the recipient on the leg (see *Touching,* above). This present is typically given while the sender is standing on a horizontal tree branch, but infrequently the sender presents while clinging to a vertical branch. The present type I is of relatively short duration, about 2 seconds or less. The sender and recipient were usually separated by 0.5–2.5 m. The maximum distance was estimated at 4.5 m. The sender gave shrill squeals in only 6.5% of the cases and did not vocalize in the remaining 93.5%.

The preceding description is based on 123 examples, 104 of which were from the CW group, 8 from the ST group, one from the BN group, and 10 from groups of undetermined identity.

The sender was an adult male once; an adult female 19 times; a subadult male 5 times; an approximate adult 5 times; an old juvenile 11 times; a juvenile 60 times; a young juvenile 15 times; an old infant 6 times; and an individual of undetermined age and sex once. The recipients were comprised of the following age-sex classes: adult male 86; adult female 11; subadult

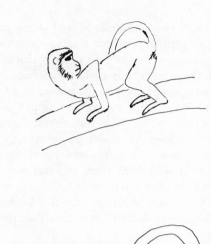

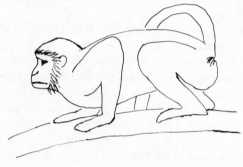

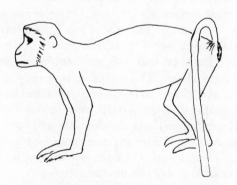

Fig. 4. Three types of present in *C. b. tephrosceles*. Top to bottom: a juvenile giving the present type I, an adult male giving the present type II, an adult female giving a sexual present.

male 20; approximate adult once; juvenile 3; and young juvenile 2. Clearly, adult males are the major recipients of this display, which is given primarily by adult females and immature monkeys, especially juveniles.

Immediately prior to the present type I, the sender approached the recipient in 68 cases; the recipient approached the sender in 12 cases; the sender harassed the recipient while the latter was copulating in 4 cases; the recipient chased the sender in 4 cases; the recipient threatened the sender with a stare in 3 cases; the recipient harassed the sender by persistently following him in one case. The preceding event was not noted in 31 cases. Following the present type I the sender moved away from the recipient in 85 cases; the recipient moved away in 2; neither moved away in 20; the recipient chased the sender in 2. The subsequent event was not noted in 14 cases. In all cases the situation appeared to be agonistic, and I conclude that the present type I is an appeasement gesture.

A detailed analysis of the data from the CW group supports this latter conclusion and further suggests that the present type I is given by subordinate monkeys to dominant ones (compare table 10 with 9 and with the weighted mean group composition in table 7). Furthermore, as SAM attained sexual and physical maturity there was an apparent change in the frequency with which specific animals were presented to. In the period prior to SAM's maturation, male CW received 66% of the presents, but in the second half of the study he received only 23%. The number of presents type I that SAM received increased from zero in the first half to 44% in the second half (table 10). This is consistent with the changes in supplantations which were described and discussed above, and it also seems clearly related to SAM's rise in the hierarchy.

Present type II (fig. 4). In this gesture the presenter flexes all 4 limbs nearly equally so that his body is lowered, but neither his hind nor his forequarters are elevated. His ventral surface actually touches or nearly touches the tree branch on which he is standing. The sender's tail is typically elevated and diverted over his back and to one side. His posterior is oriented toward the recipient. He does not, however, look over his shoulder toward the recipient, but instead looks straight ahead. The recipient is usually seated and sometimes places his hands and/or feet on the sender's hips, tail base and/or dorsal sacrum. Exceptions to this general pattern were infrequent. For example, the sender's tail was once flopped over the shoulder and onto the back of the recipient. In another case the sender extended one foot along the side and then onto the back of the recipient. Once the sender bent his head down and looked between his arms and legs toward the recipient.

The distance between the sender and recipient is usually much less than in the type I present. In the 48 examples of the type II present the sender and the recipient were in physical contact 31 times, were within 0.3 m 7 times, 0.5 m apart once, one m apart once, and their distance was not noted in 8 cases.

There were no vocalizations during 41 of the type II presents, the presence or absence of vocalizations was not noted in 6 cases, and in only one case did the sender chist while presenting.

This present is primarily the activity of adult males, and of the 48 examples the sender was an adult male 44 times, a subadult male twice, an adult female once, and a juvenile once. The recipient was an adult male 46 times, an adult female once, and a juvenile once.

During the present type II the recipient placed one or both hands and/or feet on and gripped the posterior of the sender in 20 cases; handled the sender's perineum in 3 other cases; neither handled nor gripped the sender's posterior in 24 cases; and in one case his action was not noted. In one case the recipient also placed his muzzle next to the perineum of the sender.

Immediately prior to the present type II the sender approached the recipient in 34 cases and the recipient approached the sender in 9. The approach was not noted in 5 cases. In general, there was no obvious interaction or stimulus evoking the type II present. However, on at least 3 occasions the present followed the harassment of the sender (an adult male) as he was copulating with an adult female, the harassment being done by the recipient to be (another adult male). In another case the present type II occurred immediately after an intergroup supplantation. On two occasions the sender gave a branch-shake display prior to presenting, and once an adult male was presented to after he had given a branch-shake display. In yet another case the potential recipient rushed toward the potential sender as he was chasing an adult female. This adult male ceased chasing the adult female and ran toward the approaching adult male and presented to him.

Grooming followed 46% of the type II presents. The recipient groomed the sender in 14 cases, the sender groomed the recipient in 7 other cases, and in another example the sender and recipient groomed one another in turns.

Analysis of the data for the present type II in the CW group clearly reveals that the dominant males present to the subordinate males (compare table 11 with 9). Furthermore, this display and dominance status are both positively correlated with frequency of copulation (see section on heterosexual behavior). These data suggest that the type II present is a display used primarily to reinforce the dominant position of high-ranking males. In addition, it may also enhance the bond between adult males within the group, especially when given at times of imminent physical aggression between the males. In other words, the present type II may, at times, act as a substitute for inter-male physical aggression.

High-Intensity Aggression (*C. b. tephrosceles*)

Branch-shake, branch-bounce, and leaping about. These 3 displays are given in similar stimulus situations, and the branch-shake and leaping-about displays are sometimes given in sequence. All 3 probably involve similar motivational states.

During the branch-shake display the arms and legs are flexed and extended in rapid succession while the displayer remains fixed on one branch, often near the end of the branch. The number of flexion-extension sets varies from one to more than 6. This flexion-extension results in the shaking of the branch and creates noise, particularly if the branch is foliated. Usually the monkey is on a horizontal branch while giving this display, but infrequently the display is given from a vertical branch. On one occasion an adult male grasped a branch on which he was seated and shook it by rocking back and forth while retaining his seated position. Of the 48 branch-shake displays described, 28 were without vocalizations, in 9 the displayer chisted, in 4 he gave wheets, in 2 barks, in one case rapid-quavers, and in 4 cases it was not noted whether he vocalized.

In the branch-bounce display the monkey squats bipedally and bounces up and down on horizontal and limber branches. The hands are free and the feet usually maintain contact with the tree branch. Infrequently the feet actually leave the branch as the displayer hops up and down in one place. In one exceptional case the displayer gripped a vertical branch with both hands and bounced his feet off this branch in rapid succession. In the 7 examples of branch-bouncing, wheets were given twice, there were no vocalizations in 3 cases, and the presence or absence of vocalizations was not noted in 2 cases. I consider the branch-bounce display to be a variation of branch-shaking.

The leaping-about display consists of a form of rapid locomotion in which the displayer essentially bounds from branch to branch. His limbs are flexed and extended much more vigorously than in typical rapid climbing, and this motion shakes every branch he hits. The displayer's back is arched convexly and he also appears to piloerect, both of which tend to increase his apparent body size. The displayer travels as far as 20 m during the leaping-about display. In some cases he drops vertically from branch to branch, creating a great noise in the process. During a total of 17 examples of leaping about, the displayer gave chists 8 times, wheets twice, no vocalizations 4 times, and in 3 cases it was not noted if he vocalized.

In the total 72 displays of branch-shaking, branch-bouncing and leaping about, 10 were given toward me, 4 toward adult males, 3 toward adult females, one toward a juvenile, two toward a young juvenile, one by an adult male after an adult female presented sexually to him, 12 by adult males after being harassed during copulation, one while a male was contesting for leadership of a group progression (see chapter 4), one after a male had been groomed by an adult female, one in apparent response to the branch-shake display of another, two during play, 9 during intergroup agonistic encounters, 3 toward another proximal group of badius, 2 toward a large raptor, 2 in response to the alarm calls of *Cercopithecus ascanius*, one in response to the loud noise of a falling tree branch, and in 17 cases neither the stimulus nor the recipient was determined.

Forty-six of the 72 displays were observed in the CW group. During the

period of August 1970 through 1 August 1971, 20 were given by adult male
CW, 2 by adult male LB, 2 by adult male ND, and one by an adult male
whose identity was not determined. In the second half of the study (2 August
1971 through March 1972), while SAM was attaining sexual and physical
maturity, adult male CW gave only 9 of these displays, SAM gave 9, adult
male LB one, adult male ND one, and one was given by a young juvenile in
play. The remaining 26 displays were observed in at least 4 other groups of
badius. Twenty-two of these were given by at least 10 adult males, one by an
approximate adult, one by an adult female, one by a subadult female, and
one by a young juvenile in play.

The data clearly suggest that these displays are threat gestures, given
primarily by high-ranking adult males. In the CW group the performance of
these displays by the dominant adult male (CW) diminished as SAM reached
maturity and rose in the hierarchy.

Branch-drop.   Only 4 cases of this threat display were observed during the
entire study. In all examples the displayer broke loose a dead tree branch and
dropped it toward me. In one case the displayer used both hands to shake
loose a dead branch that was lodged in the fork of a tree. In another, the
displayer grabbed a dead branch, drew it back, and then threw it forward
and toward me in an underhand manner. In addition to these 4 cases,
adult male CW twice tried to break loose a dead branch in what I presume
were attempted branch-drop displays. In both cases he vigorously pulled on
and bit a dead branch that was loosely attached to the tree. Adult males
performed this display in all 6 cases (male CW 3; male ND 1; and unidentified adult male 2). Chist calls were given during 4 displays, wheets during 2,
and wheets and chists during one. Five of these displays were directed toward
me and one toward another badius group.

Gape (fig. 5).   Nineteen examples of this threat gesture were noted: 11 from
the CW group, 2 from the ST group, and 6 from groups of uncertain identity.
In this gesture the mouth is opened and very slightly puckered into a gape and
held that way for about 1–2 seconds. The lips are not retracted and the mouth
corners are not uplifted. In all cases the teeth were unclenched. In 15 cases
the teeth were not exposed; in 2 cases they were slightly exposed; and in 2
other cases they were exposed. The aperture of the open mouth was judged to
vary from about ⅓ to fully open. As to the position of the displayer's body, the
forequarters were extended toward the recipient on 8 occasions. In 2 cases the
sender lunged or leapt toward the recipient, and in the remaining 9 cases the
position of the sender's body was not noted. In addition, the sender grabbed
or slapped toward the recipient without making contact on 4 occasions;
grabbed and held the recipient once; and gently poked the recipient with one
hand on one occasion. In all cases the sender stared toward the recipient. In
18 cases there were no vocalizations. Once the displayer, adult male ND, gave

2 wah!s toward the recipient, adult male CW, who then chased ND. Three
times the recipient and the displayer were in physical contact. In 3 other cases
the distance between them varied from 0.5 to 1 m, and in 13 cases the
distance was not noted.

The displayer was an adult male in 7 examples, an adult female 10 times,
an approximate adult once, and a juvenile once. Twice the recipient was an
adult male; 6 times, a juvenile; twice, 2 juveniles simultaneously received the
display; once, an approximate adult and a juvenile were simultaneously
displayed to; 6 times a young juvenile was the recipient, and twice, an old
infant.

Immediately prior to the gape display the recipient harassed the displayer
on 2 occasions as he copulated; there was an agonistic encounter of undetermined nature in 3 cases; the displayer approached the recipient in 2; the
recipient approached the displayer in 2; the recipient dangled and gamboled
in front of the displayer in 5 (in 3 of these cases the displayer was an adult
female with a young, dark infant); the recipient played with the displayer
once; the recipient groomed the displayer twice; the recipient chased the
displayer once; and in one example both recipient and displayer fed immediately prior to the display.

Immediately after the gape display, the recipient moved away in 8 cases;
the recipient chased the displayer twice; they both departed once; the
displayer repeated the gape once; once the recipient gave a type I present;
displayer and recipient remained together after one display; they embraced
once; the recipient suckled from the displayer once; the displayer groomed
the recipient once; and in 2 cases the subsequent event was not noted.

Although the stimulus situations evoking the gape were quite variable and
ranged from very mild to very intense aggressive encounters, it seems reasonable to conclude that the gape is a defensive threat whose function is to
maintain distance between the displayer and the recipient.

Two encounters were observed which I consider to have involved variations
of the gape. In one of these an adult male rapidly approached another adult
male and opened his mouth widely with his unclenched teeth exposed next
to the shoulder of the other male, who was seated. He held this open-mouth
position for at least 3 sec., during which the recipient male lowered his
head and gave chist calls. The recipient then groomed the displayer.

In the second variation a young juvenile approached and sat in contact
with an adult female who was seated. The adult female immediately turned
and made a biting motion toward the young juvenile, who showed no response. This biting motion appeared as an extremely rapid variation of the
gape.

Stare with extended forequarters (fig. 5 and pl. 9). This aggressive threat
gesture was observed only 16 times (13 in the CW group, 2 in the ST group
and one in an unidentified group). In this display the limbs are flexed, but

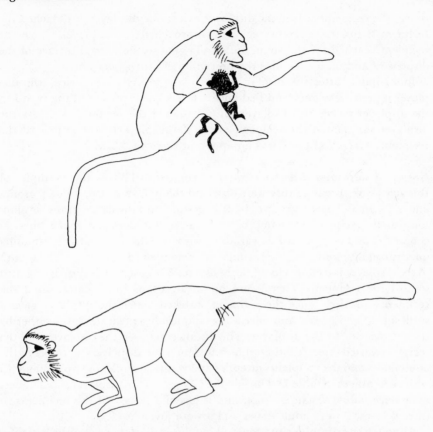

Fig. 5. *Colobus badius tephrosceles*. An adult female, holding a newborn infant, gapes and slaps toward another monkey. An adult male stares with extended forequarters.

the arms more so than the legs, resulting in a forward extension and slight lowering of the forequarters and head. The head is slightly lower than the shoulders, and the displayer stares toward the recipient. The tail is held in a normal position, and only once was it arched over the back. In 10 examples the mouth was held closed; once the lips were slightly parted, but the teeth were not exposed; and in 5 cases the mouth position was not noted. There were no vocalizations in 13 displays, wheets in one, rapid quavers in 2, and wahls in one. The distance between the displayer and the recipient was not noted in 12 cases but varied from 1 to 3.5 m in the other 4.

Adult male CW gave this display 5 times, adult male ND twice, adult male LB 3 times, and unidentified adult males and females 3 times each. The recipient was adult male CW twice; SAM once; adult female BT once; an unidentified adult male and female once each; an approximate adult once; juveniles 3 times; young juveniles 5 times; and an unidentified monkey once.

Immediately prior to this display the recipient harassed the displayer 4 times; the recipient chased the displayer twice; the displayer approached the recipient 3 times; once the recipient approached the displayer; once the recipient reached for the young infant clinging to the ventral surface of the displayer; and on 5 occasions the preceding event was not noted.

Immediately after the display the recipient moved away twice; the displayer moved away once and fled once; the recipient presented the type I to the displayer twice; the displayer lunged toward the recipient 4 times; the displayer supplanted the recipient once; the displayer grabbed toward the recipient twice; and there was no apparent event in 3 cases.

Grab and slap toward without physical contact.   Thirty-nine examples of this aggressive threat gesture were observed (33 in CW group, 4 in ST group, and 2 in an unknown group). In this gesture the displayer grabs or slaps toward the recipient with one or two hands but does not make physical contact. The body posture is variable, with the displayer sitting, standing quadrupedally, squatting, or staring with extended forequarters. The mouth of the displayer is usually closed, although on 3 occasions the displayer gaped toward the recipient. There were no vocalizations in 33 cases; in 2 the recipient squealed; once the recipient shrieked; once the displayer gave a shrill squeal and a wah! and once a low-amplitude squeal was given either by the recipient or by the displayer. The distance between the displayer and the recipient varied from less than 0.5 m to 1 m. The displayers were primarily adult males and the recipients juveniles and young juveniles (compare table 12 with percentages in table 7). Outside the CW group the displayer was an adult male once, adult female 3 times, and a juvenile twice. Of the recipients in these 6 cases, 2 were adult males and 4 were juveniles.

Although the stimulus situation evoking the grab or slap was quite variable, when one considers only those cases in the CW group where the recipient was a juvenile or young juvenile (75.8% of the total cases) it is apparent that harassment of the displayer by the recipient is the chief stimulus. In all but 3 of these 25 examples the recipient either harassed an adult male (see below) or tried to touch a young and dark infant that was clinging to an adult female immediately prior to the grab or slap by the harassed male or female. In 2 of the remaining 3 cases the displayer used this gesture to supplant the recipient from food, and in the final case a young juvenile gave a branch-shake display (probably in play) next to an adult male, who then slapped toward the young juvenile. Usually the recipient flees from the displayer after being grabbed or slapped toward.

The data for the CW group indicate that dominant monkeys usually slap or grab toward their subordinates (table 12). In the one exceptional case a young juvenile or old infant climbed into the lap of adult male CW and then began to squeal and squirm about, whereupon an adult female with another young juvenile clinging to her ventral surface approached and slapped toward male CW. The young juvenile then fled out of CW's lap and CW leapt away and chisted.

Lunge toward or chase.   In a lunge, one animal rushes toward another in a fast run. A chase is virtually the same motor pattern, but is of longer duration and covers a greater distance. Lunges and chases are both distinguished from supplantation by the more rapid approach of the displayer. Eighty-four social interactions involved chases or lunges. Fifty-nine of these were in the CW group (table 13), 8 in the ST group, 2 in the BN group, and 15 in other groups of unknown identity. There were no vocalizations in 50 cases, but there were some in 34. In most of these 34 encounters it was not determined who vocalized; but it was definitely the recipient that squealed in 4 cases, screamed in one, and gave a wah! in one, and the chaser that gave chist calls in 2 cases, wheets in one, and barks in one. The chase or lunge resulted in definite physical contact between the chaser and chasee in only 5 encounters.

The great majority of lunges and chases were performed by adult males and usually against other adult males and immature monkeys (compare table 13 with percentages in table 7). In groups other than the CW group the relationship was much the same, with 68% of the chasers being adult males and 28% unidentified as to age or sex. In these other groups, juveniles were the recipients in 24% of the cases, adult males in 24%, adult females in 12%, young juveniles in 4%, and the recipient was unidentified in 36%. Nine of the 84 encounters were multipartite, involving either more than one chaser or more than one chasee.

Immediately prior to chases or lunges in the CW group, the recipient harassed the chaser as he copulated in 19 cases; the preceding event was not determined in 24 cases; the recipient harassed the chaser in a nonsexual context (see below) in 7 cases; in 5 cases either the chasee or chaser was involved in an aggressive encounter with at least one other monkey prior to the chase; in one case the chase was apparently precipitated by the approach of another badius group; and in 2 cases the chasee stared toward the chaser in a threatening manner prior to the event. The recipient fled in all cases.

The data for the CW group demonstrate that dominant monkeys usually chase or lunge toward subordinates. In one notable exception, adult males LB, ND, and SAM joined forces and simultaneously chased the dominant male CW. This was in September 1971, approximately one month after SAM had performed his first heterosexual mount—a time when the dominance hierarchy among the adult males of the CW group seemed unstable. In February 1972 SAM chased adult male CW unassisted by the others, further indicating his rise in the hierarchy.

Aggressive physical contact.   Physical contact was made in only 21 aggressive encounters (15 in the CW group and 6 in at least 3 other groups). The nature and frequency of these contacts were as follows: push with one or both hands 4 times; push and slap with one or both hands once; shake with one or both hands 3 times; slap once; grab 4 times; jab with one hand twice; kick once; maul once; bite 3 times; and once the contact was not clearly seen. None of these contacts resulted in any visible wound. There were no vocalizations in 4 cases; the recipient sqwacked in 2 cases, screamed in one, squealed in 7, gave wah!s in 3, gave rapid quavers and wah!s in one; and it was not noted if vocalizations occurred in 3 cases.

The aggressor was usually an adult male or an adult female, and the recipient an immature monkey. In the CW group the aggressor was: adult male CW once; adult male LB 8 times; adult male ND once; an adult female 4 times; and a juvenile once. In other groups the aggressor was an adult male 3 times; an approximate adult twice; and undetermined once. The recipient in the CW group was adult male LB once; a subadult once; an adult female twice; a juvenile 4 times; a young juvenile 4 times; and an old infant 3 times. In other groups the recipient was an adult male once; a juvenile twice; a young juvenile twice; and undetermined once.

Aggressive physical contact is a relatively rare event in the social life of red colobus, and when it does occur the aggressor is usually considerably larger than the recipient. This may be why there was never any retaliation in these encounters, and it may also be the means whereby serious physical aggression is avoided.

Multipartite Aggressive Encounters

The majority of agonistic interactions involving more than 2 monkeys occurred in the intergroup conflicts. Multipartite aggressive encounters within the group were prevalent only during the harassment of a copulating pair by adult males and immature monkeys. Outside of the sexual-harassment situation, participation by more than 2 badius in intragroup aggression was extremely rare.

Redirected aggression. Only 2 cases of redirected aggression were observed. In one of these an adult male who was examining the perineum of an adult female suddenly left her and chased another adult male. What was presumed to be the same male soon returned to the female and began to examine her perineum again. However, she soon fled from him, and he then chased a young juvenile, caught and shook it. I concluded that this male redirected his aggression away from the other adult male and/or possibly the adult female and toward the young juvenile.

In the second case an adult male chased a subadult male of the same group during an aggressive encounter with another group. This seemed to be a case in which the adult male redirected his aggression away from the other group and toward a subadult male in his own group.

Solicitation for aid. Solicitation for aid refers to a social interaction in which an animal that is being attacked or threatened by another solicits the aid of one or more other animals against its aggressor. The soliciting monkey usually effects this by looking rapidly back and forth between its aggressor and the monkey whose aid it is soliciting, or it runs to the side of the monkey that it is soliciting. Most of the badius multipartite aggressive encounters probably involved some solicitation of this kind, though we have only 2 clearly observed examples of it. In one case a juvenile approached and pulled the tail of a young juvenile. The young juvenile squealed and looked toward a third badius. The first juvenile then looked in the same direction and fled immediately. The second case involved a subadult who fled from the scene of a vocal agonism to the side of adult male ND. ND looked toward the subadult's aggressor and gave chist calls, and the aggression terminated. Although solicitation for aid undoubtedly occurred many more times than this, the impression remains that among badius this kind of social interaction is much less frequent than among some of the other cercopithecids, such as *Cercopithecus aethiops*.

Intervention in aggression by adult males.   This refers to a social interaction in which an adult male runs to the scene of an aggressive encounter involving other monkeys and by so doing terminates the aggression. Again, this may have occurred in many of the multipartite agonistic encounters, but only 2 cases were clearly observed. In one case adult male ND was grooming adult female BT, but when a fight accompanied by screams and chasing broke out in the group he left her immediately and ran toward the encounter, which then terminated. The second case was more complicated and seemed to involve 2 interventions. Adult male CW ran toward the scene of a vocal agonism and then chased adult female BT, whereupon adult male ND ran toward them. CW then ceased chasing BT and ran toward ND and presented type II to him, thus terminating the aggressive encounter.

Harassment during copulation.   Two types of harassment occurred during copulation. One type was strictly vocal and involved no physical contact during the actual harassment. This vocal harassment was performed only by adult males and is considered to be a form of intrasexual selective pressure because it often interrupted the copulation before its completion. The vocalizations were usually given at distances of 5 m or more from the copulating pair and included wahls, rapid quavers, and screams. Vocal harassment occurred in 14 of 108 bouts of copulation in the CW group.[1] Only two cases of such harassment occurred during the first half of the study (August 1970 to 1 August 1971), whereas 9 occurred in the second half (2 August 1971 through March 1972) (table 14). The increase in the second part of the study again correlates with an apparent instability in the dominance hierarchy of the CW group brought about by the maturation and rise in the hierarchy of SAM.

Only 2 vocal harassments by adult males were observed in a total of 66 copulation bouts in at least 5 other groups.

The second type of harassment was performed by immature monkeys and often involved physical contact. It sometimes occurred simultaneously with the vocal harassment of adult males but was obviously more common (table

14). On occasion, two immature monkeys harassed a copulating pair together. The nature of these harassments was quite variable, but in general the harasser would run toward the copulating pair and then bounce round about them and slap toward or actually grab the adult male. Physical contact was made in 15 of the 23 harassments involving immature monkeys in the CW group and in 2 of the 6 harassments in other badius groups. The contact always occurred while the copulation was in progress. During this physical contact the harasser did one of a number of things to the adult male mounter, including: hitting and pulling his head; slapping him in the face; grabbing his head and then his hips; muzzling and handling his perineum; climbing onto the mountee's back and then, while holding the mounter's head with both hands, placing its own muzzle in contact with the mounter's muzzle; thrusting its head into the mounter's chest and effecting dismount; bounding off the mounter's back and then grabbing the mounter's head with both hands and twisting its own body so as to pull the mounter off the female; and in the most mild cases merely reaching out with one hand and touching the mounter. Considering the average composition of the CW group it would appear that young juveniles tend to perform this kind of harassment slightly more often than older juveniles (compare tables 7 and 14).

Although both types of harassment could effectively disrupt the copulation, a chi-square test demonstrates that there is no significant difference in the number of harassments occurring in successful and unsuccessful bouts of copulation. Successful bouts were those in which there was a distinct pause at the termination of pelvic-thrusting, presumably when ejaculation occurred (see section on heterosexual behavior). Of the 108 copulation bouts observed in the CW group, 27 were successful, 64 were unsuccessful, and the results were undetermined in 17 other bouts. Considering only successful and unsuccessful bouts it was found that there was no harassment in 81.5% of the successful and in 70.3% of the unsuccessful bouts, and that there was harassment in 18.5% of the successful and in 29.7% of the unsuccessful bouts. These results are not significantly different ($\chi^2 = 0.713$, $p > 0.30$; Siegel 1956 pp. 107-9). Similarly, the number of bouts in which adult males harassed the copulating pair did not differ significantly as between successful and unsuccessful bouts ($\chi^2 = 0.017$). The significance of these harassments may lie not so much in their disruption of copulations in general, but rather in their ability to decrease a low-ranking male's chances of achieving a successful copulation, if it is reasoned that only the more dominant animals could persist through the harassment to a successful copulation. Thus, this harassment may produce differential reproduction among the males, i.e., sexual selection.

Relations between Adult Male and Immature *C. b. tephrosceles*

Evidence has been presented which suggests that within social groups of *b. tephrosceles* there is a kind of subgroup consisting of the adult males. This subgroup is characterized by stable membership over a relatively long period (several years) and a rigid dominance hierarchy, expressed as priority of access and correlated with copulation success (see below) and stable roles in the stereotyped present type II display. The cohesion of this adult male subgroup is most apparent in its united aggression against neighboring groups of red colobus.

Two kinds of social interactions may be directly related to the achievement of membership in this subgroup and to the expulsion of young males from the group. One of these is the harassment of adult males by immature monkeys in nonsexual situations, and the other is the harassment of young males approaching puberty by the adult males. The 2 kinds of harassment are dependent on the age of the immature monkey. Only as he approaches subadulthood does a male become the object of harassment by adult males.

Harassment of adult males by immatures in nonsexual situations. The behavior patterns employed in this kind of harassment were extremely varied, but in all cases the essence of the social interaction consisted of an immature monkey making a definite approach to an adult male, who was often resting, and then pestering him one way or another, such as slapping toward him or pulling his tail, and then fleeing from him. The adult male usually ignored this harassment or made a low-intensity threat gesture toward his harasser.

Eighty cases of this harassment were observed in the CW group and 14 more in at least 3 other groups. In 12 cases the harasser merely approached the recipient and then turned and fled immediately; 4 times it reached toward the recipient without making physical contact; once it extended its head toward him as if to smell him; 3 times it only approached and sat next to the male; twice it slapped toward him without making contact; once it touched him on his perineum; once it hung from his tail and one leg; twice it touched him on the back; twice it muzzled his genitals; 6 times it pulled his tail; in two other cases it presented type I before pulling his tail; twice it rolled into his lap; twice it climbed onto him; once it presented type I before grabbing him; 3 times it touched him first and then presented type I; once it laid in front of him and then presented type I; 3 times it reached toward him without making contact and then presented type I; 42 times it approached and only presented the type I before fleeing; and 4 times it dangled beneath him.

In 30 cases the male showed no response to the harassment; 10 times he slapped toward the harasser without making contact; 5 times he merely looked toward the harasser; 4 times he lunged; once he made an incipient lunge; 3 times he grabbed the harasser; once he grabbed and shook the harasser; twice he pushed it away; 3 times he grabbed the harasser's hips, and on one of these occasions he also muzzled its perineum; once he touched its perineum; once he yawned; once he gaped; once he held and gently bit the harasser; once he chased it; once he followed it; and once he groomed it. In 28 cases the adult male's response was not noted.

There were no vocalizations in 59 cases; the harasser squealed or shrieked in 35 cases; and the adult male gave wheets in one case and chists in one other.

In 78 cases the harasser fled or climbed away from the adult male; in 14 cases it remained near him; and in 2 cases the harasser's final act was not noted.

In the CW group the harasser was a subadult male in 5 cases; a subadult of undetermined sex in 3; an old juvenile in 8; a juvenile in 42; a young juvenile in 15; an old infant in 6; and a young, dark infant once. There was a positive correlation between dominance status and the frequency with which an adult male was harassed. The most dominant male was harassed most. In the CW group, adult male CW was harassed on 30 occasions, LB 24, ND 12, SAM 13, and in 2 cases the male's identity was not determined.[2]

The harasser's behavior in this type of social interaction clearly has elements of both approach and flight. On the one hand he seems attracted to adult males, and on the other his tendency is to flee from them. It is not clear how this kind of encounter relates to the adult-male subgroup, partly because I was unable to determine the sex of the majority of the harassers. If they all proved to be males, then it is plausible that this kind of harassment is a means of testing the aggressiveness of the recipient adult male. In addition, it may be that any kind of interaction with the adult males, even harassment, is to the advantage of immature monkeys in gaining future admission to the adult-male subgroup, insofar as all social interactions increase familiarity between the participants.

Greater familiarity with the dominant male may mean a better chance of being admitted to the adult-male subgroup. The advantage to males of being admitted to this subgroup is that there seems only one other alternative: exclusion from the total group.

Harassment of young males by adult males.   In this type of social interaction an adult male persistently follows a young male, who moves away from him giving shrill squeals or shrieks. This differs from a spatial supplantation in that it covers a longer period of time and a greater distance. Adult males usually follow the young males for at least 30 seconds and sometimes for several minutes. Although the route followed in such encounters is circuitous, it usually covers at least 30 m.

Only 12 cases were clearly observed in the CW group. The recipient gave shrill squeals or shrieks in 7 of these cases and no vocalizations in 3; in 2 cases it was not determined who vocalized. The harassing male gave no calls in 4 cases, wheets in 6, and chists in one; it was not determined if he called in 2 other instances.

Most of the harassing was done by the dominant male. Male CW harassed in 7 cases, male LB in one, SAM in 2; and in 2 cases the male was not identified. SAM was the recipient in 6 cases, an unidentified male in 4, an old juvenile male once, and a juvenile once. SAM was harassed 6 times by male CW between 20 April 1971 and 2 June 1971, while he was a young subadult. His physical and sexual maturation were rapid after this period. On 2 August 1971 SAM not only performed his first sexual mount but also his first harassment of a juvenile. In September 1971 he also harassed an old juvenile male.

This sample surely represents a gross underestimate of the actual frequency of this kind of harassment. With more examples, its role in the integration of young males into the adult-male subgroup or their expulsion from the group might become apparent. At present it is uncertain how this operates. SAM's case was the most clearly observed, and yet it offers no understanding of the process. There was no apparent interaction which demarcated the period when his role as the harassed terminated and his role as the harasser began. The change was clearly correlated with his physical and sexual maturation, but there was no social encounter which made it apparent why he should remain in the group and rise in the dominance hierarchy rather than being expelled from the group.[3] His maturation and rise in the dominance hierarchy clearly disrupted the stable and exclusive copulating role of male CW, but this change too was not accompanied by any prominent social interaction, such as a contest of physical aggression. It all seemed to come about rather undramatically. As soon as SAM began copulating in August 1971, so did males LB and ND, who until that time had performed virtually no sexual behavior (see section on heterosexual behavior). The manner in which a young male is either admitted into the adult-male subgroup or expelled from the group, and the relation of this process to harassment, probably involves some very subtle behavior, which was missed in this study.

Miscellaneous Information on Aggression in *C. b. tephrosceles*

Two issues are considered here. The first concerns the frequency of aggression and the spatial relations between dominant and subordinate red colobus while feeding. The wood of dead trees and/or possibly the insects living in this dead wood seemed to constitute a choice food item for the red colobus (see chapter 5). When feeding on a dead tree, the entire group would cluster around it and gnaw on the trunk. Frequently monkeys would break loose pieces of wood, carry them several meters away, and then feed on them. The frequency of aggression definitely increased when the monkeys of the CW group were feeding on dead trunks. This was undoubtedly because of the concentrated nature of the choice food source and, consequently, the close proximity of feeding animals. Several of these conflicts involved biting and pushing. On one occasion 2 adult males of the ST group attacked me when I approached them as they were feeding on a dead tree trunk. One of them actually leaped to the ground and ran toward me, stopping only 2–3 m from me. Not all monkeys exhibited this kind of aggression during these feeding bouts, and, in general, it was the adult males who were most aggressive. Several adult females and immature monkeys often fed on the dead wood while in physical contact and without any aggression. In marked contrast to the aggression around dead trees was the apparent tolerance for close proximity in many other feeding situations. SAM and adult male LB were both seen feeding peacefully while in physical contact with adult females and immature monkeys. Adult males ND and CW also fed together in contact without any aggression. One of the most remarkable cases concerned adult males CW and LB. CW approached LB, who was seated and feeding. CW stood over and in contact with LB and then reached over LB's shoulder and ate leaves off the branch which LB had pulled toward himself and was also feeding on. LB showed no response to this but continued feeding. CW ate more leaves from the same branch and then soon left, without the display of any aggressive or submissive gestures.

The second issue concerns the absence of an eyelid display in *b. tephrosceles* in spite of the fact that they have pinkish eyelids which contrast with their dark faces. Several other species having pink eyelids and dark faces, such as baboons, vervets, and some mangabeys, display the eyelids as part of an aggressive gesture. Eyelid displays have not, to my knowledge, been described for any colobinae, and they may be restricted to the cercopithecinae.

Aggression in Other Subspecies of Red Colobus

Aggressive encounters involving vocalizations were common in 3 of the other 4 subspecies studied: *b. temminckii, b. badius,* and *b. preussi.* Rarely, however, were these encounters clearly observed. No aggression was seen during the brief survey of *b. rufomitratus.*

The gape, as described above for *b. tephrosceles,* was also observed in the other 3 subspecies and appeared to be of more common occurrence in their repertoire than in that of *b. tephrosceles.* It was given by adult males toward adult males; adult females toward adult females; adult males toward adult females; an adult female toward a young juvenile who tried to suckle from her; and an adult female and a subadult toward me. In these other subspecies the gape also appeared to be an aggressive threat gesture, which was usually responded to with flight, but in one case the recipient (*b. temminckii*) lowered its hindquarters and assumed a crouched position.

Slapping and grabbing toward without physical contact, and chases and lunges, were observed in all 3 subspecies.

Staring with extended forequarters (mouth closed) was seen only in *b. temminckii* and was given by a subadult toward me. Only one case of harassment during copulation was observed, and this occurred during an intergroup aggressive conflict of *b. temminckii.* This single case may have been atypical because in the 3 other bouts of copulation observed in this subspecies there was no harassment. Only in *b. temminckii* was the yawn clearly used as a threat gesture. The single example was given by an adult male during an intergroup conflict. He opened his mouth widely and exposed his canines.

Spatial supplantation was seen only in *b. badius,* where an adult male supplanted an adult female.

Multipartite aggressive encounters seemed to occur quite frequently in all 3 of the other subspecies, but only one case was clearly observed. This concerned an adult male *b. preussi* who intervened in a conflict and, apparently, terminated it.

Comparison with Other Colobinae

There is no information available on aggressive behavior from the other studies of red colobus.

The aggressive behavior of *C. guereza* includes spatial supplantations, chases, leaping-about displays, and grimaces, as for badius (pers. observ.). Branch-shaking, gape, stare with extended forequarters, and touching have also been observed in guereza (J. F. Oates, pers. comm.). Guereza give at least 2 displays which seem to have no counterpart in badius. One of these has been described by Schenkel and Schenkel-Hulliger (1967) as "chewing and lip-smacking," and they consider it to be an appeasement gesture. The mouth is slowly opened and closed with a resonant clicking sound being produced as the mouth is opened (pers. observ.). This display has been directed by both adult males and adult females toward me. The second is a penile display given by adult males during mildly stressful situations; usually toward the observer. The penis is erected as the displayer sits with his legs extended (J. F. Oates, pers. comm. and pers. observ.). There is apparently, a dominance hierarchy among the males of some populations, although such hierarchies do not involve females (Schenkel and Schenkel-Hulliger 1967). In contrast, Ullrich (1961) did not observe a dominance hierarchy in his study group. Like the badius, juvenile guereza sometimes harass copulating males (J. F. Oates, pers. comm.).

Some of the gestures and displays given by *Presbytis entellus* during aggressive encounters appear to be similar, if not identical, to those of badius, including: grimace, stare, lunge, chase, hit or slap, bite, rapid glance, leaping-about display, present, and touching (Jay 1965). There are a number of displays in the repertoire of entellus which seem to have no counterpart in badius. These are: slap ground, crouch and then stand suddenly, head bob, bite air, head toss, and move tongue in and out of mouth (Jay 1965). The movement of the tongue in and out of the mouth seems to be absent from some populations (Yoshiba 1968). Immature male entellus from the age of 10 months to 4 years demonstrate an interest in adult males. They run to an adult male, touch him, embrace him from a frontal position, or mount him. All 3 of these patterns occur alone or in combination. Immature females do not engage in this type of interaction (Jay 1965). This interest in adult males by immature male entellus is very reminiscent of the harassment of adult males by immature badius. Harassment of subadult males by adult males is apparently absent in entellus. In some populations of entellus, copulations

are harassed by subordinate males (Jay 1965), but in others it is more typically
the adult females who harass the copulating pair (Yoshiba 1968). Similar
variation exists in the prevalence of dominance hierarchies. In some entellus
populations, dominance hierarchies of the adult males are conspicuous,
stable, and linear (Jay 1965, Vogel 1971) ; whereas in others, dominance
relations are restricted to a single adult male and subadult males because of
the group composition (Yoshiba 1968); and in some other areas dominance
relations seem to be absent among the adult males (Vogel 1971). In Jay's
(1965) study, the dominance relations of adult males were not strongly
correlated with their frequency of copulation, in contrast to the situation with
badius. The one shift in dominance rank order among adult males which
occurred within an entellus group was not accompanied by aggressive fighting
(Jay 1965), apparently like the single case observed in my badius group.
Dominance relations among the adult females of some entellus populations
are nonlinear and bidirectional, and dominance interactions among the
females appear to be of much more frequent occurrence than among female
badius (Jay 1965). However, dominance hierarchies have not been detected
among females of the entellus at Dharwar (Yoshiba 1968). Adult female
entellus sometimes form aggressive alliances against other monkeys. Such
alliances were rarely observed in badius.

Leaping-about displays have been described for adult male *Presbytis johnii*
and are often given during intergroup confrontations (Tanaka 1965 and
Poirier 1968c). Dominance hierarchies have been described for both males
and females of this species, but there seem to be no special relationship or
harassment between adult and immature males (Poirier 1969).

GROOMING BEHAVIOR

Only social grooming in *C. b. tephrosceles* is considered here. In this social
interaction, one monkey uses its fingers to examine the fur of another and
removes particles from the fur by direct application of the mouth. Solicitation
for grooming usually consists of the potential groomee presenting the part of
the body to be groomed to the potential groomer. Typically only one monkey
grooms another, and only on rare occasions do 2 monkeys groom the same
groomee simultaneously. Simultaneous mutual grooming was not observed.

All grooming encounters that were observed in the CW group on all
observation days between November 1970 and March 1972 inclusive are
considered in this analysis. As stated in the introduction to this chapter,
grooming bouts were distinguished from one another on the basis of partner
identity, direction of grooming, and a time interval of at least one hour
between grooming bouts involving the same partners and the same direction
of grooming. In many cases I was unable specifically to identify one or both
members of a grooming pair, and in such cases only the age-sex category was
noted. Distinction of subadults from adult females was sometimes difficult,
especially with poor observation conditions; when identification was uncer-

tain, the monkey was scored as an approximate adult. Although all 3 adult males were easily distinguished in November 1970, adult females were not recognizable until later in the study. Even after they were distinguished, adult females still posed greater problems of recognition than adult males. Consequently, the grooming scores of the 4 distinct adult females increased with time after I was initially able to recognize them. Specific individuals in other age classes were usually not distinguishable. My data on grooming lack information on the duration of grooming bouts because of observation problems and the conflict with systematic sampling of activity, which was done at regular time intervals (see chapter 5). Consequently, bouts of short and long duration are lumped together; but, of course, bouts of long duration have a greater likelihood of being observed and scored than do those of short duration. The bias of my opportunistic sampling method is in favor of long-duration grooming bouts. In general, it appears that adult males groom other monkeys for shorter periods than do adult females. My sampling method is therefore biased against scoring adult males as groomers.

Individuality in grooming roles was very pronounced in the CW group (table 15). For example, adult male LB was seen to groom another monkey on only one occasion, whereas adult male ND was frequently a groomer and adult male CW was intermediate to these 2 in his role as a groomer. Some of the intermonthly variation might be attributable to differences in sample size (number of hours of observation and number of grooming bouts), but much of it remains to be explained. Consider, for example, the last 3 months of the study, in which the hours of observation and number of grooming bouts scored were much the same. Adult male LB was relatively constant in his scores both as groomer and groomee, whereas males CW and ND showed much variation in their grooming roles during these 3 months.

Individuality in grooming roles was not apparent among the adult females, but females BT and PGCW may have received slightly more grooming than females GCW and KT (tables 15 and 17). The relation between grooming roles and dominance status among adult males is not particularly striking (compare table 9 with 15, 16, and 17). However, the lowest-ranking adult male (ND) did groom much more frequently than the other adult males. Furthermore, with the increasing maturation of SAM (first recognized in April 1971) there was a striking change in the grooming relations within the group. SAM became the frequent recipient of grooming and, as of September 1971 (after his first copulation), was usually groomed more than the other adult males (tables 15–17).

In spite of the marked individuality in grooming roles, it is apparent that adult females groom more than do adult males and that adult males receive more grooming than do adult females (compare average group composition in table 7 with tables 15–17). Analysis of the total grooming bouts for the CW group reveals that adult males were the groomers in 81 cases and adult females in 440. In contrast, adult males were the groomees in 221 bouts and adult females in 173. Casting these data into a two-by-two matrix and testing them with a chi-square test corrected for continuity (Siegel 1956, pp. 107–10) reveals that the differences between males and females are highly significant ($\chi^2 = 164.9$, $p < 0.001$). These results are further substantiated by consulting tables of the cumulative binomial probability distribution, which show that the probability of a discrepant distribution at least as great as the one attained here is less than 0.00001, for both the data of groomer and groomee (Mosteller, et al. 1961 and Harvard Computation Laboratory's Tables of the Cumulative Binomial Probability Distribution 1955).

It is also evident from tables 15–17 that adult males groom less than they are groomed and that adult females groom more than they are groomed. The groomer and groomee data for adult males CW, LB, and ND were lumped together and compared by means of Wilcoxon's signed-ranks test for 2 groups (groomer and groomee) arranged as paired observations (Sokal and Rohlf 1969, pp. 400–401). The results of this test were highly significant and confirmed the conclusion that males are groomed more than they groom others ($T_S = 50$, $p < 0.0098$, 2-tailed). The data for adult females were tested similarly and were also highly significant, supporting the conclusion that they groom more frequently than they are groomed ($T_S = 20$, $p < 0.0096$, 2-tailed).

The distinction in grooming roles of adult males and adult females may be related to the dominance of males over females and to the fact that mothers are the primary groomers of their own infants and young juveniles.

Comparison of table 7 with tables 15–17 indicates that members of the juvenile class participate in grooming bouts slightly less than might be expected on the basis of their numbers in the group. They appear to groom about as frequently as they are groomed. Young juveniles are involved in grooming bouts much as might be expected on the basis of their percent composition of the group membership and seem to be groomed slightly more than they groom. Infants are groomed much more than they groom (only 2 cases), and the incidence of their being groomed is generally higher than expected from their numbers in the group.

The frequency of participation by a female in grooming bouts does not seem to be affected by whether or not she possesses a young, dark infant. Neither does her sexual state affect her grooming frequency. Lumping the data for the 3 recognizable adult females (GCW, BT, and PGCW) who were observed to copulate during this study, and segregating these data into 2 classes (months in which the particular female copulated and months in which she did not copulate), results in a two-by-two matrix in which the females were groomers 80 times and groomees 23 times in months during which they copulated, and were groomers 84 times and groomees 33 times in months when they did not copulate. These differences were not significant when tested with a chi-square test corrected for continuity ($\chi^2 = 0.71$, $p > 0.30$, 2-tailed) and support the conclusion that the distribution of an adult female's grooming activities between groomer and groomee are not

affected by whether or not she copulates. This conclusion is further substantiated by Spearman rank correlation coefficients which were computed separately for the same 3 adult females and which compared the number of times
she was mounted by an adult male per hour of observation with the number
of bouts in which she was a groomer per unit time, and also with the number
of bouts in which she was a groomee per unit time, for each month that she
was recognizable. Thus, for each of the 3 females 2 coefficients were computed; one compared her monthly rate of sexual mounts with her rate as a
groomer, and the other compared her monthly rate of sexual mounts with her
rate as a groomee. None of the 6 coefficients was significant (r_s ranged from
0.21 to 0.83, with $p > 0.05$); leading to the conclusion that a female's sexual
behavior does not affect her role in or frequency of grooming bouts.

Comparison with Other Species, Particularly Colobinae

There have been no published accounts of grooming relations of other red
colobus groups, populations, or subspecies.

Only 2 studies of *C. guereza* have dealt with grooming in any detail (Leskes
and Acheson 1971, and Oates 1974). Both give similar results and demonstrate that the majority of grooming is performed by adult and subadult
females, much like that described for badius. On the basis of membership
composition and the grooming relations of two groups of guereza, Oates
concludes that adult females not only groom more but also receive more
grooming than expected. The latter point differs from the badius I studied.
However, adult guereza females did groom about 24% more than they were
groomed, which resembles the trend in badius. Juvenile and infant guereza
were rarely groomers but were groomed about as much as one would expect
on the basis of their numerical representation in the group. Juvenile badius,
in contrast, are frequent groomers, although their participation in grooming
bouts is slightly less than expected. The only adult male guereza in Oates's
group rarely groomed but was groomed much as expected, thus differing
from adult male badius, who are groomed more than expected.

Grooming seems to be an extremely important social behavior in groups of
Presbytis entellus (Jay 1965) and also in *Presbytis johnii* (Poirier 1969). Like
the badius, *P. johnii* adult females are the chief groomers, and they groom
longer than do adult males. Poirier (1968b) also reports that female *P. johnii*
seldom groom their own infant, in apparent contrast to badius and guereza.

The grooming relations described for the CW group of badius bear some
similarities with certain of the cercopithecinae; for example, adult female
baboons do most of the grooming, and adult male baboons groom for shorter
periods than do females (Hall and DeVore 1965). However, in other ways they
are quite different. For example, grooming is stated to be directly related to
dominance status in baboons, with the most dominant animals receiving the
most grooming; also, estrous female baboons are groomed about 4 times
more than are non-estrous females (Hall and DeVore 1965). In vervets it has

been shown that females with newborn infants receive much more grooming than in the prepartum period and after the infant is one month old (Struhsaker 1971). None of these relationships were found among the red colobus.

HETEROSEXUAL BEHAVIOR

Description of Postures for *C. b. tephrosceles*

In a typical heterosexual mount the adult male approaches an adult female, grabs her tail base with one hand and handles her perineum and clitoris with the other. He often places his muzzle near her perineum either for visual and/or olfactory inspection (pl. 10). She then assumes the posture called a sexual present, in which her arms and legs are only slightly flexed, her tail is diverted to one side, but not elevated, and she looks straight ahead and not back toward the male (fig. 4). He mounts her from behind, grabbing her hips or sides slightly anterior to the hips with his hands while keeping both feet on the supporting tree branch. He then delivers pelvic thrusts. Vocalizations are usually absent. Ejaculation, assumed to coincide with a pronounced pause which terminates the thrusting, appears to occur only after a series of incomplete mounts (mounts without a pause). Heterosexual mounts are often harassed by immature monkeys and less frequently by adult males.

The preceding generalized description is based on observations of 160 mounts in the CW group and 72 additional mounts in at least 6 other groups. The 160 mounts observed in the CW group were distributed among 108 bouts of heterosexual mounting; bouts between the same partners were separated by at least 10 minutes. In these 108 bouts the adult male initiated the interaction by approaching the adult female on 32 (29.6%) occasions, while the female initially approached the male on only 5 occasions (4.6%). The initiation of the bout was not determined in 71 cases. Before mounting, the adult male grabbed the female's tail base in 26.8% of these 108 bouts, handled her perineum and/or clitoris in 22.2%, and muzzled and/or closely examined her perineum in 21.3% of the bouts. These premounting behaviors are not mutually exclusive, and the total number of bouts in which one or more of them occurred was 41 (37.9%). These frequencies are minimal because their presence or absence was often not noted. Furthermore, although usually given prior to the first mount of a copulation bout, these behaviors were sometimes given between incomplete mounts. They were never given after a complete mount, i.e., a mount terminating with a pause. All 4 adult males in the CW group performed these premounting behaviors. Grabbing the tail base, and handling and muzzling a female's perineum and clitoris, by adult males were observed on numerous other occasions when they were not followed by mounting.

In the CW group only 24 examples of the female's sexual present were adequately observed and described. These 24 examples were performed by at least 6 different females. In 8.7% of these 24 presents the female's limbs were described as flexed; in 47.8% slightly flexed; in 17.4% not flexed; and their position was not noted in 26.1% of the cases. The female looked straight ahead in 78.3% of the presents; looked to one side in 4.3%; looked over one shoulder toward the adult male in 13.0%; and her head position was not noted in 4.3% of the cases. Her tail was diverted to one side but not elevated in 34.8% of the presents; it was held in the normal downward position in 17.4%; elevated and diverted in 4.3%; and the tail position was not noted in 43.5% of the examples. The posterior quarters of the presenting female were never elevated; were lowered only once; and were probably in the normal position during the remaining cases, although this was usually not noted. In only one case did the presenting female lie on a tree branch with her arms and legs greatly flexed. In an additional 16 presents described for at least 6 other badius groups, the pattern was essentially the same as that described for the females of the CW group. However, in 4 of these cases the female's posterior was elevated, a pattern not observed in the CW group. And in 3 other cases the presenting female lay on her ventral surface.

During the actual mount the male's feet remained on the supporting tree branch in 59 (36.9%) of the 160 mounts; their position was not noted in 99 cases. In one of 2 exceptional cases, the male once kept only one foot on the tree branch while the other dangled free, and in the second case one foot remained on the branch and the other gripped an ankle of the female.

There were no vocalizations during 132 (82.5%) of the 160 mounts observed in the CW group; their presence or absence was not noted in 15% of the cases; adult male CW gave low-amplitude wheet calls during 3 (1.9%) incomplete mounts; and adult male LB once barked during an incomplete mount.

Only 27 of the 160 mounts observed in the CW group terminated with a pause and were thus considered to be complete. At least 40.7% of these complete mounts were preceded by and occurred in the same bout with incomplete mounts. The number of incomplete mounts preceding the complete mount ranged from 1 to 5. Typically only one incomplete mount preceded the complete mount. However, this surely represents a minimal number and probably an underestimate, because conditions of observation were often such that I could not be confident that incomplete mounts did not precede the complete mount. These results I think justify the conclusion that male *b. tephrosceles* are multiple mounters who achieve ejaculation only after a series of mounts in the same bout, like rhesus monkeys (Altmann 1962), and in contrast to single mounters, such as vervet monkeys, who ejaculate on the majority of mounts (Struhsaker 1967a).

Mounting bouts were not typically accompanied by grooming between the 2 sexual partners. Grooming was associated with only 22 (20.4%) of the 108 bouts of mounting observed in the CW group. In 19 of these the female groomed the male; in 2 the female and male groomed reciprocally but not simultaneously; and in only one case did the male groom the female without reciprocation.

Harassment during copulation by adult males and immature monkeys has already been dealt with in the section on agonistic behavior. Vocal harassment by adult males occurred in 13% and harassment by immatures in 21.3% of the 108 copulation bouts in the CW group.

Description of Postures for Other Red Colobus Subspecies

Only 5 copulation bouts, which included 6 mounts, were observed among *b. temminckii*. Since none of the 6 mounts included an ejaculatory pause, all were judged to be incomplete. The only occasion when 2 mounts occurred in rapid succession between the same partners was during an intergroup conflict and may have been an atypical bout. This was also the only bout in which the copulating pair was harassed by other monkeys. Although the sample is small, it appears that the males of this subspecies grip the females with their hands in a more anterior position than do other red colobus subspecies and most other cercopithecidae. The male's hands are placed well anterior to the female's hips, and in one case they gripped the female's thoracic region. In another example the male also pressed his arms against the female's sides and hips while his hands gripped her sides more anteriorly. The position of the male's feet during mounts was variable; either they were placed on the supporting tree branch, or they were gripping the female's ankles, or they were alternately grabbing at her ankles and then moving back to the branch. Twice the female stood quadrupedally during the mount; once she lay ventrally on a tree branch; and in 3 cases her stance was not clearly noted. The adult male muzzled the female's perineum once before mounting her, and in another case he muzzled her back after an incomplete mount. No vocalizations were given in any of the bouts.

In the study of *b. badius*, 14 bouts of copulation were observed, which included 15 mounts. There was definitely no ejaculatory pause in 3 of these mounts, but the presence or absence of a pause could not be determined for the other 12 cases. The direct evidence for multiple mounting in this subspecies is not strong, with only one observed case in which 2 mounts between the same partners occurred within one minute. The indirect evidence for multiple mounting is, however, quite compelling. First, there is the notable absence of observations of ejaculatory pauses, which suggests that several mounts precede ejaculation. Second, the quaver vocalization given by females during mounts and briefly just after the dismount often occurred in temporal clusters from what seemed to be one source. This suggests that females are mounted several times in rapid succession, as occurs in multiple mounting. In only 2 examples was premounting manipulation of the female by the male observed. In these cases he grabbed her tail base with one hand and then muzzled her perineum. The position of the male's hands and feet during mounts was usually not noted. However, in 3 cases his hands gripped the female's sides just anterior to her hips, and in 6 mounts the male's feet remained on the supporting tree branch, which is much more like the posture of male *b. tephrosceles* than like *b. temminckii.* In one mount the male held his mouth open in a puckered shape, with his teeth concealed. The female's stance was not noted in 11 cases, but in most of these she was probably standing in an approximately normal quadrupedal position. This normal position was clearly noted in only one mount. In one case the female was semiprone in a half-lying and half-sitting posture. In another, all 4 limbs were greatly flexed, bringing her into a prone position with her ventral surface in contact with the tree branch. The female's head position was not noted in 13 cases, but once she looked over her shoulder toward the male mounting her, and in another case she lowered her head between her arms and looked between her arms and legs toward the mounted male. Female *b. badius* were not always receptive to the males. Once an adult male delivered pelvic thrusts into the dorsal sacrum of an adult female while she was seated. Quaver vocalizations were given by the female in 12 mounts, not given in only 2 mounts, and their presence was not noted in one case. Copulation almost certainly occurs at night in this subspecies, because the quavers associated with copulation were heard well after sunset on several occasions. These vocalizations may also be of importance in inducing what appears to be contagious copulation, a kind of social facilitation in mounting. The evidence supporting this hypothesis is based on the fact that very often the quaver bouts from one copulating pair were immediately followed by similar bouts from other sources. However, this is a very tentative hypothesis and needs verification.

Fourteen copulation bouts comprised of 18 mounts were observed in *b. preussi.* Four of these bouts consisted of 2 mounts each. The mounts within a bout were separated from one another by 2.5 minutes or less. Ejaculatory pauses did not occur in 2 mounts, but their presence or absence could not be determined for 16 mounts. Ejaculate was clearly observed on the female's perineum after only one of these mounts. It would appear that male *b. preussi* are also multiple mounters. Premounting manipulation of the female by the male was not observed. The position of the male's hands during the mount was never adequately described, but they gripped the female in the general vicinity of her hips. His feet gripped her ankles in 2 mounts, and in another only one foot gripped an ankle, while the other foot remained on a tree branch. The position of his feet was not noted in 15 cases, but it appears that *b. preussi* grip the female's ankles much more than do males of *b. tephrosceles*. The stance assumed by the female while being mounted was poorly described, but in general she was in an approximately normal quadrupedal position. In at least 2 cases, however, her limbs were greatly flexed, bringing her into a prone posture with her ventral surface in contact with the supporting tree branch. Females gave quaver vocalizations during 10 mounts; sqwacks and quavers in 3; iks in one; yowls in one; no vocalizations in 2 mounts; and the presence or absence of vocalizations was not noted in one case. In contrast to all other subspecies of red colobus studied, the males of *b. preussi* frequently vocalized during mounts. In 4 cases they gave barks; yelps in 7; and it was not determined if they vocalized in 7 other mounts. Although premounting manipulation of the female by the male was not observed, on 2 occasions males muzzled the perineum of adult females having huge perineal swellings. These muzzlings were not followed by mounts or attempted mounts.

Relation between Female's Perineal Swelling and Sexual Behavior in
C. b. tephrosceles

The perineal swellings of adult females were referred to in chapter 2, where it was indicated that the size of this structure shows considerable variability between the different subspecies. In *b. tephrosceles* the swelling undergoes changes in size and coloration for each individual female. Its maximum size in *b. tephrosceles* was visually estimated at about 5 cm deep, 5 cm wide, and 7.5 cm long. Typically, swellings were much smaller than this. The color of the female's genital skin varied from dusky grey to crimson. The swellings varied in color from pink to bright pink, but unswollen perineal skin could also be of these colors. Swellings were never seen in females with newborn infants. If this structure is, in fact, a sexual swelling, which is under estrogenic control like that of baboons and chimpanzees, then one would expect to find correlations between the development of the perineal swelling and sexual behavior. Demonstration of such correlations and determination of swelling duration and cycle length are dependent on nearly continuous and accurate records. Unfortunately, the nature of my study in no way permitted continuous records of recognizable females. The majority of observations of the CW group were made during the first week of each month, which, combined with the observation conditions, precluded continuous and highly accurate records of the sexual behavior and state of the perineal skin of recognizable females. However, in spite of these deficiencies, let us consider the available data.

In the 160 heterosexual mounts observed in the CW group the female's perineal skin was swollen in 26.9% of the cases; not swollen in 31.9%; and the state of the perineal skin was not determined in 41.2% of the mounts. I believe that during the first few months of this study I was not so perceptive of the small swellings of *b. tephrosceles* females as I later was. Having just studied *b. preussi,* in which the females develop huge swellings, I was probably disregarding the small lumps of female *b. tephrosceles.* This is partly verified when the data are analyzed in 2 periods: August 1970 through May 1971 and June 1971 through March 1972. In the first half of the study, swellings were detected in only 14.9% of the females who were mounted as compared to 31.9% in the second half. Likewise, more females who were mounted were scored as unswollen in the first half (55.3%) than in the second (22.1%). It would also seem that, because I was more acutely aware of the existence of the small swellings, I became more conservative in the second half, where I scored 46% of the females who were mounted as having perineums of an undetermined status. This compares with 29.8% in the same category during the first half. These data suggest that swollen females tend to be mounted slightly more often than unswollen females.

Records for individual females are restricted to 3 adult females (GCW, PGCW, and BT) and to a subadult female (M) who was recognizable in the CW group for only 5 months. The correspondence between swelling and the occurrence of copulation for a female is fairly good (table 18), but a refined analysis is not warranted in view of the deficiencies in the data and the sampling method described above. The size and color of a particular female's swelling obviously varies from month to month, but whether the swelling is on a monthly cycle like that of many other primate estrous cycles cannot be determined from these data. Most of the sampling was done at approximately monthly intervals, which prohibits the distinction between the case where a female is continuously swollen for 30 days from one in which she cycles once or more than once during the 30-day interval separating the samples. The cycle length in the swelling of female *b. tephrosceles* remains undetermined.

The case of female GCW suggests that the swelling is under hormonal control, much like that in baboons. She was last swollen in mid-December 1971 and showed no indication of further swelling until that phase of the study was terminated in mid-March 1972. Furthermore, she did not copulate again after mid-December 1971, even though the males of the CW group were sexually very active. When GCW was next observed at the end of August 1972, she had a clinging infant, which had the approximate coloration of an adult and which I judged to be about 4 months old. This would place its birth at sometime in April or May 1972. If she were impregnated in mid-December 1971, this would give an approximate gestation period of about 4.5–5.5 months, which seems reasonable considering the 5.4-month gestation given for *Presbytis* by Napier and Napier (1967). I conclude from this that swelling does not occur in pregnant female *b. tephrosceles*. Further evidence for the hormonal control of swelling is provided by another example from female GCW's record. On 1 October 1971 she was judged to have a bright pink swelling about 1.5 cm deep, but which by the next day had obviously increased in size and was estimated to be about 2.5 cm deep. On 3 October her swelling was virtually the same, and, in addition, ejaculate was visible on her vagina. On the 4th and 5th the swelling was still bright pink but had diminished in size, suggesting the onset of detumescence. This pattern resembles that of baboons, in which sexual swellings can rapidly increase and decrease in size during the period of a few days. However, it is obvious that more data are required for an understanding of the physiology of this swelling in *b. tephrosceles*.

The preceding observations indicate that the color of the perineal skin of a female *b. tephrosceles* is not a particularly good indicator of her physiological state. Pink perineums occur in females with newborn infants as well as in sexually receptive females. Perineal swelling indicates sexually receptivity and probably high levels of estrogen, whereas an unswollen sexual skin tells nothing of these 2 parameters.

I suggest either that the early states of estrus in *b. tephrosceles* are not indicated by any swelling of the perineal skin or that such swelling as may develop is so small as to go undetected by most observers. This would account for some of those cases in which unswollen females copulated. The baboons provide another analogy for this case in that the tumescence of their sexual skin develops over a period of several days during which they are sexually receptive even when the swelling is exceedingly small.

Relation between Female's Perineal Swelling and Sexual Behavior in Other Subspecies

The great majority of female *b. temminckii* lacked perineal swellings, and their genital region was either dusky in color or mottled with dark mahogany and grey. Only 6 females were seen with perineal swellings. Two were described as having slight swellings of crimson color; 2 had conspicuous pink swellings; and one had a large crimson swelling. None of these swellings were as large as those of *b. badius* or *b. preussi*. All but one of the females having clinging infants lacked swellings. The swelling of the one exceptional female with a clinging infant was small and pink. The rarity of swellings in this subspecies might be a seasonal phenomenon, as suggested by the relative abundance of infants and the infrequency of copulation seen during my brief survey.

Females with perineal swellings were commonly seen during the study of *b. badius*. Some of the swellings were very large and estimated to be 13 cm deep and 14 cm wide, matching Pocock's (1935) description for *b. waldroni*. These swellings were usually pink or bright pink in color, but a few were dusky in appearance. Unswollen females had grey perineums. None of the females having clinging infants were swollen. Only once did a large young juvenile suckle from a female with a small, bright pink swelling. In the 15 heterosexual mounts recorded for this subspecies the female had a large swelling in 3 cases; a huge swelling in 5 cases; and in 7 mounts the female's perineum was not clearly observed. It is apparent that female perineal swellings are associated with sexual receptivity and are probably under estrogenic control.

The swellings of female *b. preussi* are even larger than those of *b. badius*, and some were estimated to be equivalent to ¼ to ⅓ the body volume of the female. These swellings were also pink. Swellings were absent from females with clinging infants or young juveniles. In the 18 heterosexual mounts observed, the female had a small swelling in 2; a large swelling in 4; a huge swelling in 5; was simply noted as being swollen in 4; and her perineal condition was not observed in 3 mounts. As for *b. badius*, there seems little doubt that the female swellings are directly related to estrus and estrogens.

Seasonality in Heterosexual Behavior of the CW Group of *C. b. tephrosceles*

There was considerable variation in the amount of heterosexual behavior observed in the CW group from one month to the next (table 19). However, when one examines the distribution of the 160 mounts recorded for the CW group there is no apparent seasonal trend. Some of the high and low monthly frequencies are due to small samples, e.g., September 1970 and July 1971 (table 19). But consideration of months of similar sample size still reveals great intermonthly variation, e.g., May and August 1971 (table 19). It is also apparent that heterosexual behavior is not related to the seasons of rainfall. Compare, for example, August 1971 (115 mm of rainfall) with February 1972 (41 mm). Mounting was seen with approximately equal frequency in these 2 months. Furthermore, some samples during months with similar rainfall were very different in mounting frequencies, e.g., May 1971 (114 mm) and January 1972 (117 mm) (see table 19). Comparison of the same months in subsequent years also supports the conclusion that sexual behavior is not on a seasonal basis in the Kibale Forest. The monthly frequencies of mounting in the period of October 1970 through March 1971 are very different from those of the same months a year later.

The lack of seasonality is probably related to the general lack of seasonality in food availability and to the fact that birth intervals for each particular female are not on a 12-month basis (see below). Much of the inter-monthly variation in mounting behavior is explained by the variation in the estrous activity of 1 or 2 individual females. For example, in March 1971, female GCW was involved in at least 7 of the 10 mounts scored for the group; and again in January 1972, female BT accounted for at least 8 of the 13 mounts. Thus, it appears that much of the intermonthly variation in the heterosexual behavior of the CW group can be accounted for by the relatively small number of adult females in the group, the lack of seasonality in sexual behavior, and an interbirth interval of more than 12 months. These conclusions are further supported by examination of individual records of sexual behavior (table 18). Female GCW was sexually active only during the first 10 months of the 13 months she was recognizable. In contrast, female BT was sexually active only in the last 3 months of the 12 months she was recognizable.

Further support for the aseasonality of copulation in *b. tephrosceles* comes from the observation of newborn infants in or near compartment 30 of the Kibale Forest each month, except for 2, from May 1970 through March 1972. None were seen in August and November 1971, probably because observations were restricted to the CW group in these months.

Seasonality in Heterosexual Behavior of Other Badius Subspecies

The short duration of the surveys of the 3 West African subspecies prohibits an analysis of intermonthly and seasonal variation in their sexual behavior.

However, the indirect evidence provided by the relative frequency of mounting and the incidence of newborn infants and swollen adult females permits some tentative and reasonable hypotheses to be made.

The very few observations of copulation and of swollen females during the study of *b. temminckii*, combined with the relatively frequent observation of neonatal infants, suggest that this subspecies has a copulation and birth season (see preceding sections and table 3). Furthermore, this subspecies experiences a more pronounced seasonality in rainfall and vegetational change than does any of the other subspecies studied, with the possible exception of *b. rufomitratus* along the Tana River. Marked seasonality may exert selective pressures for the maintenance of a copulation and birth season in *b. temminckii*.

The indirect evidence for both *b. badius* and *b. preussi* suggests the absence of seasonality in copulations and births (see preceding sections and tables 4 and 5). Heterosexual mountings, swollen adult females, and infants of all ages including those with neonatal pelage were commonly seen in the same social groups for both subspecies. The tropical rain forests where these 2 subspecies were studied show much less seasonality in rainfall and probably less general seasonality in vegetation than does the savanna woodland of *b. temminckii* in Senegal and Gambia. In this respect, the habitats of *b. badius* and *b. preussi* are much like that of *b. tephrosceles* in the Kibale Forest.

During the extremely short survey of *b. rufomitratus*, only one newborn infant was seen and no heterosexual behavior was observed. The nature of the habitat, seasonality of rainfall, and probable seasonality in vegetation resemble those of *b. temminckii*, and I would therefore surmise that *b. rufomitratus* too has a strong tendency toward seasonality in sexual behavior and births.

Relation between Mounting Success and Dominance in the CW Group

Comparison of tables 9 and 20 reveals the close correspondance between priority of access (dominance) and the frequency of participation in heterosexual mounts. The dominant adult male did most of the mounting and copulation. Accordingly, there is also a good correlation between mounting success and the other correlates of dominance, e.g., the receiving of the present type I display and the sending of the present type II.

Prior to the physical and sexual maturation of SAM, adult male CW performed at least 84% of all heterosexual mounts, with males LB and ND doing very little (table 20). At this time there appeared to be a very rigid hierarchy in terms of priority to estrous females. It is also recalled that very few harassments of copulation by adult males occurred during this first part of the study (table 14). In the second part of the study, when SAM reached maturity and began copulating, the rigid priority of access to females broke down. Not only did SAM mount females, but the frequency with which males LB and ND mounted increased considerably when compared to their performance in the period prior to SAM's maturation. Accordingly, CW did proportionally less of the mounting after 1 August 1971 than he did before (42.7% in table 20). Correlated with this apparent breakdown in the copulating hierarchy was an increase in the frequency of harassment of copulations by male ND (table 14). Subsequent study in 1972 and 1973 indicated that the copulation hierarchy had again stabilized in the CW group, but with SAM now the chief copulator and the other 3 males doing very little.

Miscellaneous Information on Sexual Behavior.

It is noteworthy that no homosexual mounting was observed in any of the subspecies of red colobus studied; nor was any other behavior seen which might be termed homosexual. The only possible exception occurred when one adult male *b. preussi* muzzled the perineum of another adult male.

Masturbation has been observed only once during my entire study of red colobus. On 4 August 1973 adult male ND held his erect penis with one hand and several times in rapid succession moved his hand along the shaft of the penis in what appeared to be a forceful manner. He stared at his penis throughout. Ejaculation did not result.

Comparison with Other Colobinae

There have been no publications, until the present study, dealing with sexual behavior in red colobus monkeys. The perineal swelling of females was described by Pocock (1935) for *b. waldroni* and by Kuhn (1967) for *b. badius*. Kuhn (1972) also observed that this swelling varied in size for captive individual female *Procolobus kirkii* (the Zanzibar red colobus). It appears that Napier and Napier (1967) are incorrect in concluding that the perineal swelling of female red colobus is unrelated to the phases of the sexual cycle.

Female olive colobus (*Procolobus verus*) also have perineal swellings (Hill 1952). However, the cyclical nature of these swellings is poorly understood. Booth (1957) reports that it is reduced in pregnant and lactating females, whereas Napier and Napier (1967) conclude it has no relation to the menstrual cycle. There are no descriptions of sexual behavior for this species.

No other colobinae have any external, physical manifestation of estrus.

The only information on sexual behavior of free-ranging *Colobus guereza* comes from the study of Oates (1974). Oates reports that he observed only 19 mounts in 10 bouts of copulation during 888 hours of observation. He believes that his observations were so few because the 4 females in his main study group were either lactating or pregnant during most of his study. The male's feet gripped the female's ankles in at least one of these mounts. An "ejaculatory" pause was observed only once, which suggests to me that guereza may be multiple mounters like badius. Because there was only one fully adult male in Oates's main study group, the relation between dominance and copulation success could not be examined.

The descriptions of sexual behavior in the Asiatic colobinae are generally inadequate. Photographs show that the male's feet remain on the substrate during copulation in *Presbytis entellus* (Jay 1965 and Sugiyama 1966). In most groups and subspecies of entellus studied to date, the female initiates copulation by approaching and soliciting the male in a manner totally unlike anything observed in badius. The solicitation consists of simultaneously shaking her head, dropping her tail to the ground, and presenting to the male (Jay 1965, Sugiyama 1966, and Vogel 1971). In some areas, however, such as in the Kumaon Hills, it is the male who initiates copulation, and the preceding solicitation is not given by the females (Vogel 1971). It cannot be determined from the reports of entellus sexual behavior whether this species is typically a multiple mounter or not. Jay (1965), however, states that "consortships last only a few hours," which suggests to me that perhaps entellus is a multiple mounter. Heterosexual behavior appears to occur throughout the year in entellus (Jay 1965, and Sugiyama et al. 1965). In the Dharwar area of southern India, estrus can be induced by the death of the female's infant, such as occurs when the single adult male of the group is forcefully replaced by another adult male, who then kills infants in the group. Jay (1965) presents data supporting the conclusion that female entellus have a 30-day estrous cycle, with an estrous period of 1–7 days duration (see Jay 1965; fig. 7–10). The correlation between dominance and copulation success in groups of entellus having more than one adult male is not particularly striking. Although the females may have a tendency to solicit the most dominant males most frequently, in one particular group it was the lowest-ranking subadult male who was ranked second in copulation (Jay 1965). Excluding this exceptional male yields a slightly better correlation between dominance and copulation success, but one which was still not significant when I tested it with a Spearman rank correlation coefficient ($r_s = 0.61$, $N = 7$, $p > 0.05$).

Bernstein (1968) observed only 18 copulations among *Presbytis cristatus*. In this species the female solicits copulation with the male in a manner similar to that described for entellus. The males are also multiple mounters, like badius. Masturbation by a male was observed only once (Bernstein 1968).

Rudran (1973*a*) reports a peak mating season for the Polonnaruwa, Ceylon, population of *Presbytis s. senex*, which is coincident with the rainy season and the period of maximum food availability. He found no mating season among *P. senex monticola*, living in a less seasonal environment.

There is virtually no information on the sexual behavior of *Presbytis johnii*, Poirier (1970) having observed only one attempted copulation in 1250 hours of observation.

BIRTHS

C. b. tephrosceles

The gestation period for red colobus is not precisely known. From copulation records and the approximate date of birth, female GCW was estimated to have a gestation period of 4.5–5.5 months (see preceding section and table 18). This estimation is weakened by the fact that no observations were made of the CW group in the period of 17 March through 30 August 1972. Consequently, the date on which GCW gave birth (April-May 1972) was estimated on the basis of her infant's appearance when first seen on 2 September 1972. This estimate was, in turn, based on records of color change for 6 infants only (see below). BT was the only other female in the CW group for whom a prepartum record of copulation was available, thus permitting an estimate of gestation. She was actively copulating until the end of my observations in March 1972 (table 18) and gave birth between 5th and 29th September 1972, soon after the resumption of observations at the end of August 1972. Assuming she was not pregnant while copulating in early March and that she conceived at that time, we can estimate her gestation period as 6-6.5 months. It is quite possible, however, that she copulated again after the observations were terminated in early March 1972 and that conception occurred at a later date, which would give a shorter gestation period. Even though BT was clearly observed on 4 September 1972, there was no visible sign of her pregnancy.

Birth intervals for specific females were estimated from 2 kinds of data. The first was the interval between the time when a particular female was first recognizable without an infant and the time her infant was born. The second was the interval between the time she gave birth as a recognizable individual and the time she was last observed in the study. So far, no individual female has given birth to more than one infant during the course of my observations. Obviously, these 2 kinds of data give only the minimal time intervals between births. The data are valuable, however, because they clearly show that individual females do not give birth annually and that birth intervals are longer than for the majority of cercopithecinae species. The minimal estimates of birth intervals for specific females in the CW group are as follows: GCW > 14 and > 15 months; BT > 18 and > 8 months; KT > 27 or > 28 months; PGCW > 13 months; and SK > 27 months.[4]

The births of *b. tephrosceles* in and near compartment 30 of the Kibale Forest do not appear to be closely linked with particular seasons. As previously mentioned, infants with neonatal coloration were seen in nearly every month from May 1970 through March 1972. In August and November 1971, when no newborn infants were seen, observations were restricted to the CW group. The birth dates of 6 infants born into the CW group were determined accurately to within a few days or weeks on the basis of direct observation. The dates of two other births in the CW group were determined by back-dating on the basis of the infant's appearance at first sighting and on the basis of the ontogeny of pelage coloration in the 6 infants whose birth dates were known with greater accuracy. The 6 relatively precise birth dates are: 2-3 November 1970; 6 March to 5 April 1971 (KT's infant); 10-17 April 1971 (SK's infant); 5-29 September 1972 (BT's infant); April 1973 (DTK's infant); and May 1973 (ETT's infant). The other 2 births were estimated to occur between April and May 1972 (GCW's infant) and between July and August 1972. The seasonal distribution of these births is not particularly

striking, but there is a suggestion of a tendency toward 2 periods of birth
peaks; one from March through May and the other from July or August
through November. Except for the month of July and the first week or so of
August, the other months are typically ones of heavy rainfall. Obviously,
more data are required to substantiate this suggestion. Among such data are
regular monthly records on many individually recognizable females, because
the distinct neonatal coloration is maintained for 2-3 months after birth.
Consequently, births occurring near the end of one of the 2 annual peak
periods of birth could be confused with births occurring at the beginning of
the subsequent period.

Other Red Colobus Subspecies

As discussed in the section on heterosexual behavior, the data suggest a trend
toward seasonality in the births of *b. temminckii,* but not for *b. badius* or
b. preussi.

Comparison with Other Colobinae

None of the other studies of red colobus present data on births. Similarly,
data on gestation period, birth interval and birth seasons are not yet available
for the other African colobinae.

   Napier and Napier (1967) report a gestation period of 166 days (5.4
months) for *Nasalis larvatus* and 168 days for *Presbytis* sp.

   Among different populations of *Presbytis entellus* there is considerable
variation in the monthly distribution of births; some are very seasonal and
others are not (Jay 1965, Yoshiba 1968, and Vogel 1971). There is no
prominent birth season in the Dharwar population of South India, but there
seems to be considerable synchrony in the time of births within a particular
social group. Different groups have their infants at different times (Yoshiba
1968). This is undoubtedly related to the male replacement, and consequent
infant killing, which is followed by estrus in the mothers of the infants. Most
studies of entellus have been of relatively short duration and do not permit
direct determination of gestation period or birth interval. However, on the
basis of indirect evidence, Jay (1965) estimates the birth interval to be about
20-24 months. Yoshiba (1968) indicates that the birth interval is at least this
long in some of the Dharwar entellus, among which young monkeys have been
seen to be nursing from their mothers at 20 and 24 months of age. These birth
intervals are similar to those estimated for *b. tephrosceles.*

   Rudran (1973a) found a birth peak coincident with the dry season among
the *Presbytis s. senex* of Polonnaruwa, Ceylon, but births did occur throughout the year. In the more uniform climate of Horton Plains, there was no
apparent birth season among the *P. senex monticola,* although within specific social groups there was a tendency toward birth synchrony. Rudran
(1973) estimates the gestation period of *P. senex* at 6.5 to 7.25 months, which
may be somewhat longer than that of badius. The interbirth interval at

Polonnaruwa is about 22–25 months and similar to that of the badius, but this interval is much shorter at Horton Plains, being about 16–17 months.

The data on births for *Presbytis johnii* suggest that they may occur in 2 annual periods (Poirier 1968*b*). Birth intervals for this species were estimated to be 18–24 months, but supporting data are not available (Poirier 1968*b*).

INFANT ONTOGENY

C. b. tephrosceles

As described in chapter 2, newborn infants have black, silky pelage on their back and sides and a grey ventral surface. The coat is totally lacking in red or brown, making infants very distinct from older monkeys. The muzzle, ears, palms, and soles are usually pink. The birth date of 6 infants in the CW group was determined rather accurately on the basis of direct observations. The age at which the infant first acquired a tinge of brown color on its cap was determined for only 4 of these 6 (one had not yet acquired the brown cap at the time of writing). These ages in months are: 3, 1.25–2.25, 1, and 2.5–3, with a mean of 2.1 months. The infant's cap becomes a rusty red and the rest of its body color assumes the adult color at about 3.5 months, as determined from 4 infants (4, 3.5–4.5, 3.5, and 1–4 months).[5] Three infants were reclassified as young juveniles at 9.5–10.5, 9–10, and 8–8.5 months of age. Two young juveniles became juveniles at 18–19 and about 22.5 months of age, respectively.

Recognition of the individual infants in the CW group was, with only one exception, based on the adult female they clung to and associated with, and on their age relative to other infants in the group. The small size of the CW group, combined with the facts that newborn infants are not handled or carried by other members of the group and that infants of exactly the same color phase were never present in the group simultaneously, makes reasonable the assumption that the infant carried by a particular adult female is the same from day to day and month to month. The validity of this method for distinguishing infants was further supported by observations of the infant of female BT, who developed a distinct notch on its tail. This infant never clung to, suckled from, or remained in close association with any other adult female than BT during the 8 months it was under observation.

While being carried by the mother, infants always clung to her ventral surface. No infant or juvenile of any other age-class ever rode on its mother's back.

Newborn infants were often embraced with one or both hands by their mother as they clung to her ventral surface. During the first month the mother often placed one hand briefly to the clinging infant's back as she climbed with it. Mothers were seen to hold the tail base of their infant with one hand and muzzle its perineum. Neonates were frequently and usually groomed only by their mothers.

The method of sampling the CW group for 5-day periods at monthly intervals precludes precise dating of the locomotor and social ontogeny of these infants. However, at least the general ontogeny can be reconstructed. Infants did not begin to climb away from their mothers until 2 to 3.5 months of age, and even then they did not move much more than a meter from her. Solitary play also began at about this age. Juveniles and young juveniles often approached a mother with a neonate and attempted to touch the infant. The mother responded to these attempts by slapping toward the approacher, staring with extended forequarters, gaping, grimacing, poking or jabbing the approacher with one hand, and shaking the approacher. The approacher employed various behavior patterns in its attempt to contact the newborn infant. It gave the present type I and a variety of patterns usually restricted to play situations, such as gamboling in front of the mother, or giving her the open-mouth play face, or dangling beneath her by its arms or legs, or giving shrill squeals as it bounced in front of the mother and infant. In all cases the approacher was repelled by the mother. So-called "aunty" behavior is clearly not practiced or tolerated among *b. tephrosceles*. The age at which infants first made physical contact with monkeys other than their mother varied from about one month to 3.5 months, although this latter figure seems too great and may be an artifact of the sampling technique.

Infants did not commonly participate in play until they were about 3.5 to 5.5 months old. These play groups usually consisted of 2 monkeys and involved gamboling (hopping and bouncing about), grappling with one another's arms, and mouthing one another about the sides of the neck and on the shoulders. The open-mouth gesture was common in play. In this gesture the mouth is held about ½ to ¾ open with the unclenched teeth only slightly exposed. The mouth is held in this position for several seconds as the monkeys grapple with one another. Young monkeys sometimes hang from and climb up the tail of older monkeys, who usually show no response to this.

Weaning behavior was observed very infrequently and consisted of the female's pushing the young monkey away from her nipples or gently biting it about its head as it suckled from her. The youngest monkey to whom these weaning behaviors were directed was about 4–4.5 months old, but this infant was still suckling from its mother at 8–8.5 months of age, just before she disappeared. At least 3 infants were still occasionally carried by their mothers at 11.5, 10.5, and 14.5 months of age, respectively. Nursing continued well beyond the first year for at least 3 infants; $\geqslant 25.5$, $\geqslant 26.5$, and $\geqslant 14.5$ months, respectively.

Comparison with Other Colobinae

Neonate *Colobus guereza* have completely white coats and pink faces and ears, in sharp contrast to the black and white adult coloration. These newborn infants are highly attractive to other group members, particularly adult females other than the mother and, to a lesser extent, immature monkeys. The neonate is frequently handled and carried by these other monkeys during its first month of life, but this attention decreases in frequency as the infant ages, being virtually absent when it is 4 months old (Oates 1974). Similar attention to infants has also been described for guereza by Leskes and Acheson (1971). Social play can first occur during the 5th week of life in guereza, which is much earlier than for badius. The play encounters are very much like those of badius, with the open-mouth gesture, grappling, and mouthing being common components (Oates 1974).

Presbytis entellus are also born with a distinctive color and, like the guereza, receive considerable attention, handling and carrying by the other females in the group, particularly adult females. This infant transferal occurs as early as the first day of life (Jay 1965). At about 2.5 months of age, infant entellus first begin to climb away from their mother and other females, and at 6 months of age they frequently play with other infants. The neonatal color is lost at about 3–5 months of age. This rate of development appears similar to that of badius. However, it seems that weaning behavior is much more conspicuous in entellus than in badius and that weaning in some populations is completed earlier than in badius. Jay (1965) estimated that weaning was complete at about 12–15 months of age, but Yoshiba (1968) reports some entellus in south India as still suckling at 20 and 24 months, which is like the badius.

Presbytis johnii has a distinct neonatal coat, lasting until about 2.5 to 3.5 months (Poirier 1968b). The newborn infants of this species are also highly attractive to other group members and are handled and carried by them, but Poirier (1968b) thinks this occurs less in johnii than in entellus. Infants are carried only in the ventral position, and mothers rarely groom their own infants. Weaning begins at 7–9 weeks and is completed at 12 months (Poirier 1968b). Infant johnii would appear to become independent of their mothers much earlier than either entellus or badius.

The infants of *Presbytis senex* are born with distinctive coats (Rudran 1973). Their ontogeny of coat color is much like that of infant badius. The first pronounced change in coat color occurs at 1–2 months and they assume the approximate color of adults at 3–4 months. In contrast, they are weaned at a much earlier age than infant badius; no suckling occurs after 7 months of age.

The newborn *Presbytis cristatus* is also distinctive in color and is handled and carried by other group members (Bernstein 1968).

Among colobinae it appears that the red colobus of the Kibale Forest are exceptional in that their newborn infants are not handled or carried by other members of the group. J. F. Oates (pers. comm.) has also pointed out that the red colobus is the only species of colobinae, except for some populations of entellus, which is not territorial. He goes on to argue that social groups which defend territories against neighboring groups of conspecifics may have a greater need for closer intragroup social bonds. One way of achieving these bonds would be through the extensive attention given newborn infants by other group members. Further support for this hypothesis is gained from the scant and indirect evidence presented in chapter 2, which suggests that the immature animals are the most mobile element in red colobus societies and most often change their membership from one group to another. It seems plausible that the less attention infants receive from other members of the group the less strong are their ties to the group during the juvenile stage of life. Data on the relative mobility of immature monkeys in other colobinae would further test this hypothesis.

MISCELLANEOUS SOCIAL BEHAVIOR

Embrace.   Only 9 cases of embracing were observed in the study of red colobus, and all 9 occurred in the CW group of *b. tephrosceles.* In 6 cases the 2 participants sat facing one another. Three of these involved mutual embracing with the arms, and in the other 3 cases only one of the participants embraced the other. In 2 cases the participants sat side by side with only one of them embracing the other. Once adult male CW approached adult female BT and sat beside her with his ventral surface in contact with her right side, and then he embraced her with both arms—one across her ventral surface and one around her back. The initiator of the embrace was adult male CW 5 times; adult male SAM once; an approximate adult once; and a juvenile twice. The recipient of the embrace was adult male CW once; adult female BT twice; adult female GCW once; adult female PGCW once; unidentified adult females twice; and juveniles twice. The initiator of the embrace approached the recipient in 7 cases; the recipient approached the initiator once; and in one example the preceding event was not observed. After the embrace the recipient groomed the initiator in 5 cases; the initiator groomed the recipient once; the initiator moved away in one case; and both fed side by side in another. In only 2 of these 9 examples of embrace was there any indication of even low-intensity aggression. Once a juvenile stared with extended forequarters toward another juvenile who was playing with a young juvenile. The staring juvenile then approached and embraced the other juvenile, and afterwards the initiator moved away. In another case adult male ND supplanted an adult female over food, whereupon adult male CW approached and embraced the same female, and she groomed him afterwards. In general, however, there was no apparent stimulus situation evoking the embrace, and the social significance of this gesture remains obscure.

Although not described, Jay (1965) lists the embrace as a submissive gesture in the repertoire of *Presbytis entellus.* Sugiyama et al. (1965) present a photograph of 2 subadult female entellus embracing, which resembles some of the ventral-ventral embraces of badius. They conclude that "embracing is conducted in order to release the high tension which arises among monkeys" during aggressive encounters.

Adult male suckles from adult female.   Two cases were observed of adult male *b. preussi* suckling from adult females. In one case the adult male suckled from an unswollen female continuously for about 3–5 minutes, and she occasionally groomed him as he suckled. In a different group and several kilometers from the site of the preceding example another adult male suckled from an unswollen adult female, alternating between her left and right nipples at least twice. He continued this for about 1.5 minutes, and then she groomed him when he ceased suckling from her.

I have found no evidence of adult males suckling from adult females in any other cercopithecid species.

INTRAGROUP DISPERSION

Spread of the CW Group

The maximum spread of a group, expressed as the maximum linear distance over which members of the same group are dispersed, is often difficult to determine with precision under the observation conditions of the rain forest. Unless one can plot the location of all members of the group at any one moment, estimates of the maximum dispersion are usually minimal approximations. The spread or dispersion of the CW group was not sampled in a systematic manner. Only extreme patterns of dispersion and general tendencies were noted. Typically, the CW group was cohesive, having a maximum linear spread of 50 m or less. Very infrequently, the group subdivided into 2 segments. The maximum distance between such segments was once 70 m. The most extreme case concerned adult female BT and her young juvenile, who were estimated to be 70 m from the nearest other member of the CW group and thus gave the group a maximum linear dispersion of at least 100 m. This was on 3 June 1973, and female BT was not seen again after 10 June 1973, apparently having died or emigrated from the CW group. At the other extreme, the CW group was often aggregated in the crown of one tree. The maximum linear spread in one such case was estimated to be only 20 m. In another example 14 of the total 19 monkeys present in the CW group were concentrated into an area of only 7 m in diameter. The relatively small size of the CW group compared to that of other red colobus groups in the Kibale Forest probably accounts for its compact nature. Larger groups, such as the ST group, seemed to be typically dispersed over greater distances (circa. 75–100 m). The role of adult males in coordinating group movements and cohesion with their vocalizations is discussed in chapter 4.

Interindividual Distance in the CW Group

Patterns of intragroup dispersion were more precisely determined through a series of samples in which the distance between a specific individual and its neighbors was estimated. Samples were collected during 10-minute periods at half-hour intervals which began at 10 minutes and 40 minutes past the hour. I attempted to collect 20 such samples for each of 7 recognizable monkeys in the CW group during the period of 2 August through 17 September 1971. In fact, the sample size varied from 19 to 28 for these 7 monkeys (table

21). The CW group remained constant in number (19 monkeys) and composition during the sample period (see table 7). In each 10-minute sample all monkeys within 5 m of the sample monkey were scored. Their approximate distance from the sample monkey was tallied only when the spatial arrangement between them persisted for at least 5 seconds. If no monkeys were within 5 m the closest neighbor within 10 m was tallied. No more than one sample was collected for each of the 7 monkeys during any one 10-minute sample period.

Lumping the data from all 7 monkeys and all 155 samples indicates that red colobus usually have no other red colobus within 2.5 m of them, and that when they do have neighbors within 2.5 m, there is usually only one (fig. 6). Rarely are there more than 2 neighbors within 2.5 m of the sample monkey. There are, however, pronounced individual differences in the number of neighbors a monkey has and its spatial relations with them (table 21).

It would appear that adult males have fewer neighbors than do adult females and that the neighbors of adult males are farther from them than are the neighbors of females (table 21). Some of these differences between the sexes can be readily explained. For example, female KT was in possession of a 5-6-month-old infant at the time of the sample, and this greatly increased the frequency with which she had a neighbor within 2.5 m and also contributed to the short distance from her neighbors. Female BT was accompanied by a young juvenile at this time, who had an affect on her sample similar to that of the infant on KT's sample. Female GCW frequently had neighbors nearby; probably because she was in estrus at the time of the sample (see table 18). The differences in interindividual distances and numbers of neighbors among the adult males is less easy to explain. ND's relative separation from other monkeys may be related to his low dominance status.

Additional insight into the basis of individual differences in spacing is gained from examination of a matrix comparing the observed with the expected frequency with which particular individuals or age-sex classes are within 2.5 m of the different sample monkeys (table 22).[6] Clearly, certain monkeys tended to be near one another more than others. Female KT was within 2.5 m of her old infant much more than expected by chance, but was within 2.5 m of other individuals much less than expected. Her apparent gregariousness was really only a function of her infant and was not the result of her being particularly close to other adults. A similar situation existed for female BT and her young juvenile. Female GCW, who was in estrus, contrasts markedly with these other 2 females, for she had 3 adult males (CW, LB, and SAM) as her frequent and near neighbors. Interestingly enough, ND, the most subordinate male and the one who did the least copulating, was never within 2.5 m of GCW. The cohesiveness of adult males, which has been discussed earlier in this chapter and in the preceding chapter, is also reflected in the matrix shown in table 22, with males CW, LB, and ND all being within

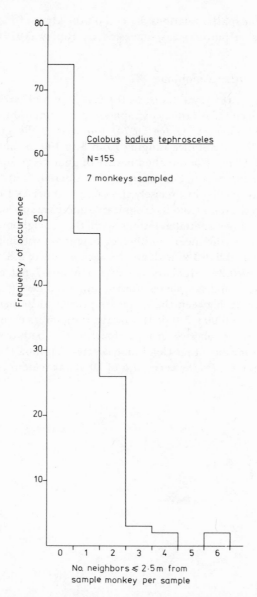

Fig. 6. Frequency histogram summarizing the number of neighbors within 2.5 m of the sample monkey per sample for 7 individuals of the CW group of *C. b. tephrosceles.*

2.5 m of one another more than expected by chance. There is no obvious explanation for the apparent affinity between adult male LB and one or more of the juveniles. It is clear from this analysis that intragroup spacing is dependent on a variety of social and physiological factors, but, in addition, it is also possible that spatial relations depend on the kinds of food being eaten and their dispersion patterns, as suggested by Ripley (1970) for *Presbytis entellus*.

Comparison with Other Colobinae

C. Guereza living in the same forest as the CW group of red colobus maintained much more cohesive groups in a spatial sense than did the red colobus. In a group of 13 guereza, Oates (1974) found that the group could be dispersed over a linear distance ranging from 3 to 100 m, but generally the spread was about 17 m. The bunched nature of guereza groups is even more apparent when interindividual distances are considered. Oates collected data on interindividual spacing in precisely the same manner as I did for the red colobus. He collected data from 9 recognizable guereza, and in virtually every index they show closer spatial relations with their neighbors than do the badius. For example, the mean number of neighbors within 2.5 m of all 9 guereza ranged from 0.6 to 3.9, with an average for all 9 of 2.8. This compares with an average of 0.86 neighbors within 2.5 m for 7 red colobus. When tested with a Wilcoxon 2-sample test for the unpaired case (Sokal and Rohlf 1969), this difference between the 2 species proved to be highly significant ($C = 53$, $0.01 > p > 0.002$, 2-tailed). Clearly, the guereza in the Kibale Forest live in spatially more cohesive groups than do the badius. There are no quantitative data for any other Colobinae species. Yoshiba (1968) notes that the troop dispersion for *Presbytis entellus* of Dharwar seldom exceeds 45 m in diameter.

4 Vocalizations

In the following consideration of the vocal repertoires of red colobus monkeys, most attention is given to *badius tephrosceles* because more data are available for this subspecies. A repertoire is described for each subspecies, and details of the stimulus situations are given for each call. Whenever possible, comparisons are made between call types of the different subspecies. Only for *b. tephrosceles* has it been possible to explore in depth the relations between vocalizations, social roles, and social behavior.

The vocal repertoires were constructed initially on the basis of my aural impressions. That is, the vocalizations were categorized in the field on the basis of the way they sounded to me. Tape recordings made in the field were accompanied by narration describing the stimulus situation of each recording, and the onomatopoeic name of each call was recorded. Subsequent spectrographic analysis in the laboratory revealed that sometimes I had lumped calls together which in fact had rather different physical properties. These cases are considered in detail in the appropriate sections. However, in general, calls that were initially classed together but later proved to have different spectrographic properties were considered as variations of a more general category. The main attribute of a vocal repertoire of this kind is that it attempts to provide an operational definition for separating calls which allows one to accumulate quantitative information on who gives what calls, when, and how often. Such quantification is essential to understanding the function of the calls. Vocal repertoires permit detailed comparisons, not only between different age-sex classes within a population, but also between populations and subspecies and thereby provide insight into phylogenetic relations. Thus, the detailed descriptions of the vocalizations are a necessary prerequisite for understanding the behavioral and evolutionary aspects of vocalizations.

Methods

All recordings were made in the field using a Nagra III recorder and a Sennheiser MKH 804 unidirectional microphone. The tape speed was usually 7½ ips, but occasionally 3¾ ips. Tapes were analyzed with a Kay-Sonagraph 6061B 85-16,000 Hz Spectrum Analyzer. The wide band filter was used, having a resolution of 300 Hz. Measurements of the sonagrams were made with a frequency and time scale generated by the internal calibration signals of the sonagraph.

Definition of Terms

The following terms are used to describe the physical properties of the vocalizations. A *unit* is the basic element of a sound, distinguished as a continuous tracing on the spectrogram. A *bout* is a cluster of one or more units and is separated from other similar clusters by a time interval approximately twice or more than that between units within a bout. *Duration* is the length of the unit in seconds. It is measured from the beginning to the end of the continuous tracing on the spectrogram. Whenever echoes were apparent in the spectrogram, they were eliminated in the measurement of duration. *Tonal sound* refers to sound having energy clearly separated into relatively narrow-frequency bands when analyzed with the wide-band filter. Harmonic sound is included here. *Nontonal sound* has energy that is continuously spread over a relatively wide-frequency range at the same instant. Noise is another term for this type of sound. *Band* refers to tonal energy that is developed at a higher frequency than the fundamental frequency of the unit. In truly harmonic sounds it is equivalent to an overtone. Band is used in those cases where the other frequencies were not harmonically related to the fundamental, or where the relationship to the fundamental was ambiguous. In other words, a band is a relatively emphasized portion of the frequency spectrum which does not appear to be an overtone. *Mean lower frequency* represents an average of all the measurements of the lower frequency of a particular sample of units. For example, if two units of a particular call type had lower frequencies of 2 kHz and 3 kHz respectively, the mean lower frequency for this sample of 2 would be 2.5 kHz. The lower frequency was considered to be the lowest frequency on the sonagram with a relatively dark tracing. Faint tracings occurring at frequencies lower than this were not measured. *Mean upper frequency* represents an average of all the measurements of the upper frequency of a particular sample of units. As for the lower frequency, the highest frequency of the spectrogram with a relatively dark tracing was measured. Two examples are presented here to further clarify these definitions. The bark in pl. 11 B is 0.09 sec. in duration and is nontonal, with energy ranging from 0.7 to 4.8 kHz. The yelp in pl. 11 P is 0.04 sec. in duration and is tonal with a fundamental ranging in frequency from 0.7 to 0.9 kHz and 3 overtones (bands) ranging up to 3.6 kHz.

The vocalizations were first classified on the basis of aural similarities and differences. Sonagrams were then made and measured. In general, the sample size of the units measured for each call reflects its relative frequency of occurrence in nature. In cases where a call is stated to be infrequently heard in association with a particular stimulus, it is implied that the stimulus was rare and not the response to it unless otherwise noted.

VOCAL REPERTOIRE OF *C. b. tephrosceles*

Twenty-five sounds have been described for the subspecies *C. b. tephrosceles* on the basis of aural impressions. Twenty of these were tape recorded and analyzed spectrographically. Some of the 25 may represent variations of only one call type, and a more conservative estimate of the number of sounds is 16.

Bark

The bark was given under the following circumstances: during intra- and intergroup agonistic encounters; in response to the loud noise of a falling tree branch; during both intra- and intergroup vocal exchanges when several individuals would call in apparent response to one another's calls; during group progressions; toward *Stephanoaëtus coronatus* (crowned hawk-eagle); in response to the noise of a fleeing and snorting giant forest hog (*Hylochoerus meinertzhageni*); in response to the rapid quaver calls from another badius group; once toward humans; once by an adult male after he had chased another adult; once by an adult male as he performed an incomplete copulation and again immediately after he dismounted; once by an adult male after he had been harassed by others during his copulation; and in several cases there was no apparent stimulus. This call was given primarily, if not exclusively, by adult males, and would appear to function as an alarm and a threat, and in both intra- and intergroup spacing.

Thirty-four barks were recorded and analyzed from 5 or 6 males in 3 different groups (pl. 11 and table 23). Thirteen were tonal and 21 nontonal in structure. The number of bands developed ranged from 1 to 6. The mean duration was 0.097 sec., but there was individual variation. Adult male LB, for example, provided 19 barks, which ranged in duration from 0.08 to 0.13 sec. and had a coefficient of variation (standard deviation/mean x 100, Sokal and Rohlf 1969) of 14%.

Some barks of *b. tephrosceles* greatly resemble some of the chirps of *b. temminckii* (pl. 18), although the latter would appear to be slightly longer in duration (table 28). The chirps of *b. badius* are less similar (pl. 16), but surely homologous to *b. tephrosceles* barks. The bark and chirp of *b. preussi* are also clearly homologous to *b. tephrosceles* barks, but the former appear to be longer in duration than the latter (pls. 20 and 21 and table 30).

Yelp

This relatively infrequent call is probably only a variation of the bark. It did, however, sound distinct enough to the observer to be classified separately. It was given after a rapid quaver on 2 occasions; in response to a rapid quaver from another group once; during intergroup agonistic encounters once; and once by adult female KT as adult female BT wrapped one arm around KT's neck and then groomed her.

Only one yelp was recorded (pl. 11 P). It was 0.05 sec. in duration, with a fundamental between 0.5 and 1.0 kHz. Three overtones were developed.

This yelp was quite different from that of *b. badius* (pl. 16), but more closely resembled the short-duration yelp of *b. preussi* (pl. 19 and table 30). The yelp of *b. rufomitratus* was of longer duration.

Chist[1]

One of the two most common calls in the repertoire of *b. tephrosceles,* the chist was give primarily by adult males under a wide variety of situations. Unless otherwise stated, the following descriptions refer to cases in which adult males were the vocalizers. Both adult males and females chisted toward humans. Once an adult male also shook a dead branch free and dropped it toward me as he chisted. It was given in response to the loud noises of falling tree branches and loud thunder claps; toward a fleeing duiker (*Cephalophus harveyi*); toward a fleeing and snorting giant forest hog (*Hylochoerus mein-ertzhageni*). It was also given in response to: the barks of bushbuck (*Trage-laphus scriptus*); the loud noise created by a bushbuck as it kicked a stone or log; the chatter calls of 2 species of squirrels (*Heliosciurus rufobrachium* and an unidentified side-striped species); the snuffling and roaring calls of nearby *Colobus guereza;* the chirps and pyows of nearby *Cercopithecus mitis;* the chirps, twitters, and hacks of nearby *Cercopithecus ascanius;* and the hoots and barks of chimpanzees that were directed toward the observer. It was also given toward low-flying crowned hawk-eagles (*Stephanoaëtus coronatus*).

Chists occurred in the following intragroup agonistic situations: during the leaping-about display and branch-shake display; by an adult male after chasing another adult; by an adult male as he chased another adult male; by an adult male as another adult male leaped toward and supplanted him; by one adult male in response to the agonistic encounter of 2 other adult males; by one adult female toward another as they had an aggressive interaction; by one adult male as he persistently followed a subadult male; by one adult male after a juvenile had approached him and then fled giving shrill squeals; by one adult male as he approached and presented type II to another adult male (once both males chisted during such an approach); by a subadult male as a female presented type I to him; and by an adult male as he chased a juvenile. In general, agonistic encounters have a high likelihood of being accompanied by chists.

Chists were sometimes associated with sexual interactions. A female once chisted as she fled from an adult male after an incomplete mount; occasion-ally an adult male chisted as he followed an adult female after an incomplete mount; infrequently an adult male chisted after a complete copulation; an adult male chisted after being harassed by others during copulation, and in response to the rapid quaver and wah! calls (see below). Less frequently chists were related to grooming encounters. Adult males sometimes chisted while being groomed by an adult female, after being groomed while climbing away from the female, or, occasionally, when the grooming female left the male.

Chists were also given by 2 adult males as an adult female fell from one branch to another and eventually to the ground. After lying on the ground in a semiconscious state for several hours, she died, from some undetermined cause. But it was only when she fell that the males chisted and not while she was on the ground. An adult female chisted as she climbed to the ground and retrieved her infant that had fallen within 8 m of me. Chists were also heard on another occasion as an older infant fell to the ground.

Chists by adult males were clearly related to the onset of group progressions. They were also given during progressions and on some occasions seemed involved in determining the direction of the progression. In examples observed in the CW group, the two adult males at the far ends of the group would chist almost continuously for several minutes as they moved in opposite directions from one another. Eventually the group would move in the direction of one of them, and the other would then join them. There were also cases in which obvious vocal exchanges between adult males within a group were not clearly related to group progressions. As in the cases in which only one adult male chisted in the absence of any obvious stimulus, the significance is not known.

Chists were also involved in intergroup relations. They were exchanged between adult males of different groups and given in response to wah! and rapid quaver calls in nearby groups. They were common during intergroup aggressive encounters and may also be related to the reunion of the group following such encounters.

On three occasions solitary males chisted toward the observer. One adult male was with a mixed group of *Colobus guereza* and *Cercopithecus ascanius*. One subadult male was with a group of *C. guereza* and another subadult male was with two *C. guereza*.

One of the more peculiar situations occurred when adult male CW chisted continuously as he closely followed an adult female *Cercocebus albigena,* who had a large perineal swelling; a super sexual stimulus?

Chists were also heard at night at the following times: 2245, 0230, 0430, and 0500 hours. At least one of these nights was moonless. On 2 of these nights rapid quavers accompanied the chists.

While an animal is giving the chist, the mouth is typically ½–¾ open, with the unclenched teeth exposed. Once, however, an adult female only retracted her lips, exposing her teeth, which were not quite clenched as she chisted toward me.

The chist would appear to function as an alarm and threat and in intragroup (cohesion) and intergroup (dispersal) spacing. There are, however, several cases in which functional explanations are not apparent, e.g., during grooming and sexual encounters.

One hundred and fifty-six chists were analyzed from at least 8 different adult males in 5 different groups (pls. 11 and 12 and table 24). Of these, 102 (65.4%) were tonal and 54 (34.6%) were nontonal. The number of bands developed ranged from 0 to 9. There was considerable intraindividual variation (see, e.g., pl. 11 G), but certain trends seemed consistent for particular individuals. Of 35 chists analyzed for male CW, 86% were tonal. Only 43.5% of the 23 chists analyzed for male ND were tonal, and 61.5% of male LB's 26 chists were tonal.

In contrast, an analysis of duration of chist calls failed to reveal a consistent pattern. Interindividual comparison of the samples in table 24 revealed only one significant difference. The sample for male CW taken on 16 February 1971 had chist phrases significantly longer in duration than those in the sample for male LB collected on 20 June 1971 ($p < 0.01$, two-tailed Median Test from Siegel 1956). All other interindividual comparisons from the duration data in table 24 revealed no significant differences at the 0.05 level. There were, however, two cases of significant differences in duration between samples for a given individual. The chists collected on 16 February 1971 for male CW had significantly longer durations than those recorded from him on 20 June 1971 ($p < 0.05$, Median Test). Similarly, the chists of male ND on 31 December 1971 were significantly longer than those given by him on 22 December 1970 ($p <_! 0.05$, Median Test). Comparisons of all other intraindividual samples in table 24 revealed no further differences in duration.

The one case of interindividual difference in duration of chists may contribute to individual distinctiveness in chist calls, but in general it appears that duration is not a character used to distinguish chists of particular individuals. The 2 cases of intraindividual differences could be accounted for by the circumstances under which the calls were given, i.e., longer chists might be given in one kind of situation than in others.

Interindividual distinctiveness in chist calls would appear to be based on characters other than duration, such as tonality and interphrase patterning. For example, adult male CW frequently preceded his chists with a wheet (see pl. 12 B, C, and H, I). Males LB, ND, and SAM, the only other adult males in the CW group, rarely called in this sequence.

Compared with the bark, the chist appears of slightly shorter duration and is more often tonal in structure. Infrequently, calls were heard which sounded intermediate to barks and chists, and the few recordings available of these would support the hypothesis that barks and chists represent points in a graded signal system (pl. 11 M, N, and O and see following call type). However, in terms of frequency of occurrence, these intermediate types were so uncommon that one rarely had difficulty classifying barks and chists.

The 3 western subspecies studied apparently lack a call that is structurally similar to the chist. The chirp appears to be the most closely related, but in all 3 western subspecies this call descends in frequency from its beginning to end (pls. 14, 16, and 21), whereas the chist ascends (pl. 11). The chist of *b. rufomitratus* males are very much like those of *b. tephrosceles*.

Bark-chist and Chist-bark

These infrequent calls are poorly understood but appear to be calls intermediate to the bark and chist. The small sample available for analysis consists of 2 bark-chists recorded from adult male LB (pl. 11 M) and 7 chist-barks recorded from males LB, CW, and ND (pl. 11 N and O). The 2 bark-chists were nontonal and were 0.09 and 0.10 sec. long. Five of the chist-barks were tonal and 2 were nontonal (the last 2 of pl. 11 O). They ranged in duration from 0.06 to 0.11 sec., with a mean of 0.08 sec. Their structural relation to barks and chists are most apparent from visual inspection of pl. 11.

Nyow

This call may be a variation of the bark, but the only sonagram available (pl. 11 Q) indicates an internal pulsation that looks different from the bark. The one example also appears to have nonharmonically related bands. The segregation is based primarily on my subjective judgment that it sounded different. This distinction was also made in the field notes. The nyow was given toward the observer and interspersed with chists. It was also given in response to the loud noise of a falling tree branch and thunder clap; in a vocal exchange between groups; and in response to rapid quavers in another, nearby group.

The single nyow recorded for *b. tephrosceles* was like that of *b. temminckii* nyows, which also have internal pulsing, nonharmonically related bands, and similar duration (pl. 15 A). It is also similar to *b. badius* nyows, which have internal pulsing and nonharmonically related bands but which are of longer duration (pl. 16 A, B, and F). Those of *b. preussi* have nonharmonically related bands and similar duration (pl. 20 F) but lack internal pulsing. Nyows of *b. rufomitratus* closely resemble those of *b. tephrosceles*.

Squeal-bark

Only 2 recordings were made of this call type (pl. 12 A). They were from two different adult males in two different groups. Both occurred in bouts of barks by these males. They sounded like a call intermediate to a bark and squeal. This may have been partly because one of the examples had both tonal and nontonal portions (second one in pl. 12 A). But structurally they are not obvious intermediates. Both examples were 0.19 sec. long.

Wheet[2]

This along with the chist was one of the two most common calls in the repertoire of *b. tephrosceles*. With only one exception, it was given exclusively by subadult and adult males. In the one exception an adult female gave a wheet, which, if not identical, was extremely similar to those of males. The stimulus in this case was not determined.

Like the chist, the wheet was given in a wide variety of situations. It was given toward humans; in response to barks of bushbuck, to the sounds of fleeing duiker, to the twitters of *Cercopithecus ascanius;* and toward crowned hawk-eagles.

It was perhaps most common during intragroup agonistic encounters, where it accompanied leaping-about displays, branch-dropping displays, and branch-shaking displays; it was given after chasing another adult; while chasing an old juvenile; toward agonistic encounters involving other group members; as an adult female presented type I to the vocalizer; by an adult male as he presented type II to another adult male; by an adult male as he supplanted others from food or spatial positions; by an adult male as he persistently followed a subadult male and also when reaching down slowly with one hand toward a subadult dangling beneath him.

In association with sexual interactions, wheets were given by an adult male: as he approached an adult female and then copulated with her, but not during the actual copulation; when following an adult female after an incomplete mount; between incomplete mounts each of which was accompanied with harassment by juveniles; and by an adult male as he embraced an adult female and then investigated her perineum.

Wheets were given by males as they approached adult females to groom or be groomed by them and when they were actually being groomed by them.

Wheets were given during group progressions and were related not only to the onset and cohesion of such progressions but also to their direction. As described in the section on chists, wheets were sometimes given as 2 adult males chisted and moved in opposite directions, one of them ultimately being joined by the other and the rest of the group.

Wheets were prevalent in intergroup encounters. They were given while animals stared at members of other groups; in response to the chists, wah!s and rapid quavers of other nearby groups; during intergroup aggressive encounters; and in apparent vocal exchanges between males of adjacent groups.

When adult males jumped onto branches that broke under their weight and then leaped onto another, they sometimes wheeted. A solitary subadult male who was with 2 *Colobus guereza* gave a wheet toward me. There were several other occasions in which there was no apparent stimulus.

The wheet appears to function as an alarm, a threat, and in intra- and intergroup spacing. As with the chist, there were situations in which a functional explanation is not apparent, e.g., during sexual and grooming encounters.

During this call, the mouth was held ⅔ to full open, with the unclenched teeth exposed.

Limited data suggest that the wheet develops ontogenetically from the shrill squeal of juveniles and may bear a relationship to the prolonged squeal of infants and juveniles (see below and pl. 12). One particular subadult male (SAM of the CW group) was heard to give only shrill squeals until 1 June 1971, when he gave a squeal that was distinctly intermediate to the shrill squeal and the wheet. This was given as he leaped from tree to tree in a leaping-about display. He again gave shrill squeals on 3 June 1971, but on 15 July 1971 he distinctly gave wheets and also the first chists that had been heard from him. Thereafter he was never heard to give another shrill squeal. In one other case an adult male gave typical wheets as a crowned hawk-eagle (*Stephanoaëtus coronatus*) flew in and perched near the group. He then began climbing toward the eagle, giving prolonged squeals which were very much like those of infants and juveniles. The eagle then flew off.

Sonagrams were analyzed of 38 wheets from 3 or 4 different males in 3 different groups (table 23 and pl. 12). Thirty were from adult male CW. All except one were tonal. Duration of this call type was extremely variable; for example, the 30 calls measured for male CW ranged in duration from 0.08 to 2.09 sec., with a coefficient of variation of 66.84%. Overtones were developed in 26 of the 38 calls, ranging in number from 1 to 3. Usually only one overtone was present. Adult male CW had a mean number of 1.39 overtones in the 23 of his calls which had any overtone; that is, 6 had no overtone and one was nontonal. Wheets were very often followed immediately by barks or chists from the same vocalizer (pl. 12 B, C; D, E, F; and H, I).

Structurally, there is no call resembling the wheet in the repertoire of any of the 3 western subspecies studied. The wheet of *b. rufomitratus* is virtually identical to that of *b. tephrosceles*.

Quavering Squeal

Only 4 recordings are available of this infrequent call (pl. 13 A–D). All were from one adult male and associated with chists which he gave in the same calling bout. The structural similarity to the wheet is striking. In fact, it is probably a variation of this call, differing primarily in the greater degree and frequency of pitch modulation. It is this modulation which lends the quavering quality to the call. The duration of these 4 calls ranged from 0.34–0.98 sec. with a mean of 0.56 sec., much like that of the wheets.

Squeals and Gasp of Dying Infant

A dark infant male, estimated as less than 4 months old, was once found on the ground in an extremely weak and relatively immobile condition. A group of badius was within audible range of the infant but was obviously moving away. After I had waited 25 minutes and it was apparent that no monkey was going to retrieve this infant, I took him back to my house, where I made extensive recordings of his cries. The cries recorded sounded like those he had given while lying on the ground in the forest. They were classified into two types; the squeal and the gasp (pl. 12 K and L). Forty calls were analyzed spectrographically — 31 squeals and 9 gasps. The squeals ranged in duration from 0.07 to 0.8 sec. with a mode of 0.18 sec. and a median of 0.17 sec. The 9 gasps were all much shorter in duration, with a range of < 0.01 to 0.08 sec. and a median and mode of 0.03 sec. The gasp occurred at the end of a squeal (pl. 12 L). Both call types were of low amplitude and high-quality recordings were only possible because the monkey was in hand. The squeals were all tonal and the gasps all nontonal. More than 12 harmonics were developed in one squeal, and in some cases the energy exceeded 16 kHz.

The tonality and high frequencies of this call resemble the wheets and the prolonged squeal of juveniles (pl. 12 M, and see below). They may, in fact, be ontogenetically related.

Prolonged Squeal

This call was given primarily, if not exclusively, by infants and juveniles. It occurred in apparent weaning and contact deprivation situations, such as when an adult female (presumed mother) pushed away and/or nipped a young juvenile or infant while it suckled or while it climbed along behind her. A very young dark infant (less than 4 months old) gave this call while sitting and climbing near an adult female and juvenile. A young juvenile called in this manner as it sat beside and once suckled from a dying adult female. On at least four occasions young juveniles gave prolonged squeals as they hesitated and made incipient leaps before crossing a wide gap between tree branches. Having once crossed the gap they ceased squealing and rejoined the group. One young juvenile gave this call almost continuously for 20 minutes before crossing a gap. A juvenile once gave prolonged squeals toward a large raptor, and, as described above in the section on wheets, an adult male once gave a squeal which sounded very much like the prolonged squeal toward a crowned hawk-eagle. Occasionally juveniles gave this call in the absence of any apparent stimulus. It was once heard at night (0430 hours).

While the animal is giving this call, the mouth is ½ to full open and puckered into an oval shape. The teeth are not exposed.

Only 4 prolonged squeals were available for spectrographic analysis (pl. 12 M). They were all given by one young juvenile as it climbed alone. Duration ranged from 0.28 to 0.80 sec., with a mean of 0.61 sec. All were tonal; 2 had one overtone (up to 11.8 kHz) and 2 had no overtones, like that in pl. 12 M. The possible ontogenetic relation of this call with the squeals of a dying infant and the wheet is suggested above.

No call with comparable structure to the prolonged squeal was found in the repertoire of the 3 western subspecies studied.

Shrill Squeal

This call was given primarily by immature monkeys. Only 4 times did adults give shrill squeals: once by a large adult male as he was chased by a smaller adult male; once by an adult female as she slapped toward another monkey; once by an adult female toward me; and once by an adult male when there was no apparent stimulus.

Most commonly, shrill squeals were given by immatures toward adults in aggressive or potentially aggressive situations. Toward adult males they were given in the following circumstances: by a young juvenile as it approached and then fled; by a juvenile as it harassed a copulating pair; by a juvenile as it approached and reached up with one hand toward an adult male, who then slapped at the juvenile, which then fled, returned, presented type I, and fled again; by a subadult as it presented type I while hanging upside down beneath an adult male; by a young juvenile as it sat in the lap of an adult male, when they were approached by an adult female with a clinging young juvenile who slapped toward the adult male, whereupon the young juvenile left him; by a young juvenile and, on another occasion, an old infant as they were roughly handled by an adult male; by an old juvenile as an adult male grabbed its tail base (as done when examining the perineum of adult females); by a juvenile as it ran past and very near to an adult male; by

(b)

(a)

Plate 1. *Colobus badius tephrosceles* (red colobus), adult males: (a) SAM, (b) ND.
Photos by Lysa Leland.

Plate 2. *C. b. tephrosceles,* adult female GCW.

Plate 3. *C. b. tephrosceles,* adult female with ventral clinging infant. Photo by Lysa Leland.

(a)

Plate 4. *C. b. tephrosceles* climbing: (*a*) adult female employing the slow crossed-alternating walk; (*b*) juvenile employing the gallop.

(b)

(a)

Plate 5. *C. b. tephrosceles* in mid-leap: (a) adult female with clinging infant; (b) adult male SAM. Photo b by Lysa Leland.

(b)

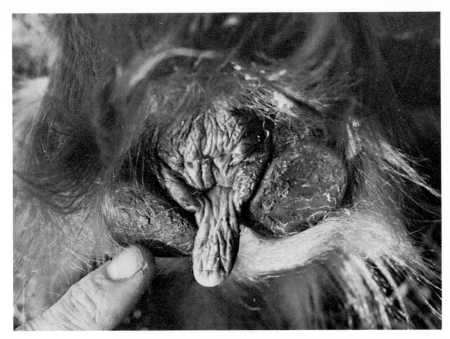

Plate 6. Genital region of adult female *C. b. tephrosceles* who died of natural causes (see chapter 5, section "Mortality"). The finger is touching her left ischial callosity. Note the large, but normal, clitoris.

Plate 7. Male infant *C. b. tephrosceles,* who died of natural causes (see chapter 5, section "Mortality"). He was estimated to be less than 4 months old at the time of death.

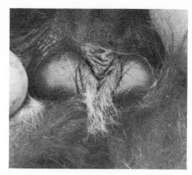

Plate 8. Perineal region of infant shown in plate 7. Note the pseudoclitoris, which is tipped with a tuft of hair.

(a)

(b)

Plate 9. *C. b. tephrosceles,* response of adult male CW to another badius group: (a) he stares toward the other group; (b) he partially extends his forequarters toward the other group just prior to moving toward them; (c) he moves toward the other group.

(c)

Plate 10. *C. b. tephrosceles,* adult male SAM manipulating and visually inspecting perineum of adult female.

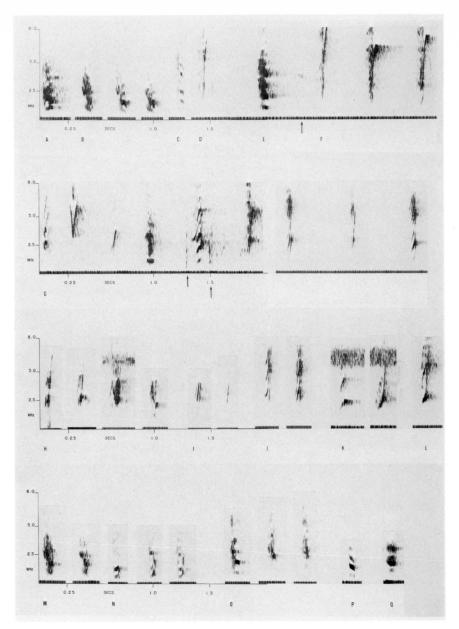

Plate 11. *C. b. tephrosceles.* A–C: barks; A: adult male ND; B: 3 from adult male LB; C: adult male CW. D–G: all from adult male CW; D: chist; E: bark (arrow indicates background noise); F: 3 chists; G: 3 chists, 1 bark, and then 2 more chists given as CW approached an adult female with whom he then copulated (the 2 arrows indicate background noise). The last 3 are chists. H: 4 chists from adult male ND. I: 2 chists from adult male LB. J: 2 chists from adult male TTT. K: 2 chists from an adult male in ST group. L: chist from an adult male of an unknown group. M: 2 bark-chists, adult male LB. N: 3 chist-barks, adult male LB. O: 3 chist-barks, adult male CW. P: yelp, adult male CW. Q: nyow. Ignore background noise at about 5–7 kHz in H and K.

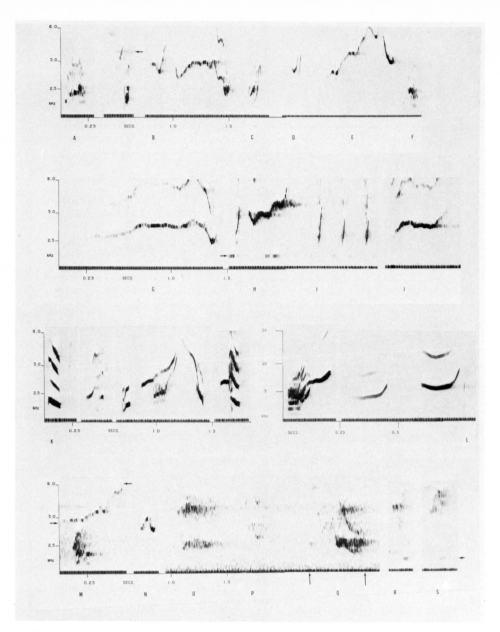

Plate 12. *C. b. tephrosceles.* A: 2 squeal-barks, from adult male LB and unknown adult male. B: wheet, followed by C: chist, from one adult male in BN group. D and E: wheet, and F: bark, from same adult male as preceding. G: wheet, adult male CW. H: wheet, followed by I: 3 chists, from one adult male in CW group. J: wheet, adult male CW as he approached an adult female with whom he then copulated. K: squeals from dying male infant, estimated as being less than 4 months old. L: gasp from same infant. M: prolonged squeal, young juvenile 11–12 months old (beginning and end are indicated by horizontal arrows) and a bark from another monkey (0.5–4.0 kHz). N: shrill squeal by a juvenile. O–Q: 3 shrill squeals (beginning and end of last squeal are indicated by vertical arrows) from one subadult male in CW group. R–S: 2 shrill squeals from one monkey, possibly subadult male of CW group (horizontal arrow indicates background noise). Horizontal arrows in A and H also indicate background noise.

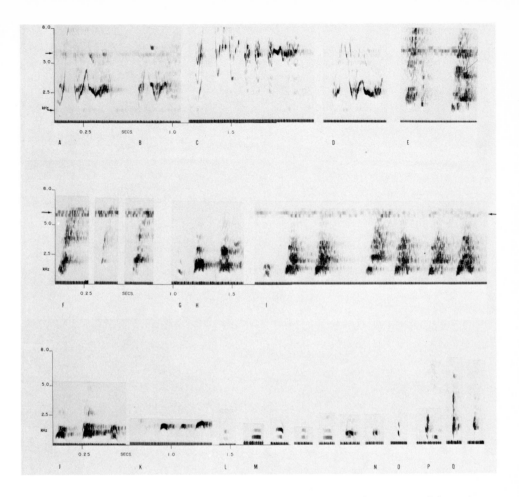

Plate 13. *C. b. tephrosceles.* A–D: 4 quavering squeals from one adult male (horizontal arrow indicates background noise). E: 2 screams (background noise at about 6 kHz and an artifact appearing as an oblique line in the second scream). F: 3 screams from one monkey. G: wah! and H: 2 rapid quavers, from one adult male (tonal types). I: 7 exhalations of a rapid quaver, probably given by an adult male as he harassed a copulating pair (first is tonal and last 6 nontonal types; horizontal arrows in F and I indicate background noise). J: 3 exhalations of a rapid quaver (first 2 nontonal, third tonal). K: 4 exhalations of a rapid quaver (the first is very faint and located directly above the K; all tonal). L: wah! probably by adult male as he harassed a copulating pair. M: 5 uh!s, possibly from 5 different monkeys. N: uh!-grunt. O: eh! P: 2-unit sneeze. Q: two single-unit sneezes.

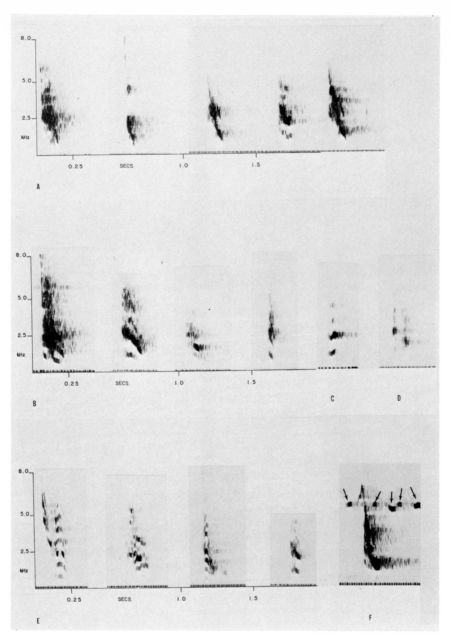

Plate 14. *C. b. temminckii*, chirps. A–B: common types. C: morning call. D: 2-unit chirp. E–F: uncommon variations. The first chirp in E was given toward the observer and the last is a morning call. The five arrows in F indicate background noise.

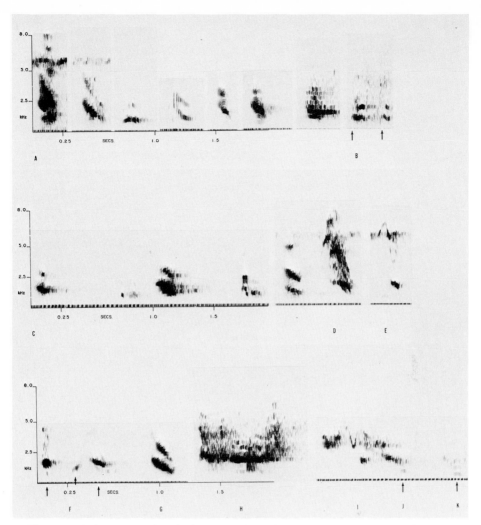

Plate 15. *C. b. temminckii.* A: 7 nyows; the third was given early in the morning, and the seventh phrase is an uncommon variation. Note the nonharmonically related bands in the first two. In the first two nyows there is notable background noise at about 5.5–6.0 kHz. B: 2-unit bark given early in the morning. The two arrows indicate the beginning of each unit. C: 5 nyows given by 3 or 4 individuals early in the morning. D: squeal or scream. E: squeal. Ignore background noise in D and E at about 6 kHz. F: A 3-unit squeal, arrows indicate each of 3 units. G: nyow from another monkey. H: scream. I: scream. J and K: 2 quavers, each indicated by an arrow.

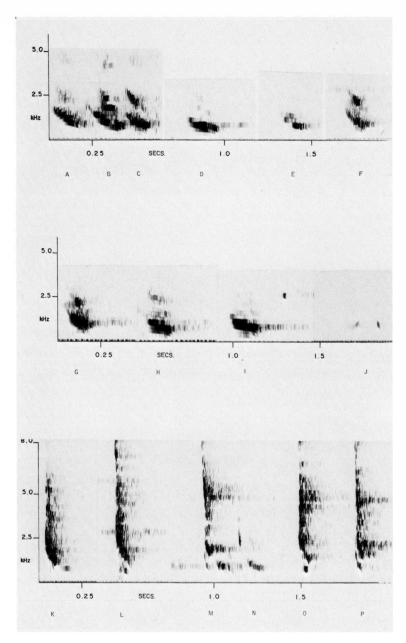

Plate 16. *C. b. badius.* A-I: 9 nyows. A-C given by 2 monkeys (note nonharmonically related bands in B), G and H by 2 adult males. J: yelp. K-P: 6 chirps, K, M, and N: 2-unit chirps.

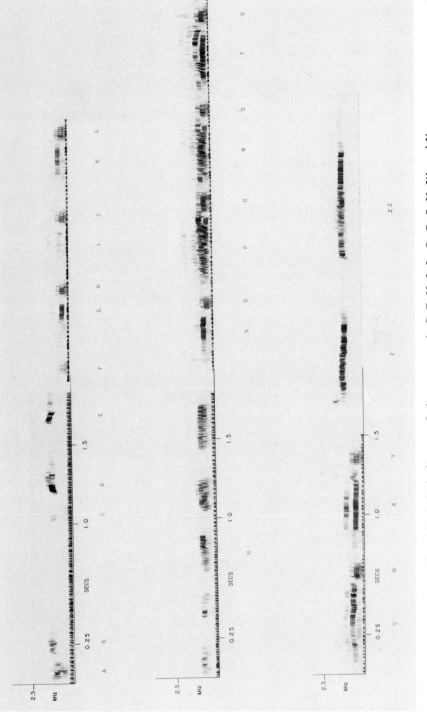

Plate 17. *C. b. badius*, copulation quavers. A, C, F, H, J, L, O, Q, S, U, W, and Y are inhalations. B, D, E, G, I, K, M, N, P, R, T, V, X, Z, and ZZ are exhalations. A–E from one monkey; F–L and N–Y from another; and M and Z–ZZ from 2 others.

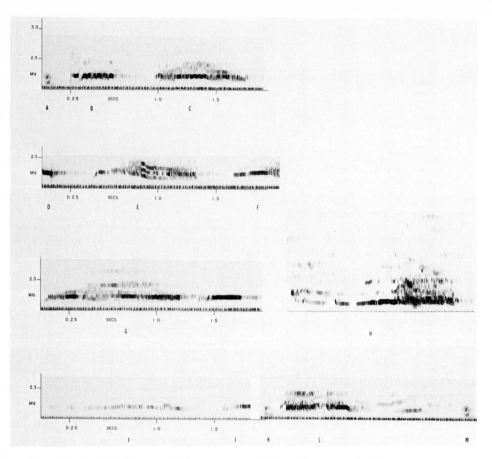

Plate 18. *C. b. badius,* copulation quavers. A, K, and M are inhalations; all others are exhalations. A brief spike of background noise reaching up to 3 kHz occurs at the beginning of B. D, F, and J are incomplete sonagrams. A–G and K–M were given by one female, and H and I–J were given by 2 other adult females.

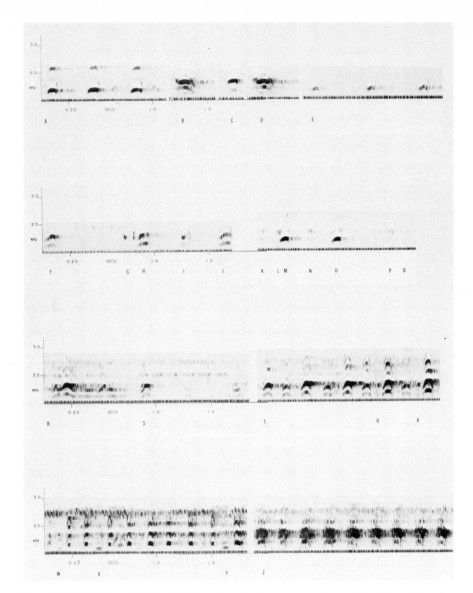

Plate 19. *C. b. preussi.* A-E: 9 yelps given during noncopulatory situations. Note the nonharmonically related bands in C–D given by 2 or 3 monkeys. E: these 3 yelps were given at the end of a rest period. F–Z: yelps and quavers during copulation. F, H, J, M, O, and S (3) are 8 yelps given by the adult male of 3 different copulating pairs. G (2 units), I, K, L, N, P, Q, R, T (9) and all of lower row (19) except W, X, and Y are exhalations of a quaver given by the female of 5 different copulating pairs. U, V, W, X, and Y are inhalations of a quaver given by a female of a copulating pair. Ignore background noise at 2.5 and 4 kHz in R–S and at 3 kHz in W–Y.

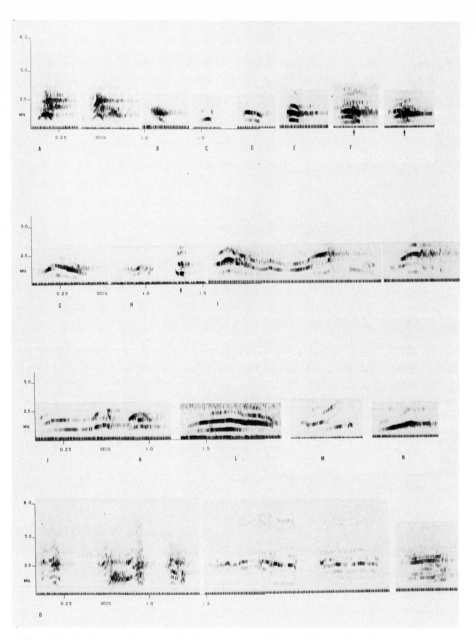

Plate 20. *C. b. preussi.* A: a single-unit bark followed by a 2-unit bark from a different monkey. B: bark after a rest period. C: bark. D: nyow, apparently related to group cohesion. E: nyow after a rest period. F: 2 nyows having nonharmonically related bands (indicated by arrows). G is a single-unit and H a 2-unit (second unit indicated by arrow) yowl from one monkey. I: 2 single-unit yowls from a different monkey. J–N: miscellaneous tonal calls. J: waa-wa, and K: wa, from one monkey. L: waa, like that given by female when pursued by male. M: wa-ah, note the nonharmonically related bands. N: whoop, possibly given during copulation. O: 6 prolonged sqwacks given by one adult male toward the observer.

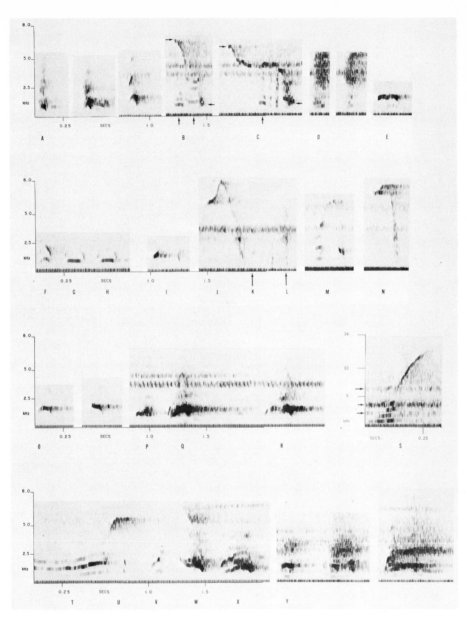

Plate 21. *C. b. preussi.* A: 3 chirps. B: sqwack-chirp (beginning and end indicated by horizontal arrows) from one and 2 yelps (indicated by vertical arrows) from another monkey. C: sqwack-chirp (beginning and end indicated by horizontal arrows) from one and one yelp (indicated by vertical arrow) from another monkey. Note that in this sqwack-chirp there are 2 vertical pulses of sound which are located immediately after the yelp from the other monkey. D: 2 sqwack-barks. E: Da-da. F: ka, G and H: 2 koos, all given by one monkey during agonism. I: squeal. J: high squeal, K and L: 2 iks, given by one monkey. M and N: squeals. O: 2 shrieks or screams. P, Q, and R: 3 screams from one monkey (possibly a female after copulation). S: high squeal (background noise indicated by 3 horizontal arrows). T–X: all from one monkey, T: OOO call. U: squeal. V–X: scream. Y: screams from one monkey. Ignore background noise at 3–4 kHz in A, B, C, D, J, K, L, M, N, P, Q, R, and Y and at about 6.5 kHz in X and Y.

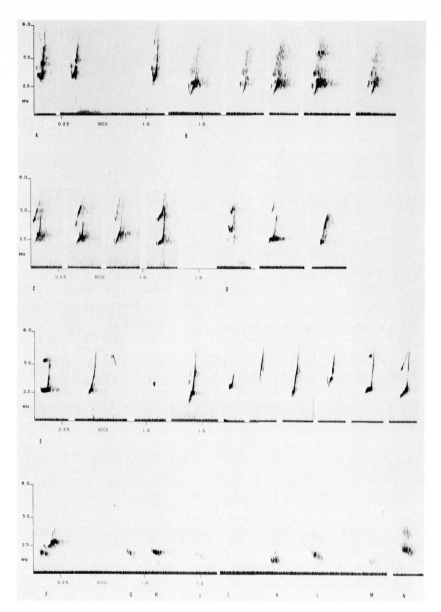

Plate 22. *C. b. rufomitratus.* A: 3 chists from one adult male. B: 5 chists from a different adult male given in response to a duiker that fled from observer. C: 4 chists from a third adult male. D: 3 chists from a fourth adult male. E: 11 chists from a fifth adult male. F: one chist and then 7 nyows (G–M) from one adult male. N: one nyow from a different adult male.

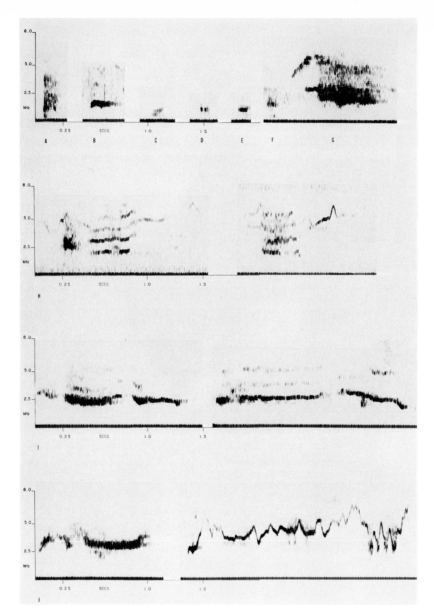

Plate 23. *C. b. rufomitratus.* A: sneeze. B: yelp. C-E: 3 grunts. F-G: 2 juvenile-like screams from one monkey. H: 2 weaning squeals from one old infant or young juvenile as it was pushed away by an adult female. I: 2 juvenile-like squeals from one monkey. J: 2 wheets from 2 adult males; the first wheet approximates a scream.

Plate 24. Lower slope habitat of compartment 30, Kibale Forest, featuring a large
Olea w. on the left.

Plate 25. Lower slope habitat of compartment 30, Kibale Forest, featuring a large *Aningeria altissima* in the center.

Plate 26. Medium slope habitat of compartment 30, Kibale Forest, featuring a large *Balanites w.* to the left of center. The exceptionally open understory here is presumably due to elephants who are attracted to the fallen fruits of this tree.

Plate 27. The interface between lower slope and valley bottom habitats of compartment 30, Kibale Forest.

Plate 28. A large juvenile *C. b. tephrosceles* foraging upside down along a branch of *Strombosia scheffleri* in the Kibale Forest.

Plate 29. A dead *Olea w.* whose wood has been fed upon by *C. b. tephrosceles* in compartment 30, Kibale Forest. Note the insect tunnel indicated by the finger and above this the tooth marks of badius.

Plate 30. Savanna woodland occupied by *C. b. temminckii* in the Forêt Classée de Narangs, Senegal, December 1969.

Plate 31. Forest-savanna mosaic occupied by *C. b. temminckii* near Somita, Gambia, January 1970. Red colobus frequently ran along the ground between the small relic patch of forest on the left and the single, isolated *Ceiba pentandra* on the right.

Plate 32. Part of a group of *C. b. temminckii* in a *Ceiba pentandra* at Niadio, Senegal, December 1969.

Plate 33. A view along the census route in compartment 13 of the Kibale Forest. This area was felled in 1968 and then selectively treated with arboricide in 1968-69. The photograph was taken in 1971.

Plate 34. Captive adult *Colobus guereza occidentalis* (black and white colobus). New York Zoological Society photo.

(a)

Plate 35. *Cercopithecus ascanius schmidti* (red-tailed monkey): (*a*) in the wild; (*b*) juvenile male in captivity, approximately 17 months old. Photo *a* by Peter Marler.

(b)

Plate 36. *Cercopithecus mitis stuhlmanni* (blue monkey). Photo by Peter Marler.

Plate 37. *Cercocebus albigena johnstoni* (gray-cheeked mangabey). Photo by Peter Marler.

a young juvenile as it followed and sat near an adult male; by a subadult male as he was persistently followed by an adult male; by an old juvenile as an adult male chased it; and by a juvenile as it leapt back and forth beneath an adult male who stared with extended forequarters and slapped toward the juvenile. Many of these situations were clearly cases in which the immatures were approaching and then withdrawing from the males in a kind of harassment.

Immatures gave shrill squeals toward adult females during apparent weaning or contact deprivation situations and as they approached and obviously tried to contact a young infant held by the female. For example, a young juvenile gave shrill squeals while seated next to an adult female, and another young juvenile did so as an adult female pushed and jabbed it away with one hand. An old juvenile gave them as it presented type I and closely approached an adult female with a newborn infant; it repeated this 3 times in rapid succession and each time the female raised one hand toward it as though to strike. A young juvenile shrill-squealed as it bounced and dangled in front of an adult female with a young dark infant. She grabbed and gaped toward the young juvenile. In one exceptional case a juvenile gave shrill squeals as an adult female bit its tail, and when it stepped aside she fed where it had been feeding.

Less frequently, immatures gave shrill squeals toward other immatures. For example, a young juvenile gave them as a juvenile grabbed its tail. Only once did a young juvenile give shrill squeals toward me. These calls were also heard during intergroup agonistic encounters, but the identity of the vocalizers was not determined. Occasionally, immatures were seen running through the trees giving shrill squeals in the absence of any apparent social interaction or other stimulus.

The mouth position was variable during this call. Once it was described as being ¾ open with the unclenched teeth exposed. In 3 other cases the mouth was puckered and about ⅓-½ open, and the teeth were not exposed.

Twelve shrill squeals were analyzed spectrographically—6 from one subadult male and 6 from one juvenile of the CW group (pl. 12 N-S). Two were nontonal (pl. 12 R and S) and 10 were tonal (pl. 12 N-Q). Even some of the tonal calls were rather noisy in that each band was spread over a wide frequency range at any given moment; see, for example, pl. 12 O and Q. This noisiness may be an important character distinguishing this call type from others such as the prolonged squeal. Duration of shrill squeals ranged from 0.08 to 0.54 sec., with a median of 0.17 and modes of 0.11 and 0.20 sec.; all shorter than the prolonged squeal. One overtone was developed in 5 of these shrill squeals, but none exceeded 8 kHz. The possible ontogenetic relation of the shrill squeal to the wheet has been discussed, but the structural affinity is not particularly striking. The structural similarity of the shrill squeal to the scream is, however, more apparent (pl.12 O and Q versus pl. 13 E and F), and there may be an ontogenetic relation here as well.

Some of the squeals given by *b. temminckii* (pl. 19 E) and screams by *b. preussi* (pl. 15 Y) resemble the shrill squeals of *b. tephrosceles* and may be homologues, but the data seem too few to warrant much speculation. No recordings are available of *b. badius* squeals. Similar squeals from *b. rufomitratus* were of longer duration and more tonal.

Scream, Sqwack, and Shriek

These terms may only refer to variations of one basic call type. Because sonagrams are available only for screams, however, the separation is based on subjective impressions of acoustical differences. All such calls were associated with agonistic encounters. The scream was given in the following situations: by an adult female as she dangled beneath an adult male who chisted toward me; by a young juvenile as an adult male grabbed and shook it; by an adult male as he was chased by one adult male and then as he chased a third adult male; by an adult male after he had given a wah! and a rapid quaver (see below); during an intergroup conflict; and toward at least 2 chimpanzees who seemed to be attacking the badius.

The sqwack was given in inter- and intragroup agonistic encounters. It was given by a juvenile as it also shrill-squealed while being persistently followed by a subadult male; and by a young juvenile as it followed, sat next to, and occasionally tried to suckle from an adult female.

Shrieks were given by 2 adult males as they approached a dying adult female as she fell (see above, section on chists); by an old juvenile as it fled from, and looked back over its shoulder toward, an adult male; by an old juvenile as it advanced and retreated from an adult female with a clinging dark infant as she slapped toward it; by a subadult male as he was chased by an adult male; by a juvenile after it approached and as it was fleeing from a subadult male who remained seated; by an old infant immediately prior to falling 12 m; and toward at least 2 attacking chimpanzees (see scream, above).

Only 6 sonagrams of the scream are available for analysis (pl. 13 E and F). All were nontonal and of relatively short duration (range 0.04–0.23 sec., mean 0.12 sec.).

Their possible ontogenetic relation with the shrill squeal has been mentioned. Examination of pl. 13 also shows a similarity between the scream and the rapid quaver. They have similar durations (see below), but the energy of the scream tends to occur more generally at higher frequencies. Also, about 24% of the rapid quavers were tonal, and their temporal arrangement within a bout was distinctive (see below).

The screams and related calls of the 3 western subspecies studied were of longer duration than those of *b. tephrosceles*. In addition, those of *b. temminckii* (pl. 15) and *b. preussi* (pl. 19) had more tonality. Screams of *b. rufomitratus* were poorly sampled and do not permit comparison.

Wah!

The wah! was associated with agonistic situations. It was most commonly given by adult males in conjunction with the rapid quaver toward a copulating pair. It was a type of vocal harassment often resulting in the disruption of the copulation. It was given less frequently among males as follows: by one adult male as he stared with extended forequarters toward another adult male, who then chased him; by an adult male as he chased and slapped toward another adult male, who fled and backed down a branch away from him; by a subadult male as he fled from an adult male who lunged toward him; by one adult male as another male grabbed and then chased him; and by an adult male as he also gave a rapid quaver and climbed over and away from another adult male. Others gave the wah! as follows: an old juvenile as it leaped aside and was supplanted from a feeding place by an adult female; an adult female as an adult male grabbed her tail base and she bolted away from him; an adult female as she moved away from a subadult male who followed her; an adult female as she approached a subadult male in a crouched and jerky manner, whereupon he stared toward her and she retreated; an adult female as she was supplanted by a subadult male; and an old juvenile as it fled upon the approach of an adult male. The wah! was also heard during intergroup agonistic encounters and during the attack by at least 2 chimpanzees mentioned earlier. An adult male once gave 4 wah!s in rapid succession as he scratched his right hand, apparently having been bitten by an insect. It was once heard at night; 0132 hours.

The wah! was usually given only once or, if repeated, only after a relatively long time interval. It was typically given with the rapid quaver — before and/or after it. Only two examples were recorded (pl. 13 G and L). They were 0.07 and 0.04 sec. long, respectively.

Structurally it closely resembles the uh! call (pl. 13 M and N), although to my ear it was distinct. The small sample does not permit further clarification of the basis for this distinction.

The wah! may be homologous to the wa! of *b. temminckii,* but there seems to be no counterpart in the repertoire of *b. badius* and *b. preussi.*

Rapid quaver[3]

Rapid quavers were given exclusively by adult males and usually toward copulating pairs as a kind of vocal harassment. This vocal harassment often resulted in the interruption of copulation. As mentioned above, it was frequently given in conjunction with the wah! Less often it was given by males in the following agonistic situations: by an adult male after he had given the rapid quaver toward a copulating pair and was being chased by the adult male of that pair; by an adult male as he gave a branch-shake toward another adult male; once by an adult male who was being chased by another adult male and, whenever the chaser ceased chasing, stared with extended forequarters and gave the rapid quaver toward the chaser, but ceased vocalizing when the chase resumed; by an adult male as another adult male grabbed him and then chased him; and once by an adult male along with a wah! as he climbed over and away from another adult male who showed no response. They were sometimes given during intergroup conflicts. Rapid quavers were heard on 10 different nights 2020, 2238, 2245, 0125, 0132, 0230, 0303, 0417, and 0500 (twice) hours. At least 2 of these nights were moonless.

Thirty-three rapid quavers were analyzed; 16 are illustrated in pl. 13 H–K. They were recorded from 3 or 4 different monkeys in 3 or 4 different groups. Eight were tonal and 25 nontonal. The average duration was 0.12 sec., but this character was quite variable, having a range of 0.03 to 0.22 sec. and a coefficient of variation of 36.66% (standard deviation/mean x 100). The lower energy in this call averaged 0.83 kHz with a range of 0.4–1.4 kHz. The upper energy ranged from 1.0 to 5.0 kHz, with a mean of 3.29 kHz. Over-tones were developed in only 3 of the 8 tonal calls. They extended up to 3.4 kHz. Typically there are 2–11 units per bout, with interunit intervals of 0.01–0.55 sec. The similarities between rapid quavers and screams are discussed in the section on screams.

The copulation quaver units of *b. badius* are more variable in duration, although generally longer than those of *b. tephrosceles* (pls. 17 and 18). The bouts are also much longer and the energy is usually more restricted in its range of frequencies in those of *b. badius* than in those of *b. tephrosceles*. The role of the vocalizer is quite different as well, for it is the female member of a copulating pair among *b. badius* that vocalizes.

The copulation quavers of *b. preussi* have similar durations and structural appearance to those of *b. tephrosceles*, but are generally more tonal (pls. 19 and 13). The energy of those from *b. preussi* is slightly lower in frequency and the bouts are much longer than those of *b. tephrosceles*. Again it is recalled that the female of a copulating pair of *b. preussi* gives the quavers, in contrast to the noncopulating adult male in *b. tephrosceles*.

Uh!

This very brief call is given primarily toward low-flying birds and presumably functions as a warning against potential avian predators. It was given toward the following avian species as they flew nearby the monkeys: black and white-casqued hornbill (*Bycanistes subcylindricus*), crowned hornbill (*Tockus alboterminatus*), harrier-hawk (*Polyboroides typus*), a small *Buteo* sp. (too small to take any but the smallest infant *badius*), pigeons (unidentified), black-billed turaco (*Tauraco schuttii*), and great blue turaco (*Corythaeola cristata*). It is certain that none of these birds prey on *badius*, and it would seem that this call is given in response to any low-flying object. For example, a harrier-hawk that was being mobbed by 2 glossy starlings soared among the *badius* at least 3 times within 2 minutes. Each time it came among the *badius* they gave uh! calls. Apparently, selective pressures have been great enough to favor this response to low-flying objects, regardless of the great number of "false alarms." It was once given toward a high soaring crowned hawk-eagle (*Stephanoaëtus coronatus*) that was calling in its typical fashion. This species is certainly a predator of *badius*. That the uh! call is primarily an alarm for avian predators is further supported by observations of the response to this call by nonvocalizing monkeys. Subadult males, adult females, juveniles, and young juveniles have all been seen to respond to the uh! call by looking toward the vocalizer and then upward in the direction from which an avian predator might strike. One badius was seen to drop to a lower branch in immediate response to this call. A juvenile once fled from the terminal end of a branch to the tree bole in response to an uh! Infants and young juveniles respond to this call by leaping into the laps of adult females who simultaneously reach out and grab them. One old infant leaped onto the back of an adult male in response to an uh! but was pushed off by him.

Much less frequently it was given during intergroup conflicts, although there is a remote possibility that I confused the uh! call with the acoustically similar wah! It was once given in apparent response to the chists from an adjacent group. On one occasion each it was given in response to the roars of *Colobus guereza,* the hacks of *Cercopithecus ascanius* and the Ka train (Marler's 1973 terminology) of *Cercopithecus mitis.* In these latter cases the badius may have, in fact, been responding to an avian predator which was evoking loud calls from these other species, because on many other occasions the uh! was not given in response to these calls.

Twenty-three uh! calls were analyzed spectrographically (pl. 13 M and N). They were recorded from perhaps as many as 5 monkeys in at least 3 groups. Duration was a fairly uniform character of this call (mean 0.07 sec., range 0.04-0.10 sec.), with a coefficient of variation of 21.6%. Twenty-two were tonal and one was nontonal. Overtones were developed in only 17 uh!s, with a mean number of 1.53. The mean lower frequency was 0.54 kHz (range 0.2-1.0 kHz) and the mean upper frequency was 0.93 kHz (range 0.6-1.3 kHz). Overtones reached up to 2.5 kHz. The uh-grunt (pl. 13 N) sounded slightly different but is, apparently, only a variation of the uh!

The uh! call is structurally distinct from other calls in the repertoire of *b. tephrosceles.* To the ear it sounds more like the wah! than any other call. The selective advantage of a distinct and relatively uniform call as a predator alarm is apparent. It is not confused with other calls and can be responded to immediately and appropriately without further information.

The unrecorded eh! call of *b. temminckii* would seem to be the counterpart to the uh! of *b. tephrosceles.* Structurally, the quaver (pl. 15 J and K) most resembles the uh!.

No counterpart to the uh! was found in the repertoire of *b. badius,* although some inhalations of the copulation quaver (e.g., pl. 19 I) and some yelps (pl. 19 S) structurally resemble it.

Eh!

This call was rarely heard and only one recording was made (pl. 13 O). It was the shortest call recorded, having a duration of only 0.02 sec. Only one other call had such a short duration, and that was one of the 5 sneezes (see below). No information is available on the function of the eh! call.

Sneeze

The sneeze was a very common sound among *b. tephrosceles*. It probably had no social significance aside from indicating the location of a monkey. While the animal was sneezing, the head and forequarters jerked forward and then returned immediately afterward to the normal upright position. The mouth was ½–¾ open and in an oval shape. The teeth were not exposed, but the tongue was sometimes partially protruded. A conspicuous spray of liquid from the nose and mouth was clearly observed during one sneeze. This description is based on eleven observations of adult males, adult females, old juveniles, and young juveniles. Sneezes were also heard at night between 2228 and 2234 hours.

The data indicate that the sneeze is, in fact, a sneeze, presumably caused by some respiratory irritation. The suggestion that it is a form of gas release (belch) does not seem feasible in view of the manner in which they sneeze. What is unusual is the very high incidence of occurrence of sneezing in *b. tephrosceles* as compared with other primates studied in the Kibale Forest and elsewhere. It was recorded from *b. temminckii*, but not with such great frequency. The sneeze of *b. rufomitratus* is like that of *b. tephrosceles*.

Five sneezes were recorded and analyzed from at least 4 animals in 3 different groups (pl. 13 P and Q). All were nontonal, with a mean duration of 0.058 sec. (range 0.02–0.11 sec.). Energy ranged from 0.08 to 6.0 kHz.

Trill

This call, which was not tape-recorded, sounded very much like the trill of *Cercopithecus mitis*. It was given only once by a young juvenile as it climbed and also gave prolonged squeals.

Flatulent-like Sound

Very infrequently heard and also not tape-recorded, this sound resembled a flatulation or possibly a belch and probably had no social significance.

General Comments

In *C. b. tephrosceles* it was primarily the adult males who approached and chisted toward humans, whereas the adult females and immature monkeys usually fled without vocalizing. This is in marked contrast to most Cercopithecines of Africa, among which it is primarily the immature monkeys and the adult females who approach and vocalize toward humans, the adult males rarely doing so.

On one occasion a dog ran through the forest barking toward the *b. tephrosceles*. Many of them remained sitting silently. Others moved away quietly. None vocalized. This was a very different response than that shown toward me by unhabituated *b. tephrosceles,* and I interpret it as a response adapted against people who are hunting with dogs, nonvocalizing monkeys being less conspicuous targets for hunters than vocalizing monkeys.

VOCAL ROLES OF *C. b. tephrosceles*

Within the CW group there were pronounced individual differences in the relative frequencies with which adult males gave particular calls. Male CW clearly gave the wheet more often than did males LB, ND, or SAM. This call type also constituted a proportionally greater part of his vocal repertoire than that of the others. Whenever possible, the identity of the vocalizer and the call type were noted (table 25). Since this particular aspect of the study was low on the list of priorities, sampling was opportunistic rather than systematic. It is possible that the data are biased toward more observable individuals or that data were only collected when little else was happening. The absolute frequency data, then, should not be given too much emphasis. However, the percentages are quite meaningful because it seems unlikely that a male is more conspicuous or observable when he is wheeting than when he is chisting. Nor does it seem likely that he will wheet rather than chist when little else is happening. Each bout scored in this study was separated from other bouts by an interval greater than that separating units within a bout. Often a bout of wheets was comprised of only one unit, whereas a bout of chists usually consisted of several units. Bouts of one call type were mutually exclusive from those of other calls. A bout of chists sometimes occurred immediately after a bout of wheets from the same male. In such cases two bouts would be scored — one wheet bout and one chist bout. For example, in table 25 it is shown that 246 vocalization bouts were scored for male CW. One hundred and forty-six (59.3%) of these bouts were comprised exclusively of wheets.

Some of the extremes in the range of percentage composition are due to small samples for certain months. Thus, in March 1972 only 5 bouts were scored in a 3-day period; no bouts were scored for male CW. There was, however, an apparent decline in the relative frequency with which male CW gave wheets. This decline was correlated with the physical and sexual maturation of male SAM. SAM first gave wheets in August 1971. In the 3 months prior to this he was intensively harassed by male CW. SAM gave many shrill squeals in this period. As of August, however, he was no longer harassed by male CW. Wheets thereafter came to constitute a large proportion of SAM's repertoire. Correlated with his maturation were other pronounced changes in the intragroup social structure, particularly with respect to sexual behavior (see chapter 3). Prior to this change, wheets constituted 68.4% (average of monthly percentages) of the bouts tabulated for male CW during a 10-month period. After this, from August 1971 through March 1972, wheets constituted only 25.5% (average of monthly percentages) of his bouts. This difference was highly significant ($p < 0.002$, 2-tailed, Mann-Whitney U-Test). Although some of the difference may be due to small samples in the last 8 months, some is, I believe, attributable to the change in male CW's social position. With the maturation of male SAM, CW was no longer the exclusive male copulator in the group. In fact, male SAM was exhibiting considerable sexual behavior, and his role in agonistic encounters suggest that as he matured he was becoming dominant to male CW, who until then had been the dominant male of the group. All of this clearly suggests that it is the dominant male in a group who gives the greatest proportion of wheets. Limited data from the BN and ST groups tend to support this hypothesis.

Males LB and ND are clearly chisters and barkers (table 25). Apparently in stimulus situations appropriate for wheets, chists, or barks, these two males have the propensity to give the latter 2 call types rather than the former one.

The apparent fact that males LB and ND gave proportionally more rapid quavers and wah!s than did the other two males is related to their infrequent participation as copulators and their frequent role as vocal harassers of male CW when he copulated.

The high proportion of shrill squeals in male SAM's sample occurred only during the first 3 months that he was sampled. During this period he was somewhat smaller in body size and not yet exhibiting sexual behavior. Male CW often followed him in a persistent manner, and this type of harassment was the primary stimulus evoking shrill squeals from male SAM.

FREQUENCY DATA ON VOCALIZATIONS OF *C. b. tephrosceles*

In an attempt better to understand the stimulus situations evoking certain call types and to contribute further to the description of these calls, quantitative data were collected on their frequency of occurrence. These data were all obtained from the CW group during 34 days in which it was followed continuously each day for at least 11½ hours.[4] On each day, all bouts of all calls were tabulated into 30-minute sample periods. The sample periods began 15 minutes before and lasted until 15 minutes after every hour and half-hour from 0645 to 1915 hours. For example, all bouts occurring between 0945 and 1015 hours were tabulated in the 1000-hour period, and all those occurring between 1015 and 1045 hours were tallied in the 1030-hour period. Bouts were defined as before, and one bout was scored for each animal calling. Thus, if 3 monkeys were giving chists simultaneously, 3 bouts would be scored. Bout length was dependent to some extent on the call type. Uh! bouts usually consisted of only one unit, lasting about 0.07 sec., whereas a chist bout typically had many units and lasted for several seconds.

All vocalizations which were obviously directed toward me were excluded. Changes in group composition during the 7-month sample period consisted of births and deaths of infants. No adult monkeys joined or left the group. Changes in group composition do not account for monthly variations in frequency of adult calls, although they obviously affect infant and young juvenile calls, such as the prolonged squeal. The two sample periods of 0700 (0645–0715) and 1900 (1845–1915) hours were biased against, because they were usually less than 30 minutes long.

The data clearly demonstrate that red colobus are extremely vocal and that the wheet, chist, and bark are the most common calls heard (table 26). Table 26 shows the extreme daily variation in the frequency with which most calls were given. For example, the number of bouts of the wheet varied from 21 to 270 per day. Such variability is not only poorly understood but compounds the problem of describing diurnal patterns in the frequency distribution of call bouts. With such variability from day to day, how does one determine the presence or absence of calling peaks? As is also discussed in the section "Temporal Distribution of Activities," in chapter 5, this extensive variability necessitates that one examine the relative changes that occur throughout each day rather than combining data for all days. For this reason, I have defined a peak of calling bouts for any given call as any frequency which is equal to or greater than 1.5 times the mean value of bouts per 30-minute sample period for each particular day.[5] For example, on 1 March 1971 there were 217 bouts of wheets during the entire 25 30-minute sample periods, or an average of 8.68 bouts per sample period. A peak of wheet bouts for this day would, by my definition, occur in any 30-minute sample period in which the number of bouts was equal to or greater than 13.02. Thus, the absolute value of peaks varies according to the number of bouts occurring on a particular day, but the relation to the mean value of each day remains the same. The number of peaks per day is much less variable from day to day than is the number of bouts (table 26). Peaks were computed only for those days in which the total absolute frequency of bouts for a given call exceeded 16. This was done because 1.5 times the mean of 16 bouts during 25 sample periods is 0.96, which means that any sample period with one bout constitutes a peak, i.e., all scores are peaks. Thus, in table 26 the number of bark peaks was computed for only 19 days, because on the other 15 days the daily frequency of bark bouts did not exceed 16.

Combining peaks for the entire sample describes the relation between calling frequency and time of day (figs. 7 and 8). Peaks clearly occur at all times of day for the 6 call types having adequate samples. Although the distribution of peaks is not uniform, there is no striking pattern. It appears that peaks of the chist and bark are more likely to occur in the morning. Wheet peaks, however, may occur somewhat more often in the afternoon. The absence of an apparent diurnal pattern for these calls may be related to the fact that they are evoked by a wide array of stimuli which are not rigidly governed by the time of the day. There does, however, appear to be a short-term cycle in peak frequency, as pointed out to me by Dr. Peter Waser. The biological significance of such a short-term cycle is not apparent. A similar cycle seems to exist for the shrill squeal and uh! calls. There is a

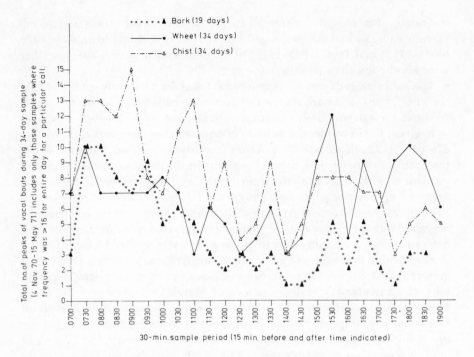

Fig. 7. Diurnal distribution of *C. b. tephrosceles* vocalization peaks.

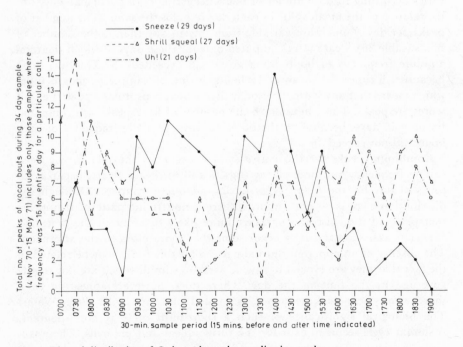

Fig. 8. Diurnal distribution of *C. b. tephrosceles* vocalization peaks.

slight suggestion of a trimodal curve for the uh! call, which may be a reflection of the flying activity of birds, evidently the primary stimuli evoking this call. A more pronounced pattern is apparent with the sneeze, although until more is known about the stimuli evoking sneezes, speculation seems unwarranted.

In chapter 5 it is suggested that, although activity is apparently related to time of day, this relationship may be affected by other factors which cause shifts in the cycle. Consequently, the relationship is of a general sort, and the precise timing of activity can be affected by weather, the attack of a predator, an intergroup conflict, etc. The same rationale is applicable to understanding the diurnal pattern of vocalizations. An analysis of intervals separating peaks of call bouts was made in an attempt to clarify the sequential aspect of vocal behavior. It is reasoned that, although the exact timing of vocalization peaks may shift one way or the other from day to day, the intervals between peaks should approach some predictable pattern. For each day the number of 30-minute sample periods separating bout peaks was tallied for each of the 6 call types having adequate samples (table 27). Clearly, the most common "interval" was contiguous. That is, given that a peak of bouts occurs in one sample period, the next peak is most likely to occur in the subsequent 30-minute sample period. This was true for all 6 calls, suggesting that peaks last longer than 30 minutes and/or that calling is contagious. (It is also possible that the most common interbout interval is less than 30 minutes, which also implies that calling is contagious.) The calling of one monkey may stimulate others to call, which would increase the chances of the peak's extending into the subsequent sample period. In fact, this exchange of calls seems to occur with the wheet, the chist, and the bark. The frequency of other intervals is inversely related to their length. If the interval between peaks were stereotyped, one would expect a relatively high frequency for one of these other interval classes. The absence of such a pattern is consistent with the earlier suggestion that, because a wide variety of stimuli evoke 3 of these calls, one does not expect a highly predictable diurnal pattern.

VOCALIZATIONS AND GROUP PROGRESSIONS

It was suggested earlier that one of the functions of the wheet, chist, and bark calls was to enhance intergroup cohesion. This function was most apparent in relation to group progressions—a progression being defined as any move by the group of 15 m or more.[6] These calls were given just before, at the onset and during such progressions, probably acting as vocal cues to initiate the move, maintain group cohesion, and possibly affect the direction of the progression. As stated earlier, further support comes from cases in which two adult males at the far ends of the group chisted as they moved in opposite directions until eventually the entire group would follow one of them.

A test of this hypothesis compares the peaks of calling with the group's movements. The results are highly significant statistically and do support the hypothesis (figs. 9 and 10).

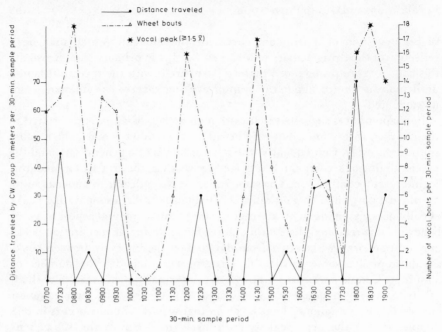

Fig. 9. Relation between progressions and vocalization peaks for the CW group of *C. b. tephrosceles* on 1 March 1971.

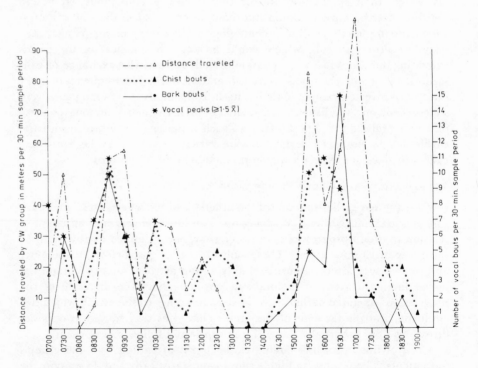

Fig. 10. Relation between progressions and vocalization peaks for the CW group of *C. b. tephrosceles* on 12 May 1971.

The hypothesis does not predict a direct correlation between the magnitude or frequency of calling and the distance traveled by the group, but rather that peaks of calling will occur just before and during group movements. Consequently, a correlation test is not appropriate. Instead, the progressions having peaks of bouts of wheets, chists, and/or barks in the preceding or the same 30-minute sample period were compared with those progressions not having such peaks in the preceding or same sample period. There were 260 of the former and 143 of the latter during the entire 34-day sample. This difference from expected values assuming the null hypothesis was tested with the χ^2 One-sample Test (Siegel 1956), with one degree of freedom, and found to be highly significant ($\chi^2 = 34$, $p < 0.001$).

A different but very similar test compares the vocal peaks of wheet, chist, and/or bark that occur in the same 30-minute sample period or the one immediately preceding the period in which a group progression occurs with those vocal peaks not so associated with group progressions. There were 376 vocal peaks that occurred in the same or preceding sample period as a group progression, and 99 that did not. Using the same statistical test as above, this difference was found to be highly significant ($\chi^2 = 162$, $p < 0.001$).

Clearly these 3 calls are related to group progressions, and the most plausible suggestion is that they function in initiating progressions and in maintaining group cohesion during progressions.

VOCAL REPERTOIRE OF *C. b. temminckii*

Of the 14 call types mentioned in the field notes, tape recordings were made of 6: chirp, nyow (bark), squeal, scream, quaver, and sneeze or cough.

Chirp

The chirp was given in the following situations: while fleeing from the observer; toward a harrier-hawk (*Polyboroides typus*) as it flew overhead; in the early morning near sunrise with no apparent stimulus; during intra- and intergroup agonistic encounters; directed toward the observer; immediately before a group progression across open ground; and upon sight of another group. Often there was no apparent stimulus evoking the call. It was sometimes given in chorus, as in the early morning when it was referred to as a morning call. When exchanged between two groups, this call was given by several members in each group. Nonvocalizing monkeys sometimes looked up toward those giving chirps even when the vocalizer was as far away as 100 m. It would appear that the chirp call has several functions, including intergroup and intragroup (group cohesion) spacing and alarm, and is somehow related to aggression, possibly as a threat.

When the monkey is seated and giving chirps, its head and forequarters jerk backward during the actual vocal emission and then return forward to the normal position. This backward jerk is repeated with each chirp. The mouth is approximately half open and the teeth are barely exposed during the chirp.

Both adult males and females gave chirps.

The chirp was apparently louder than the nyow. On at least one occasion a monkey gave a chirp call which then blended into a high-pitched quaver.

Sonagrams were made of 69 typical chirps (pl. 14) from at least 16 different individual monkeys. Measurements of these sonagrams, summarized in table 28, reveal considerable variation in duration and pitch (frequency). Fifty-eight percent were nontonal and 42% tonal in structure. In the tonal calls the average number of bands was 2.3, with a range of 1-4. In addition to these 69 sonagrams, 10 were made of atypical variations of the chirp call. These 10 were subdivided into 2 types (pl. 14)—8 of one and 2 of the other. They were all tonal in structure, having 1 to 6 bands developed over the fundamental. The first atypical variation differed most noticeably from the typical chirp in having more bands developed over the fundamental. The second differed most noticeably in having a fundamental that modulated over a much greater range of frequencies.

Nyow (bark)

The nyow is more common than the bark, which is considered a variation of it. The nyow and the bark are regarded as one call type because of structural similarity and were given under the following circumstances: upon sight of another group; immediately prior to a group progression across open ground; during intergroup agonistic conflicts; in the early morning near sunrise with no other apparent stimulus; and toward a harrier-hawk as it flew overhead. Often there was no apparent stimulus. Sometimes the nyow seemed to be given in countercalling between 2 or 3 monkeys within a group, resembling a chorus, and also between groups, involving as many as 3 individuals in each group. Other monkeys sometimes looked toward the vocalizer. It was given by adult males, possibly to the exclusion of other age-sex classes (see "General Comments" below). Nyows may function in intergroup and intragroup (group cohesion) spacing and as an alarm.

In contrast to the chirp, the nyow is given with the monkey's mouth scarcely open, the teeth not exposed, and the head and forequarters not jerked. This contrast was apparent when one particular adult male both chirped and nyowed.

Sonagrams were made of 40 common types of the nyow and of 4 uncommon variations (pl. 15 A), including one 2-unit bark (pl. 15 B). Recordings were made from at least 9 different monkeys. There was considerable variation in the duration (table 28) and structure of the common nyows (pl. 15 A–C and G). Only one of the 44 nyows was nontonal in structure. The average number of bands for the common nyow was 2.28, with a range of 0-6.

Squeal and Scream

The calls included in this category were given during intra- and intergroup agonistic encounters, including mild and intense aggression, and were also given toward the observer and while monkeys were fleeing from the observer. One particular adult male first screamed toward us and then chirped and barked. Complete sonagrams were available for only 3 single-unit squeals and 3 screams and for one 3-unit squeal (pl. 15 D-I). The separation of screams from squeals was done subjectively and may reflect the degree of tonality present in the call, a squeal having more than a scream. All 3 squeals were completely tonal, whereas the 3 screams had both tonal and nontonal components. Furthermore, squeals were shorter in duration (mean 0.21 sec.) than screams (mean 0.57 sec.), though with such a small sample this division should not be given too much emphasis. The distribution of energy was similar for squeals and screams, ranging from 0.9 to 9.5 kHz. The mean lower frequency of the fundamental was 2.79 kHz and the mean upper frequency was 6.13 kHz.

Quaver

Quavers were heard during intergroup conflicts and when another group was sighted. On one occasion a monkey quavered as another chased a third. Quavers were also directed toward the observer.

Only 2 sonagrams of the quaver were available for analysis (pl. 15 J and K). They were 0.14 and 0.13 sec. in duration and had energy distributed between 0.25 and 1.9 kHz. Both were tonal, but harmonics were developed in only one.

Sneeze or Cough

This was an abrupt and explosive sound, possibly of no social significance. Two sonagrams were available, measuring 0.02 and 0.09 sec. in duration and having frequencies ranging from 0.25-1.4 kHz. One was tonal with one harmonic developed and the other was nontonal.

Other Calls

Tape recordings were not made of the remaining calls, as listed and described below. Some may be only variations of calls described above.

Sqwack. This call was given upon the observer's approach and just before the monkey fled.

Rraugh. This was given during vocal agonistic encounters.

Whine. This call was also described as a whimper or coo. It was sometimes given in chorus and may have functioned in maintaining group cohesion, for example, during group progressions. This call was also given toward a harrier-hawk as it landed, but often there was no apparent stimulus.

Yelp. Given early in the morning, this call may have functioned in maintaining group cohesion.

Wa-ah! This was heard during a low-intensity agonistic encounter.

Eh! The eh! call was an unrepeated, but a contagious vocalization that spread through the group. It was given toward a harrier-hawk.

Woo. Pronounced frequency modulation gave this call a quaver-like effect, so that at times it seemed to be an incipient crow. Stimulus situations were not determined, although once it was given by a peripheral (straggler?) monkey about 100 m from the van of the group. The wooing apparently evoked an outburst of barks and chirps from the group van.

Wa! This abrupt call resembled an abbreviated quaver and was once given as another monkey squealed.

Ack. It was an abrupt and short call given toward the observer.

General Comments

Chirps, nyows (barks), whines, screams, squeals, and quavers sometimes occurred together in outbursts. In such situations there always seemed to be more sources of quavers and chirps than of nyows (barks), suggesting that nyows are given only by adult males, because they are typically outnumbered by adult females in heterosexual groups.

Vocalizations in response to the proximity of humans appears to depend on the age and, possibly, sex of the monkey and on the behavior of the human. For example, old juveniles and some adults (females?) fled from the author without vocalizing, but adult males invariably vocalized. Vocalizations were not given by any member of the group when a cyclist would pass beneath them on a path or road. However, if a human at the same distance were to stop and look at the monkeys, several would vocalize toward him.

VOCAL REPERTOIRE OF *C. b. badius*

Twelve call types were described in the field notes. Tape recordings were made of 5.

Chirp

This call was given in the following circumstances: in response to a falling tree branch; in apparent response to the barks of *Cercopithecus diana;* coincident with the onset of a group progression; associated with agonistic encounters, but it is uncertain whether the chirp was given by direct participants of agonism; toward a human and upon flight from a human; associated with the resumption of the group's activity following a rest period; toward *Stephanoaëtus coronatus* (crowned hawk-eagle); once in association with the copulation quaver, probably given by the mounter, who also barked (in all other copulations the mounter did not vocalize); and sometimes following a copulation quaver bout but not necessarily given by one of the copulating pair. Often the stimulus was undetermined. The chirp would appear to serve several functions. When it was given by several monkeys in an outburst, it would seem to enhance group cohesion, as when given during group progressions. It also appears to serve as an alarm and a threat call. The functional significance of this call during and after copulation remains obscure.

Sonagrams were made of 18 chirps, recorded from an undetermined number of monkeys but probably all from one group (pl. 16 and table 29). The majority were single-unit chirps, with 2-unit chirps forming a distinct minority of the sample. All were nontonal. A comparison of pls. 18 and 16 reveals the pronounced similarity between chirps of *b. temminckii* and *b. badius*. The greater variation apparent in those chirps by *b. temminckii* may be only a function of the larger sample available for this subspecies.

Nyow or Bark

The nyow, which could also be called a bark, was given under the following situations: in response to a falling tree branch; in apparent response to the barks of *Cercopithecus diana;* toward and when fleeing from humans; coincident with the onset of a group progression; in association with and after an agonistic encounter but not necessarily given by one of the direct participants; associated with the resumption of the group's activity after a rest period; toward *Stephanoaëtus coronatus;* twice in association with the copulation quaver and presumably given by the mounter, who also chirped; and sometimes after a copulation quaver bout, but not necessarily by one of the copulators. As with the chirp call, the stimulus evoking the nyow was often undetermined. It too was sometimes given in outbursts and may likewise enhance group cohesion, as when given during group progressions. In fact, the circumstances under which the nyow call was given are essentially the same as those for the chirp, suggesting similar functions. It was definitely given by adult males. Because there were fewer sources of this vocalization during outbursts of nyows and chirps, and because there are fewer adult males than females in a group, it is possible that only adult males give nyows.

Sonagrams were made of 31 nyow calls from at least 5 different monkeys and 3 different groups (pl. 16 and table 29). All were tonal with many closely spaced bands, ranging in number from 1 to 10 and averaging 5.6. Their appearance is indistinguishable from the nyows of *b. temminckii,* although those of *b. badius* may be slightly longer in duration.

Yelp

Although the yelp was sometimes associated with the resumption of the group's activity following a rest period, the stimulus evoking the yelp call was often not determined. Only two sonagrams are available for this call (pl. 16 J). These 2 yelps were tonal, measured 0.08 and 0.02 sec. in duration, and each had energy ranging from 0.75 to 1.25 kHz.

Screech, Scream, and Shriek

These 3 descriptive names were considered to represent variations of one basic call type. All were associated with agonistic encounters. An old juvenile or subadult badius once gave a scream as it ran toward a small hornbill, which then flew off. Only one recording was made. This screech measured 0.45 sec. in duration and had its tonal energy distributed between 3.4 and 4.0 kHz.

Copulation Quaver

This call was given exclusively by adult females in association with copulation. In some cases it was given only while the female was actually mounted by the male. In other cases she gave the quaver not only during the actual mount but also immediately after the dismount for a brief period. Copulation quavers were not given in all copulations but seem to have been given in most. It was a very common call during this study, heard in the Tai Reserve near Troya, as well as in the forest between Béréby and Tabou. Copulation quavers often occur in clusters and, when emanating from one adult female, suggest that *b. badius* are multiple mounters. They were also heard at night, indicating nocturnal copulation too.

While the female was giving copulation quavers, her mouth was puckered, with the teeth not exposed.

A bout of copulation quavers was sometimes terminated with a squeal. Typically a copulation quaver bout consisted of a series of short duration exhalations and inhalations, with some of the exhalations becoming relatively long in duration and sounding like a moan (pls. 17 and 18). A total of 142 exhalations and 57 inhalations were analyzed. They were recorded from at least 4 different monkeys in at least 3 different groups and comprise 16 different bouts of copulation quavers (table 29). All except 2 inhalation units were tonal in structure. Extreme inter- and intraindividual variation is apparent in the exhalations (pls. 17 and 18). Duration was a character of notable variation in the exhalations units (range of 0.06-1.89 sec.). The number of bands (possibly overtones) developed was also quite variable ($\overline{X}$ = 0.87, range = 0-6). Inhalations appear much more stereotyped and were generally shorter in duration than exhalations (table 29). The number of bands averaged 0.47, ranging from 0 to 3. The inhalations of *b. badius* appear to be the same as the quavers of *b. temminckii.*

The entire duration of copulation bouts, including intervals between units, was also quite variable. Complete recordings were made of only 2 bouts. They were 9.5 and 22.5 seconds in duration. Nearly complete recordings were made of 2 other bouts, measuring 7.25 and 30.5 sec. Three other bouts that were incompletely tape recorded, but for which estimates are available of the relative proportion the recordings represent of the total bout, had partial durations as follows: 15 sec. (approximately last ⅘ of bout); 9 sec. (approximately last ¾ of bout); and 7.5 sec. (approximately last ½ of bout).

Other Calls

The following calls were not tape recorded.

Sqwack, growl, yowl, and squeal.   All 4 of these call types were associated with agonistic encounters.

Roar-quaver.   Acoustically, this call resembled the copulation quaver. It was, however, associated with multipartite agonistic encounters.

Whine or whimper.   This was given by a young juvenile as it attempted to suckle from an adult female, i.e., placing its mouth near her nipples, but not making contact because of her averting movements. Eventually the adult female ran away from the young juvenile. This differs from the whine described for *b. temminckii,* because of the context and vocalizer.

Juvenile-squeal and juvenile-scream.   These are apparently variations of one call type. They were given just after a juvenile broke contact with an adult female and just prior to resuming the ventral clinging position on the female and suckling from her. I surmise that they are associated with the early stages of weaning.

General Comments

As with *badius temminckii,* there were cases in which *b. badius* fled from humans without vocalizing.

VOCAL REPERTOIRE OF *C. b. preussi*

A total of 18 calls were described. Fifteen of them were tape recorded.

Bark and Chirp

The bark and the chirp are apparently variations of one call type. The bark was the common form and occurred in the following circumstances: in response to a falling tree branch; upon the approach of a *Cercocebus torquatus;* in response to the adult male loud calls of *Cercocebus torquatus, Cercopithecus nictitans,* and *Cercopithecus pogonias;* after a vocal agonism, but not necessarily given by one of the direct participants; given by an adult male as he rapidly approached a vocal agonism involving others; upon the resumption of the group's activity following a rest period; coincident with the onset of a group progression; once given by an adult male as he copulated (males usually yelped during copulation); toward humans; and apparently in response to any disturbing or sudden stimulus. Barks often occur in outbursts and in some cases may be vocal exchanges between adult males within a group. I think they are given exclusively by adult males, and they seem to serve several functions, including alarm, threat, and group cohesion.

Thirty-two barks from at least 5 different monkeys in at least 3 different groups and 5 chirps from at least 3 different monkeys in 3 different groups were analyzed spectrographically (table 30 and pls. 20 and 21). The 5 chirps had a mean duration of 0.15 sec. (range 0.08-0.24), not apparently different from the barks. They too were all tonal in structure and had frequencies similar to the barks. The mean lower pitch of the chirp fundamental was 0.7 kHz, and the mean upper frequency was 2.18 kHz. The overtones had a mean upper frequency of 4.71 kHz. The chirps may have had fewer overtones (mean 1.8, range 1-2) than the barks (mean 2.5, range 0-5).

A comparison of the barks and chirps of *b. preussi* with the chirps of *b. temminckii* and *b. badius* reveals differences that may account for their acoustical distinctiveness (pls. 14, 16, 20, and 21). Those of *b. preussi* are all tonal, whereas 58% of *b. temminckii* are nontonal and all of *b. badius* are nontonal. The duration of *b. preussi* chirps approximate those of *b. temminckii* but appear to be slightly longer than those of *b. badius*. The upper frequencies attained in the chirps of *b. temminckii* and *b. badius* are higher than those of *b. preussi*.

Nyow

This call was given under virtually the same circumstances as the bark. All 34 nyows analyzed spectrographically were tonal (table 30) having 1-7 bands ($\overline{X}$ = 3.6). They were recorded from at least 6 different monkeys in at least 4 different groups. Their structural similarity to the barks is striking (pl. 20). The most apparent difference between these 2 call types is duration. A median test (Siegel 1956) shows the nyow calls to be significantly longer in duration than the barks ($0.01 > p > 0.001$). The functional significance of this difference, if any, is not apparent.

Nonharmonically related sounds occurring together in some of the nyow calls indicate that 2 different resonators are sometimes operating simultaneously (pl. 20 F).

Yelp

There are two types of yelps (pl. 19 and table 30). Those generally given during copulation (N = 91) by adult males are significantly shorter in duration than those given in noncopulatory (N = 62) situations (Median Test from Siegel 1956, $0.01 > p > 0.001$).

The yelps of shorter duration were usually given by the adult male during copulation and were simultaneous with the female's copulation quaver. Yelps did not occur in all copulations, however. This type of yelp was also given in the following situations: once an adult male yelped as he approached an adult female whom he then mounted; once an adult female groomed an adult male, but upon seeing us she fled, and the male then gave a series of short-duration yelps; once an adult male gave these yelps as he sat next to an adult female who quavered (usually done only during copulation); and once the call was associated with a chase.

The longer-duration yelps were given: after a vocal agonism but not necessarily by one of the direct participants; in association with an agonistic encounter; in response to the adult male loud calls of *Cercopithecus pogonias, C. mona, C. nictitans,* and *Cercocebus torquatus;* in response to a falling tree branch; upon the resumption of the group's activity following a rest period; and coincident with the onset of a group progression. This yelp type was sometimes given in outbursts by several monkeys, and in such cases may have been a vocal exchange between the adult males of the group. The stimuli evoking this call are much like those for the bark, chirp, and nyow. Its function may also be similar to theirs.

Both types of yelp are always tonal. The short yelp had 0–5 bands ($\overline{X}$ = 1.4) and the long yelp 0–7 bands ($\overline{X}$ = 1.6). Table 30 reveals that they are indistinguishable in the other parameters measured, with the exception of duration. These summaries are based on 62 long-duration yelps from at least 8 different monkeys in at least 4 different groups and on 91 short-duration yelps from at least 8 different monkeys in at least 4 different groups. They are readily distinguished from the yelps of *b. badius* in structure and duration (pls. 16 and 19).

Sqwack-bark

This uncommon call was given in response to a falling tree branch and toward humans. Only 2 recordings were made (pl. 21 D). They were 0.11 and 0.14 sec. in duration and were tonal, with the fundamental being between 0.6 and 1.0 kHz and overtones extending up to 5.5 kHz.

Sqwack-chirp

This call was given toward humans and at least once was definitely given by an adult female. The 2 recordings available for analysis were 0.26 and 0.58 sec. long and were tonal, with fundamentals ranging from 0.7 to 6.25 kHz and 0.9 to 6.9 kHz respectively (pl. 21 B and C). One and 5 overtones were developed respectively.

Prolonged Sqwack

This call was given by an adult male toward humans and once at the end of a copulation quaver by an unseen vocalizer (vocalizations from other monkeys did not usually follow copulation). The mouth is approximately ½ to ¾ open, with the unclenched teeth exposed. The mouth is closed exactly when the vocalization terminates. Twenty-one prolonged sqwacks were recorded and analyzed from 3 different monkeys and 3 different groups. Eighteen were from one adult male as he vocalized toward me. All were tonal, with 2–16 ($\overline{X}$ = 7.9) bands developed over the fundamental (table 30 and pl. 20 O).

Yowl

This whinelike call was given only by estrous (swollen) females, usually while seated alone. It was also given once by an adult female as she presented to an adult male before he mounted her and again immediately after the dismount. On another occasion a swollen female yowled as she approached an adult male and was then followed by him. Apparently, the yowl usually serves to signify the location of an estrous female.

During the yowl the vocalizer's mouth is ⅔ open and the unclenched teeth are exposed throughout. The ventral abdomen is pumped in and out throughout the vocalization.

Thirty-seven yowls were analyzed spectrographically (table 30 and pl. 20 G, H, and I). They were recorded from at least 5 monkeys in at least 2 groups. All were tonal and demonstrated some frequency modulation. The number of bands developed was 0-4 ($\overline{X}$ = 2.0).

Squeal and High Squeal

These two terms apply to variations of one call type, based on acoustical impressions. They were definitely heard in association with rapid yelps once and were presumably given by an estrous female. Once they were given toward humans (pl. 21 M and N) and on another occasion they occurred at the end of a bout of yowls, presumably by an estrous female.

Twelve of these calls were analyzed. They were recorded from at least 7 monkeys in at least 3 groups (table 30 and pl. 21 I, J, M, N, S, and U). All were tonal with 0-6 ($\overline{X}$ = 2.3) bands developed. The degree of frequency modulation was variable, but most pronounced in those classified as high squeals; compare, for example, pl. 21 I and J. The highest frequency attained was also in a high squeal; 13.5 kHz (pl. 21 S).

Da-da

Only recorded once, the da-da was presumably given by an estrous female. Although pl. 21 E clearly shows it as a single-unit call, it sounded like 2 units.

Copulation quaver.

The copulation quaver consists of a series of exhalations and rarely inhalations usually given by an estrous (swollen) female during copulation and immediately prior to and just after copulation. In this *b. preussi* differs from *b. badius,* among which it is given only during copulation. Once an adult female gave the copulation quaver as she sat next to an adult male who gave a rapid series of short-duration yelps. There was at least one copulation in which the female did not give the copulation quaver and another in which iks were given rather than quavers (see below).

During copulation quavers the female's mouth is ½-⅓ open and the unclenched teeth are exposed.

There are often several bouts of copulation quavers in close temporal association. When given by one female, they suggest that *b. preussi* are multiple mounters.

Copulation quaver exhalations were recorded and analyzed from at least 18 different monkeys in at least 4 different groups and totaled 315 calls (pl. 19

and table 30). Of these, 297 were tonal and 18 nontonal. The number of bands in the tonal calls averaged 2.4 (range 0–7). The vast majority were of short duration and resembled the yelps of adult males. However, longer variations did occur (pl. 19 R). The mean duration of 0.14 sec. is appreciably shorter than the homologous call for *b. badius* (0.58 sec.), although in distribution of energy they are remarkably similar (tables 29 and 30).

Only nine inhalations of the copulation quaver were recorded. They averaged about 0.04 sec. in duration and were all tonal (pl. 19). They too seemed to be of shorter duration than those of *b. badius*.

An idea of the temporal pattern of these copulation quaver bouts can be gained through an analysis of the interexhalation time interval. Two complete bouts were analyzed in this way. In the shorter of the two (about 7 sec. total duration), the mean interval between exhalations was 0.27 sec., with a range of 0.18–0.51 sec. for all 7 intervals. In the longer bout, where there were 80 intervals, the mean time between exhalations was 0.13 sec., with a range of 0.01–0.50 sec.

Measurements were made of the total duration of four copulation quaver bouts, including the interphrase intervals. They measured 24, 20.4, 13.4, and 7 sec. A fifth bout in which only about ⅘ of the bout was actually recorded lasted for 14.5 sec. Although the sample is small, it appears that bout duration is similar to the copulation quaver bouts of *b. badius*.

Ik

The stimulus situation of this call was observed only twice; once it was associated with a chase and screams, and on another occasion it was given by an adult female with a small perineal swelling as she copulated. Only 6 units were recorded, which, unfortunately, did not include this latter case. Classifying the calls given in these two situations as the same call type may not be justified.

The six iks analyzed were nontonal and had a mean duration of 0.052 sec. (range 0.03–0.12 sec.) (pl. 21 L). Their energy ranged from 0.08 to 2.8 kHz.

Miscellaneous Tonal Calls

Included in this category are 6 infrequently heard calls, most of which were tape-recorded only once or twice (pl. 20 J–N). The stimulus situations evoking them were never clearly seen, but they may all have been given by estrous females. The *waa-wa* and the *wa* were preceded by a sqwack. The *waa* sounded like a call given by a female being pursued by an adult male, to judge from other observations of encounters not tape-recorded. The *wee* (not illustrated) was once given in association with the da-da call and once was followed by copulation quavers. The *whoop* may have been given during copulation and spectrographically resembles a long-duration type exhalation of a copulation quaver. The *coo* (not illustrated) was followed by a scream from the same monkey.

All 9 units analyzed within this category exceeded 0.24 sec. in duration, averaging 0.43 sec. They were all tonal, with the energy of the fundamental ranging from 0.5 to 2.0 kHz. Some overtones reached 4.5 kHz.

Shriek, Scream, and Screech

These 3 terms represent variations of one call type based on aural impressions. Most typically they were given in agonistic encounters. In one case, screams might possibly have been given by a female after copulation. Juveniles definitely gave shrieks. Once a screech preceded a copulation quaver.

A total of 35 shrieks and screams were recorded and analyzed from at least 3 different monkeys (pl. 21 O–R and V–Y and table 30). Thirty-three were exhalations and 2 were inhalations. Thirty-two of the 33 exhalations were tonal and one was nontonal. Bands averaged 1.7 in number, ranging from zero to 7. Both inhalations were nontonal and were much shorter in duration (0.02 and 0.05 sec.) than the exhalations ($\overline{X}$ = 0.30 sec.). The inhalations had energy ranging from 0.9 to 2.4 kHz.

Structurally, these calls are like those of *b. temminckii* and *b. badius*.

OOO

Only one recording was made of this call (pl. 21 T). It was apparently given in an agonistic situation.

Ka Koo Koo

This too was recorded only once (pl. 21 F–H) from one monkey during an agonistic encounter. The ka phrase was 0.12 sec. long, and the 2 koo phrases were each 0.17 sec. long. All 3 are tonal with one overtone each. The ka reaches higher frequencies than the koos and shows more frequency modulation.

Other Calls

The remaining 3 call types were not tape-recorded.

Squeal-chist. This was heard once prior to copulation, but it was undetermined whether it came from the male or the female.

Screech-chist. Heard on one occasion only, this call was given by a swollen adult female in association with her yowls.

Yow. Another uncommon call, it was heard during an agonistic encounter.

General Comments

As with the previous 2 subspecies, some *b. preussi* fled from humans without vocalizing.

One is impressed with the basic structural similarity between yelps, barks, nyows, yowls, and copulation quavers. The main features separating these calls appear to be duration and interphrase interval and, to a lesser extent, the number of harmonics developed. The data suggest that modification of these characters around one basic fundamental frequency account for many of the different call types and their communicative distinctiveness in the vocal repertoire of *b. preussi.*

Vocal Repertoire of *C. b. rufomitratus*

During only 2 days of study (21 and 22 November 1971) 9 call types were recorded from *C. b. rufomitratus.* Very little information is available on context or function.

Chist

This call was given by adult males toward me and in response to fleeing duikers. Fifty-eight chists were analyzed from at least 7 adult males in possibly 4 groups (pl. 22 A-F). Mean duration was 0.07 sec., with a range of 0.03-0.20 sec. All but one were tonal. Overtones were present in only 41, with a mean number of 1.1. The mean lower frequency of the fundamental was 2.21 kHz, with a range of 1.40-4.75 kHz. The mean upper frequency of the fundamental was 5.85 kHz, ranging from 2.25 to 8.0 kHz. Overtones reached up to 8 kHz.

Comparison of pls. 11 and 22 reveals the striking similarity of the chists of *b. rufomitratus* and *b. tephrosceles.* They are similar in duration, frequency, and shape (both ascend from beginning to end of call).

Nyow

Recordings were made of 7 nyows from at least 2 monkeys. Some occurred in the same bout with chists. They had a mean duration of 0.10 sec., with a range of 0.07 to 0.12 sec. All were nontonal. The mean lower frequency was 1.16 kHz and the mean upper frequency was 2.5 kHz. Maximum frequency was 4.5 kHz.

The nyows of *b. rufomitratus* most closely resemble the nyows of *b. tephrosceles;* compare, for example, pls. 22 N and 11 Q.

Sneeze

Only one sneeze was recorded (pl. 23 A). It is similar in structure and duration to that of *b. tephrosceles.*

Yelp

The single yelp recorded (pl. 23 B) was much longer in duration than that of *b. tephrosceles* (0.12 versus 0.05 sec.). It occurred at the end of a bout of juvenile-like squeals during an aggressive encounter. Its appearance is not similar to any call of *b. temminckii* or *b. badius,* but it somewhat resembles the long yelp of *b. preussi.*

Grunt

Three examples of this call were recorded from one monkey (pl. 23 C–E). In duration it resembles the uh! call of *b. tephrosceles,* but differs from it in being nontonal. It also sounded different.

Scream

Two recordings of this call were made from one monkey (pl. 23 F and G). The calls sounded as though given by a juvenile. They are nontonal and somewhat resemble the screams of *b. tephrosceles,* but appear more like those of *b. temminckii* (pl. 15 H) and *b. preussi* (pl. 21 Y, 3d phrase).

Weaning Squeal

This call was given by an old infant or young juvenile as it was pushed away from and by an adult female. The 3 recordings were made from this one individual (pl. 23 H). They ranged in duration from 0.78 to 1.5 sec., mean of 1.14 sec. They are of longer duration, have more overtones, and cover a greater range of frequencies than do the prolonged squeals of *b. tephrosceles.*

Juvenile-like Squeal

This call was given at least once during an aggressive encounter. The vocalizer was not seen, but it sounded like a juvenile. Twenty of these squeals were recorded from at least 4 monkeys (pl. 17 I). Mean duration was 1.46 sec., with a range of 0.62–3.58 sec. All were tonal, with overtones developed in 19. The number of overtones ranged from 1 to 5, with a mean of 2.5. The mean lower frequency of the fundamental was 1.70 kHz, and the mean upper frequency was 3.68 kHz. Overtones reached up to 7.5 kHz.

These squeals were of longer duration and more distinctly tonal than the shrill squeals of *b. tephrosceles.*

Wheet

This call was given only by adult males toward me. Thirteen recordings were obtained from at least 4 different adult males in 4 different groups (pl. 23 J). They ranged in duration from 0.68 to 1.98 sec., with a mean of 1.34 sec. All were tonal. Overtones were developed in 6 only, with a mean number of 1.67. The mean lower frequency of the fundamental was 2.22 kHz, and the mean upper frequency was 7.25 kHz. Overtones reached 8 kHz.

These wheets are virtually identical to the wheets of *b. tephrosceles* in structure and duration. Some of them, such as the second unit in pl. 23 J, modulate more and thus resemble the quavering squeal of *b. tephrosceles.*

General Comment

C. b. rufomitratus seemed to vocalize far less than any of the other 4 subspecies studied, even in response to humans.

Interpopulational Comparisons

Detailed structural comparisons of call types from the 5 different subspecies
or populations have been made in the appropriate subsections above. Only
calls adequately sampled will be considered here. The comparisons are summarized in table 31. Screams, shrieks, and screeches bear great similarities in
all 5 subspecies studied and, in fact, greatly resemble such calls in more dis-
tantly related primates as well. This general similarity makes these call types of
little value in understanding phylogenetic relations among the varieties of red
colobus.

Greatest attention is given to those vocalizations concerned with intragroup
cohesion, because they are most likely to be involved with maintaining interspecific segregation and thus to be evolutionarily more conservative than
other call types. They are also the most common vocalizations and the ones
best sampled.

When *b. temminckii* is compared with *b. badius,* it is seen that the 2
subspecies are very similar in their bark, chirp, and nyow calls and in the
inhalation units of the quaver. One possible difference between these 2 forms
is that a copulation quaver was not found in the repertoire of *b. temminckii,*
but this could be due to an inadequate sample.

The differences between *b. temminckii* and *b. preussi* are striking in the
bark, chirp, and nyow calls. Furthermore, the yowl, copulation quaver, and
miscellaneous tonal calls of *b. preussi* were not found in the repertoire of
b. temminckii.

The bark, chirp, nyow, yelp, and copulation quaver calls of *b. badius* and
b. preussi differ from one another. In addition, the yowl and miscellaneous
tonal calls of *b. preussi* are apparently not given by *b. badius.*

The bark of *b. tephrosceles* appears similar to the chirp of *b. temminckii.*
Likewise, the nyow of these 2 forms seems to be similar. The wah! of
b. tephrosceles looks much like the wa! of *b. temminckii.* These 2 subspecies
differ in that the chist, wheet, and prolonged squeal of *b. tephrosceles* were
not found in the repertoire of *b. temminckii.*

The bark of *b. tephrosceles* resembles the chirp of *b. badius,* and their
nyows appear alike as well. However, they differ in their bark, yelp, and
quaver. The function of the latter call is also very different in these 2 subspecies. Furthermore, the chist, wheet, prolonged squeal, wah! and uh! calls
of *b. tephrosceles* were not found in the sample of *b. badius* vocalizations.
There was no copulation quaver in the *b. tephrosceles* repertoire.

Pronounced differences were found in the bark, chirp, nyow, and quaver
calls of *b. tephrosceles* and *b. preussi.* These 2 subspecies also differed in the
function of the latter call. The chist, wheet, prolonged squeal, wah! and uh!
calls of *b. tephrosceles* were not found in the repertoire of *b. preussi,* although *b. preussi* had a copulation quaver and *b. tephrosceles* did not.

The small sample of *b. rufomitratus* demonstrated striking affinities with that of *b. tephrosceles,* particularly in the chist, wheet, nyow, and sneeze calls. They differed in the yelp call. The grunt of *b. rufomitratus* was not found in the repertoire of *b. tephrosceles.* The calls of *b. rufomitratus* bore little similarity to those of the western 3 subspecies studied.

In summary, it appears that these 5 subspecies can be arranged into 3 groups based on similarity of vocal repertoires; (1) *b. temminckii* and *b. badius,* (2) *b. preussi,* and (3) *b. tephrosceles* and *b. rufomitratus.* The variety of call types appears to be richest for *b. preussi.* Although these distinctions can be made, it also seems that the 5 forms have enough similarities in their vocalizations to indicate phylogenetic affinity at least at the superspecies level.

There is considerable confusion regarding the taxonomic status which the various forms of red colobus should be given. They are usually considered subspecies of *Colobus badius* (e.g., Rahm 1970). However, Dandelot (1968) considers the red colobus to be comprised of 6 species and 11 subspecies. He aligns *b. badius* and *b. temminckii* as subspecies of *C. badius;* and *b. preussi, b. tephrosceles,* and *b. rufomitratus* along with 4 other forms as subspecies of *C. rufomitratus;* but he indicates some doubt about the placement of *b. preussi.* The data on vocalizations support Dandelot's conclusions on the close affinity of *b. badius* and *b. temminckii* on the one hand and of *b. tephrosceles* and *b. rufomitratus* on the other, but do not support the concept that *b. preussi* is particularly close to the latter two forms. Whether or not it is justifiable to divide the red colobus into 6 species cannot yet be answered. In a comparative study of *Cercopithecus* species, I concluded that vocalizations were relatively stable characters and very reliable indicators of specific affinities (Struhsaker 1970). Data on several widely separated populations and subspecies of 6 species of *Cercopithecus* revealed no intraspecific, qualitative differences in vocalizations.

These results however, do not permit one to extrapolate to the colobinae, that is, to claim that differences in the vocalizations of red colobus populations necessarily mean they represent different species. The vocalizations concerned with group cohesion and therefore presumably important in species recognition are likely to be the most conservative vocalizations from an evolutionary standpoint. The comparisons in table 31 indicate that, of the 5 populations studied, *b. preussi* is the most different relative to these calls. Perhaps it is specifically distinct from the others or is in a state of incipient speciation. It does, however, seem of equal significance that it is precisely in these calls that the widely separated populations of *b. temminckii* and *b. badius* on the one hand, and *b. tephrosceles* and *b. rufomitratus,* on the other, are most alike. This would support the hypothesis that these 4 populations are closely related and are possibly members of the same species.

It remains to be explained why these 3 groups of populations differ so in most of their vocal repertoires. The evolutionary rates and selective pressures affecting vocalizations of red colobus may be very different from those affecting *Cercopithecus* calls. One striking difference in the biology of red colobus and most *Cercopithecus* species is the degree of congeneric sympatry. Most forest *Cercopithecus* monkeys live in sympatry with at least 2 congeneric species — in some cases, with as many as 4 congeners. In contrast, red colobus live with no more than 2 congeners. Furthermore, if one accepts that black and white colobus are in a different genus from red colobus, then the difference is even more pronounced, with most red colobus forms having no sympatric congeners. Their only sympatric congener would be the olive colobus in West Africa, which too is considered by some to be in a separate genus. This means that selective pressures favoring stereotypy in vocalizations for species recognition would be less important in red colobus, having few or no sympatric congeners, than in the *Cercopithecus* monkeys, with their high degree of congeneric sympatry. Consequently, one might expect greater variation in vocalizations between populations and subspecies among red colobus than among *Cercopithecus* species. This would be possible through "random drift" even in large populations having low selective pressures for a particular character (Dobzhansky 1970). "Random drift" can also operate in small isolated populations. The disjunct and isolated nature of many of the red colobus populations and the exceedingly small size of some (e.g., *b. rufo-mitratus* and *b. preussi*) make this latter hypothesis distinctly plausible.

Whatever the evolutionary history of these differences in vocal repertoires, they do pose the interesting possibility of their being related to differences in social behavior and social organization between the different populations of red colobus. Detailed studies of forms other than *b. tephrosceles* would test this hypothesis. Some of the differences may be related to differences in physiology and morphology. For example, the female vocalizes during copulation only in the 2 populations where she develops huge perineal "sexual" swellings at the time of sexual receptivity (see chapter 3).

Comparison with Other Studies

Other studies concerned with vocalizations of African colobinae are those of Hill and Booth (1957) and Marler (1970). Both deal with red colobus, but only in Marler's study were tape recordings and quantitative measures made. He presents data on *C. b. tephrosceles* collected from Kanyawara in 1965. The area studied by him is very near my major study area. Marler suggests that the calls of this subspecies can be placed in 3 general categories — the chist, the chook, and the wheet. He concludes that variations intermediate to these 3 types support the hypothesis that the vocal repertoire of *b. tephrosceles* is "one continuously graded system," as implied by Hill and Booth (1957).

Marler's wheet call is the same as what I have described. Some of the wheets which I measured were considerably longer in duration (up to 2.09 sec.) than those measured by Marler (about ½ sec.). The results of our frequency measures are virtually identical, as are the shape and appearance of the sonagrams illustrated. Marler saw the context of this call only once, when it was given toward a crowned hawk-eagle.

The chist call of Marler's study is the same as in my description. Our measurements of frequency and duration are in agreement, and the examples illustrated in the 2 studies are obviously of the same call type. He concludes that it functions in general alarm and in maintaining group cohesion, such as in group progressions — also in agreement with my findings.

The chook call described and illustrated by Marler appears to be the same call that I term the rapid quaver. Marler expresses uncertainty regarding the context of this call but believes that it accompanies aggressive behavior. Since it took me several months before I could clearly observe the complete context of the rapid quaver, it is not surprising that Marler did not view it clearly in his 5-day study.

Professor Marler kindly gave me a demonstration tape giving samples of his 3 categories. It has proved extremely useful in verifying our comparisons. It also brought forth an apparent difference in our classifications. Among Marler's examples of chook calls are some vocalizations which I would clearly categorize as barks. Spectrographically, the barks of my system look like those Marler illustrates (1970, fig. 4) as intermediate forms to the chook and chist calls. The 34 barks which I measured were shorter in duration ($\overline{X} = 0.097$ sec.) than the 34 rapid quavers ($\overline{X} = 0.12$ sec.). A greater proportion of the rapid quavers (73%) were nontonal as compared with barks (62%). Furthermore, the inter-unit time interval is always much shorter in a bout of rapid quavers than in a bout of barks.

The variability present in the vocal repertoire of *b. tephrosceles* makes plausible Marler's suggestion that it is a graded signal system. Although structurally intermediate calls linking distinct call types can be found, these intermediates occur with such infrequency that it is rarely difficult to categorize any given sound into a particular call type. Only because of this was I able to quantify the frequency of call types. The vocal repertoire of *b. tephrosceles* may be considered to consist primarily of a graded system, but one in which certain segments of the continuum occur with much greater frequency than others. These common calls are from different and noncontiguous areas of the continuum and are thus readily distinguished from one another.

It seems significant that one call type whose distinctiveness no one would argue is the uh! call. Since this call is directed toward low-flying birds, including potential predators, it seems that natural selection has favored an unambiguous vocalization which permits an immediate and adaptive response by the other monkeys without additional information.

Marler and I clearly differ in the number of call types assigned to *b. tephrosceles*. My conservative estimate of 16 excludes possible intermediate forms, such as the bark-chist and yelp, and possible variations of one call type, such as the sqwack and shriek. I attribute most of our difference in repertoire size to the difference in the sample sizes of the 2 studies.

Because of the short duration of his study, I believe Marler was also misled in his belief that red colobus lack vocalizations used in intergroup spacing. As I have described above, the wheet, chist, and bark all seem to have intergroup spacing as one of their functions. These calls, it is true, are of lower amplitude than the roar of the sympatric black and white colobus, which also seems to have intergroup spacing as a function. However, the absence of territoriality, the extensive overlap in home ranges, and the frequent close proximity of red colobus groups alleviates the necessity of a high-amplitude call to effect intergroup spacing. Intergroup communication is extremely important among red colobus, but the calls employed are of lower amplitude than the roar of black and white colobus.

Although I reached my deductions by somewhat different means, I do agree with Marler that the vocal communication of *b. tephrosceles* is more complex than that of *Colobus guereza* and that an obvious correlate is the more complex intragroup social structure of the larger groups of red colobus.

Hill and Booth (1957) and Marler (1970) have described the pronounced differences between the vocalizations of black and white colobus (*C. guereza* and *C. polykomos*) and those of red colobus. My results are consistent with these findings. In fact, I see greater structural similarity in the vocalizations of red colobus and some *Cercopithecus* species (see Struhsaker 1970) than between red colobus and black and white colobus.

An important observation which remains to be fully explained is the great number of stimulus situations evoking certain call types, such as the wheet, chist, and bark of *b. tephrosceles,* and the several functions they appear to serve. It is suggested that these 3 calls, and possibly others, serve primarily to focus attention of other monkeys on the vocalizer and then onto the stimulus evoking his calls. Having 3 call types that serve similar functions may provide greater opportunities for individual vocal distinctiveness.

5 Ecology

Most of the data for this chapter were collected from the CW group of *b. tephrosceles* in Uganda. Ecological data from other social groups and other subspecies were collected opportunistically and not in systematic samples.

THE KIBALE FOREST

The Kibale Forest Reserve is located in Toro District of Western Uganda (fig. 11). Its total area comprises about 56,000 ha. (Kingston 1967) and runs along a north-south axis approximately 56 km long. Located near the eastern edge of the western Great Rift Valley, it is only 24 km east of the Ruwenzori Mountains and is equatorial in position (0° 13' to 0° 41'N and 30° 19' to 30° 32'E). The southern end of the reserve is but 6.5 km from the north end of the Queen Elizabeth National Park. The Kibale Forest is drained by two major rivers, the Dura and Mpanga, both of which empty into Lake George, following the gentle slope from the north of the forest (1,590 m) to the south (1,110 m). These relatively high altitudes account for the moderate temperatures of this region. Based on 52 years of records from the nearby town of Fort Portal, the daily mean minimum and mean maximum temperatures are 12.7° and 25.5° C. The mean annual rainfall of 1,475 mm is relatively well dispersed, falling, on average, 166 days per year (Kingston 1967). However, there are distinct wet and dry seasons. March through April and September through November are generally wetter than the other months (table 41).

The Kibale Forest is a mosaic of vegetation types and consequently difficult to describe in simple terms. One estimate places the amount of the reserve which is dominated by trees at only 60%. The remaining 40% of the reserve is dominated by various types of grassland, woodland-thicket, and colonizing forest (Wing and Buss 1970). The high forest in this reserve, where some of the trees attain heights of 55 m, has been classified as moist evergreen forest but closely related to moist montane forest. It has also been described as lowland tropical rain forest having affinities both with montane rain forest and mixed tropical deciduous forest. In fact, the character of the forest changes considerably as one moves from north to south, descending 480 m in elevation. The northern sector of the forest has been classified as a *Parinari* forest because of the prevalence of this conspicuous emergent tree. *Carapa,*

114

Strombosia, Aningeria, Newtonia, and *Olea welwitschii* are also common and conspicuous large trees in this part of the forest. Further south these species are less common, and *Pterygota, Piptadeniastrum,* and *Chrysophyllum albidum* become obvious elements in the forest (Kingston 1967). Near the very south end of the reserve are large tracts of nearly monotypic forest, where the dominant emergent is ironwood, *Cynometra alexandri.* For more details of the vegetation and plant lists see Kingston (1967) and Wing and Buss (1970).

Parts of this forest reserve have been occupied by humans in the not too distant past, which probably accounts for many of the grassy hilltops scattered throughout the forest. I have found old potsherds along a roadcut on Nyakatojo, a grassy hill near the Kanyawara Forest Station. The first maps

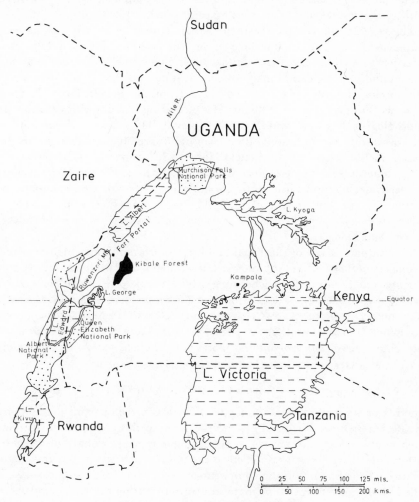

Fig. 11. Location of the Kibale Forest

made of this region, about 36 years ago, indicate that some of the grassy hills within the reserve were populated by humans (Kingston 1967). The size of many of the trees, however, clearly indicates that large parts of this forest have been relatively undisturbed by humans for at least 200–300 years. Fossil leaves found in a nearby area and estimated to be about 1,000–10,000 years old indicate that the forests and probably the climate were essentially the same then as now (Kingston 1967).

The Kibale Forest was first gazetted as a Central Forest Reserve in 1948 and remains under the administration of the Uganda Forestry Department, whose basic plan is to exploit the natural forest on a felling cycle of 70 years and to plant the grassy hills with exotic softwoods (pines and cypress) and eucalyptus (Kingston 1967). Timber exploitation has been completed in the northern third of the reserve and is progressing southward.

The vegetational complexity of the Kibale Forest, combined with its geographical position at the interface of central and east Africa, results in a diverse mammalian and avian fauna. It is like a meeting ground of the central and east African faunas—of the savanna and the forest. As many as 276 species of birds have been collected from the Kibale Forest and its environs (Friedmann and Williams 1970). During a 3-day visit, 2 amateur ornithologists (P. Hamel and D. Childs) observed 117 species in the vicinity of the Kanyawara Forest Station alone. Common large mammals living in the reserve include elephant, bushbuck, Harvey's red duiker (*Cephalophus harveyi*), blue duiker (*Cephalophus monticola*), giant forest hog (*Hylochoerus meinertzhageni*), and bush pig (*Potamochoerus porcus*). Less common are buffalo, hippopotamus, and warthog. Also present are lion, leopard, golden cat (*Felis aurata*), serval cat, spotted hyena, and a variety of viverrids (civet, genets, and mongooses).

The primate fauna of the Kibale Forest is outstanding and includes: red colobus (*Colobus badius tephrosceles* Elliot), black and white colobus (*Colobus guereza occidentalis* Rochebrune), blue monkey (*Cercopithecus mitis stuhlmanni* Matschie), red tail monkey (*Cercopithecus ascanius schmidti* Matschie), l'hoesti monkey (*Cercopithecus l. lhoesti* Sclater), grey-cheeked mangabey (*Cercocebus albigena johnstoni* Lydekker), baboon (*Papio anubis*), chimpanzee (*Pan troglodytes*), potto (*Perodicticus potto*), *Galago demidovii*, and a larger bushbaby believed to be *Galago inustus*.

THE MAJOR STUDY AREA—COMPARTMENT 30 OF THE KIBALE FOREST

The major study area of *C. b. tephrosceles* was in compartment 30 of the Kibale Forest Reserve (fig. 12). The vegetation in this compartment is relatively undisturbed and mature, representing a typical example of *Parinari* forest. A few large trees were felled in the past by pitsawyers, but this made relatively little impact on the compartment as a whole. No large trees were felled in the home range of the main study group during this study, and there was no indication of any felling in the recent past. The compartment is

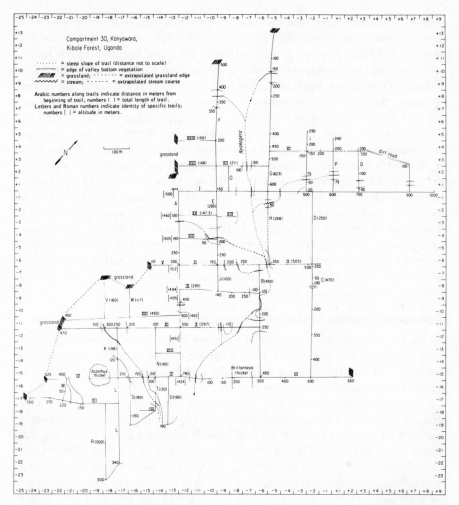

Fig. 12. Schematic map of the main study area in compartment 30 of the Kibale Forest. Solid lines indicate trails. Marginal numbers specify rows and columns, which permit specific identity of 50 m x 50 m quadrats. Quadrats are identified first by row and then by column number.

comprised of about 300 ha. and contains the head waters of the Nyakagera stream, which flows down the middle of this forest block and joins the Dura River. The study area is bounded by two grassy hills, Nyakatojo on the west and Nyamusika on the east, and more forest to the north and south. In essence, this forest compartment covers the slopes of these two hills and the valley of the Nyakagera. As one descends these hills into the valley and along a maximum drop of 165 m (1,530 m to 1,365 m), a pronounced vegetational catena is experienced. The hilltops are dominated by elephant grass (*Pennisetum purpurem*), but have recently been planted with *Pinus caribaea, P. radiata,* and *Cupressus lusitanica.* On the upper slopes, which are well drained and often have shallow soils, one finds a preponderance of *Chaetacme, Teclea, Dombeya, Diospyros,* and *Millettia,* trees of generally low stature (≤ 30 m). Further down the slope the forest becomes much taller, with some of the emergents attaining heights of 55 m (pls. 24-26). The understory is relatively open here, and large trees such as *Parinari, Aningeria, Newtonia, Mimusops, Celtis africana, Lovoa swynnertonii, Olea welwitschii,* and *Strombosia scheffleri* are common. As one approaches the valley bottom, there are fewer large trees, many breaks in the upper canopy, and an abundance of semiwoody plants in the understory, particularly *Brillantasia.* In the valley bottom one finds the waterlogged soil supporting a swamp forest; *Symphonia, Neoboutonia,* and *Mitragyna* are common trees, though widely spaced (pls. 27 and 28). There are fewer trees and species diversity is lower than on the slopes. In the understory of the valley swamp there is an abundance of semiwoody plants, including *Vernonia, Acanthus,* and *Mimulopsis solmsii.* This oversimplified description applies to that portion of compartment 30 most heavily used by the main study group of red colobus. Other parts of the compartment differ, particularly in the north end, where the hillslopes are more gentle (Oates 1974).

Vegetational Analysis of the Major Study Area

Detailed enumerations were made of the trees in the home range of the main study group of red colobus (CW group) to obtain a better vegetational description of the area and to gain some idea of the abundance and dispersion patterns of red colobus food species.

Strip enumerations were made throughout the home range of the CW group. Foot trails which had been previously cut along compass bearings and provided access to all parts of the CW group's home range were used for this enumeration (fig. 13). The clearing of these trails in no way affected the current density of the trees enumerated. All trees within 2.5 m of the trail and 10 m or more in height were identified. A height of 10 m was selected because the red colobus rarely descend below this level in the forest. The transect sampled was 2,873 m long and 5 m wide. Excluding overlap areas that occurred at trail junctions gave a total sample area of 1.43 ha. This represents about 3-4% of the total area occupied by the CW group. Trees tallied in this sample numbered 469, or about 328 trees of 10 m or more in height per hectare (table 32). Unidentifiable trees account for only 1.5% of the sample. The 10 most common trees comprise 79.4% of the sample, and the top 3 species 48.1%.

Indices of dispersion were computed for the 10 most common tree species, using the ratio of the variance/mean (Greig-Smith 1964). When this ratio is less than one, a regular or uniform dispersion is indicated; if greater than one, a contagious or aggregated dispersion. Computation of a variance and mean were possible because the tree enumeration data were collected and segregated in 50 m sections along the entire 2,873 m transect. Nine of the 10 most common trees had aggregated dispersions in this study area (table 32). Only *Celtis durandii* appears to have a uniform dispersion. This tendency toward aggregated dispersions is consistent with the impression of a vegetational catena which occurs from the ridge tops, down the slopes, across the valley, and up to the next ridge top.

Further evidence for the generality of this clumping tendency is provided by data from another method of vegetational analysis in the main study area. In this method all trees of at least 10 m height were identified within quadrats of 0.25 ha. area. A grid system was superimposed over a map of the study area, and 5 quadrats were selected on the basis of their frequent use by

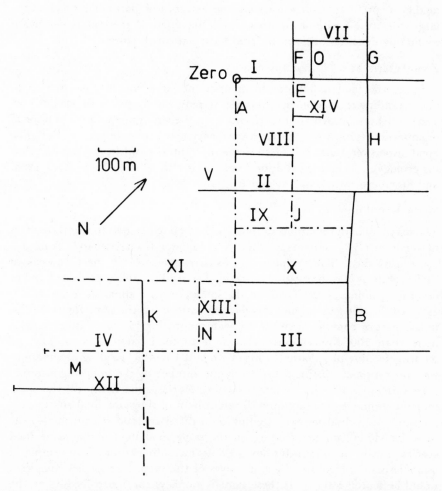

Fig. 13. Part of study area in compartment 30, Kibale Forest, showing location of strip enumerations of trees. Trails along which these strip enumerations were conducted are indicated by dash-dot lines.

the CW group of red colobus. The preponderance of certain tree species in only one of the 5 quadrats, such as *Uvariopsis,* or the great disparity in density of certain species between quadrats, such as *Diospyros,* not only supports the generalization that most tree species in this forest have aggregated patterns of dispersion, but also demonstrates the vegetational heterogeneity of this area (table 33).

Comparing the results of the strip and quadrat methods reveals similar tree and species densities. There were 328 trees per ha. and 35.7 species per ha. in the strip enumeration and 307 trees and 32.8 species per ha. using the quadrat method. Because of the vegetational catena and the aggregated dispersion of many species in this forest, it would appear that the strip method is preferable for estimating the densities of particular species in a large area (the entire study area), and the quadrat method is better for evaluating densities in a specific and local part of the forest.

FOOD HABITS OF *C. b. tephrosceles*

Study of the food habits of red colobus monkeys not only is essential to understanding their ecological relation to plants and other animals but may also contribute toward an understanding of their ranging patterns and social organization (group size, intragroup dispersion, and intergroup relations). Food dispersion, indeed, is often made the central point in hypotheses relating ecology to social organization (e.g. Crook and Gartlan 1966, Lack 1968, and Eisenberg, Muckenhirn, and Rudran 1972).

Methods and definitions

The majority of observations of food habits were collected during those days in which the CW group was systematically followed from dawn to dusk for 3–5 consecutive days. The samples were usually taken during the first week of each month. A few observations were made outside of this time period and of other red colobus groups living in the same area of compartment 30. It was assumed that the general monthly feeding pattern was the same for all badius in this part of compartment 30. In each month for 19 consecutive months, more than 100 observations were made of food eaten. The number of monthly feeding observations ranged from 104 to 468. Feeding observations were operationally distinguished from one another by the following criteria: (1) a different individual monkey feeding on the same item; (2) the same individual monkey feeding on a different item of the same food species, (3) the same individual monkey feeding on a different food species, or (4) the same individual monkey feeding on the same food item of the same food species at least one hour after any previous such observation. For example, if 5 monkeys were feeding on young leaves of the same tree species, this item would be scored 5 times. If these same 5 monkeys continued feeding on the same item for more than an hour, the item would again be scored 5 times after an hour had passed since the previous scoring. If the same 5 monkeys then began eating leaf buds of the same tree, this item would be scored 5 times regardless of the time interval between the feeding on leaf buds and on the young leaves of the same species.

The distinction of categories of food items was usually obvious. Ripeness of fruit was determined by color or, for species in which unripe and ripe fruits are of the same color, by size. The segregation of young leaves into 3 categories was based on relative size — a distinction which the badius obviously made as well, judging from the data showing preferences for different size classes of young leaves. In the first few months of the study I did not distinguish different categories of young leaves but lumped all observations into one category.

Systematic observations of food habits were analyzed on a monthly basis, and these monthly summaries were further summarized and divided into 2 periods — the first 7 months and the last 12 months of the study. This was done to provide a summary for one annual cycle without bias toward or against any foods that were preferred or available only on a seasonal basis.

General Results

Red colobus monkeys, like all members of the subfamily colobinae, are generally classified as leaf eaters. Their specialized and ruminantlike stomachs are presumed adaptations to this diet (Drawert, Kuhn, and Rapp 1962). Feeding data for *b. tephrosceles* in the Kibale Forest support this general conclusion but clearly indicate that members of this subspecies prefer the young growth of plants. Young leaves constitute about one-third of their diet while mature leaves comprise only 4–5% of their food. Sixteen to 20% of their diet consists of floral and leaf buds, and the petioles of mature leaves make up another 13–19% of their food (tables 34 and 35).

During the systematic samples of feeding behavior a total of 46 plant species plus 16 unidentified species were eaten by red colobus in compartment 30 of the Kibale Forest (tables 34 and 35). Only one species (*Strychnos mitis*) which was fed on in the systematic samples collected between September 1970 and March 1971 was not also eaten in the period of April 1971 through March 1972. However, many more species were eaten in the latter than the former period (see tables 34 and 35). The entire list of food species eaten by red colobus in the Kibale Forest was increased by only 6 species (total 52 plus 16 unidentified species) when all other observations outside the systematic samples were included. Four of these 6 observations were made in two other areas 8 and 20 km. south of the main study area. On one occasion each, badius were seen eating: the young seeds including the wings of *Pterygota mildbraedii* Engl. after gnawing through the thick pericarp, which was subsequently discarded; flowers, seeds from green pods, and small young leaves of *Cynometra alexandri* C. H. Wright; leaves and leaf buds of *Celtis mildbraedii* Engl.; green to yellow-orange fruits of *Maytenus* sp.; and, in compartment 30, the young leaves of *Pygeum africanum* Hook. *Cynometra*

alexandri, *Celtis mildbraedii*, and *Maytenus* have not been found in compartment 30. *Pterygota* is very local and generally uncommon in compartment 30, and *Pygeum* is extremely rare there. In compartment 13, about
1.5 km from compartment 30, badius ate the flowers and medium and
small-sized young leaves of *Macaranga schweinfurthii* Pax. (a swamp tree not
found in compartment 30) and *Parinari e.* fruits. Although *Parinari* was
common in compartment 30, its fruits were rarely seen there during this
study. Nonsystematic observations in compartment 30 added two more specific food items to the lists based on systematic observations. These were the
fruit of *Ficus brachylepis* (seen once) and the young leaves of *Ficus natalensis*
(seen on several occasions). This latter species was only recently identified,
and it is quite possible that young leaves of it which were eaten during the
systematic samples were classified under *Ficus* sp.

Considering the food preferences for one annual cycle (the last 12 months
of the study) it is apparent that *Celtis africana* was the most preferred species,
comprising 15.36% of the 2,898 feeding observations for that period (table 35).
It was also the most common food species in the first 7 months of the study,
but only because this period included several months in which it was fed upon
heavily (table 34). In another 7-month period, such as May–October 1971, it
ranked much lower (table 36). Perhaps of greater significance is the observation that relatively few species constitute a large proportion of the red colobus
diet. Rank ordering the species according to their percentage of the total
sample reveals that in both the first 7-month period and the last 12-month
period the top 10 species comprise about 75% of the total diet. In the 12-month
period the top 3 species alone constitute 40.8% of the diet.

Ignoring the species and considering only the part of the plant eaten shows
that in the annual cycle the petioles of mature leaves and leaflets were the most
common item fed upon (tables 34 and 35). However, if one lumps together all
5 categories of young leaves, they exceed the mature petioles. Rank ordering
the parts eaten, regardless of species, also demonstrates that few items constitute a large portion of the diet. In the 12-month period the top 10 items
comprise 87.4% of the diet, and in the first 7-month period the top 10 items
represent 83.6% of the sample. To a certain extent this is dependent on the
limited number of categories of food items recognized in this study.

Monthly Variation and Seasonality

There were considerable differences from one month to another in the diet of
red colobus. One way of demonstrating this is to summarize the monthly
variation for a selection of food species. Table 37 reveals this variation for the
top 10 food species of the last 12-month period, plus *Lovoa s.* and *Fagaropsis
a.*, which were included because they represent extremes. Most of these
species were highly variable in the percentage of feeding observations which
they comprised from month to month, as indicated by the high coefficient of
variation. In contrast, *Newtonia* showed relatively little variation and was

among the top 5 food species in every month except one when it was sixth in rank. *Markhamia* is another species which regularly constituted an important food species; like *Newtonia*, it may be considered a staple source of food for the red colobus. Although important in the total annual percentage of feeding observations, species like *Celtis africana*, and *Celtis durandii*, and *Aningeria* cannot be considered staple food species on a month-to-month basis. Even more extreme in variation are the examples provided by species like *Lovoa s.* and *Fagaropsis a.*, which are very infrequently important in the monthly diet and rank relatively low in the total feeding observations for the year.

A monthly breakdown by percentage of the most important specific food items, food species, and food items regardless of species gives further insight into monthly variations of red colobus food habits (tables 36, 38, and 39). In general it appears that patterns of food availability and utilization are extremely variable among the red colobus food species in compartment 30, and no single form seems to predominate. Certain specific items, such as *Newtonia* leaves of unknown age,[1] are important foods in every month, but much more in some months than others (fig. 14). The flowers and fruits of *Teclea* provide the best examples of seasonal foods with an annual cycle, occurring as important items only in September 1970, July, August, and September 1971. This is directly related to the pronounced seasonality and relatively high degree of synchrony in the flower and fruit phases of this species (figs. 15 and 16).[2] *Lovoa* young leaves are also seasonal foods with a 12-month cycle, being closely related to their seasonal availability (fig. 17). *Celtis africana* leaf buds and young leaves appear as important foods with a triannual cycle: September–November 1970, January–April 1971, August–September 1971, and February 1972 (figs. 18 and 19). This species is not synchronized in its phenology. The data for *Celtis durandii* floral buds and fruits suggest a bi- or triannual cycle as red colobus food (figs. 20 and 21). Certain phytophases of this species persist for long periods, and its phenology appears complex. Biannual cycles are suggested for the young leaves of *Celtis durandii* and *Markhamia* as sources of red colobus food. These cycles tend to follow the average availability curves, as determined from 10 trees of each species. Clearly, however, neither species is highly synchronized for this phytophase (figs. 22 and 23). The young leaves of *Bosqueia* are less clearly a seasonal food and provide yet another variation in the monthly pattern with which specific items are eaten. Their peaks are separated by intervals of 1–2 months (fig. 24). No phenological data are available for this species. The petioles of *Balanites* mature leaves and leaflets and *Chaetacme* large young leaves appear as important foods on an irregular basis (figs. 24 and 25). The curve for *Aningeria* floral buds as badius food has two major peaks, separated by 9 months, but also has lesser peaks interspersed (fig. 26). This is probably due to the lack of synchrony for this phytophase for *Aningeria*. In spite of year-round availability, the petioles of *Markhamia* mature leaves and leaflets

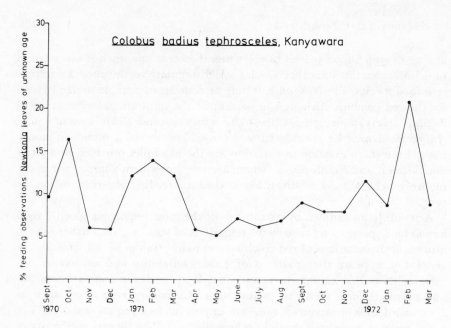

Fig. 14. *Newtonia* leaves of unknown age: percentage of monthly feeding observations.

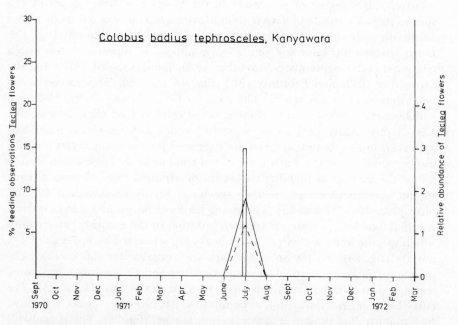

Fig. 15. *Teclea* flowers: comparison of monthly feeding observations and relative abundance, based on monthly evaluations of 5 individually marked trees. Here and in figs. 17-27 the vertical bars indicate the range in abundance scores for the trees sampled and the dotted line connects the monthly mean values of these abundance scores. The solid line represents the percentage of feeding observations.

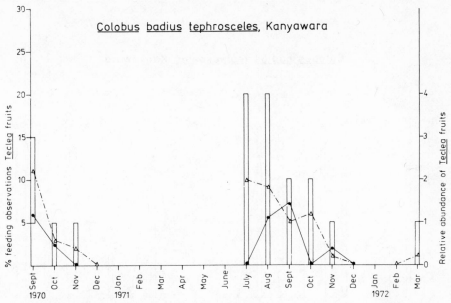

Fig. 16. *Teclea* fruits: monthly feeding observations and relative abundance scores (5 trees). Symbols as in fig. 15.

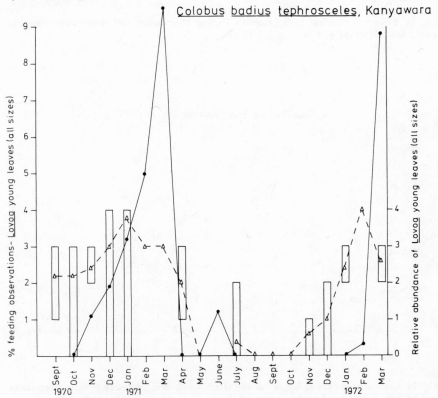

Fig. 17. *Lovoa* young leaves (all sizes): monthly feeding observations and relative abundance (5 trees). Symbols as in fig. 15.

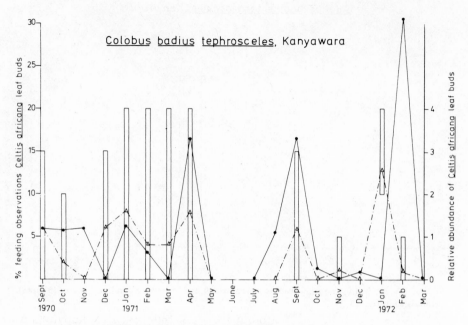

Fig. 18. *Celtis africana* leaf buds: monthly feeding observations and relative abundance (5 trees). Symbols as in fig. 15.

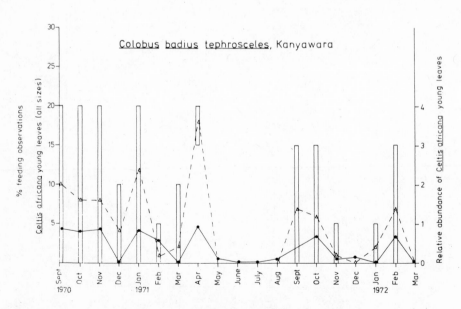

Fig. 19. *Celtis africana* young leaves (all size classes combined): monthly feeding observations and relative abundance (5 trees). Symbols as in fig. 15.

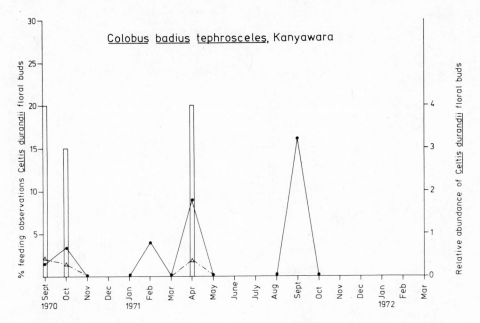

Fig. 20. *Celtis durandii* floral buds: monthly feeding observations and relative abundance (10 trees). Symbols as in fig. 15.

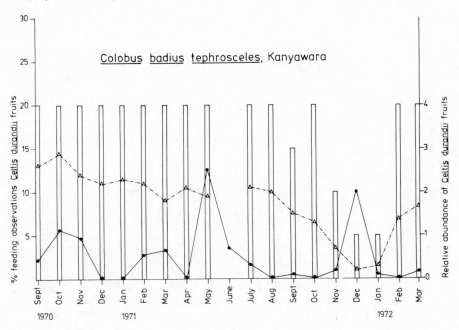

Fig. 21. *Celtis durandii* fruits (all size and age classes combined): monthly feeding observations and relative abundance (10 trees). Symbols as in fig. 15.

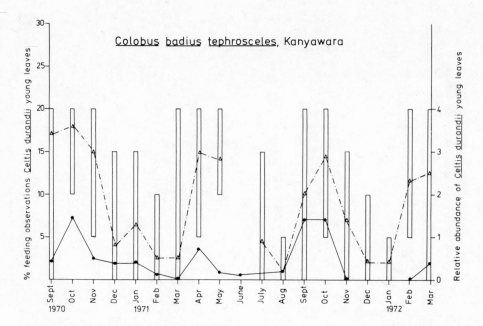

Fig. 22. *Celtis durandii* young leaves (all size classes combined): monthly feeding observations and relative abundance (10 trees). Symbols as in fig. 15.

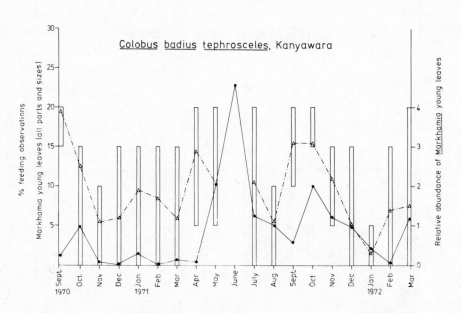

Fig. 23. All parts of *Markhamia* young leaves of all sizes: monthly feeding observations and relative abundance (10 trees). Symbols as in fig. 15.

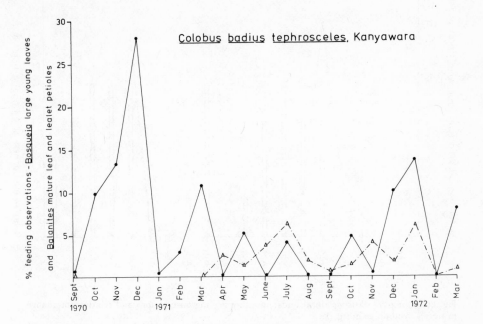

Fig. 24. *Bosqueia* large size young leaves (solid line) and *Balanites* mature leaf and leaflet petioles (dash-dot line): percentage of monthly feeding observations.

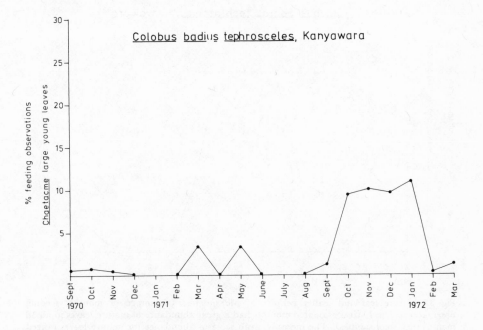

Fig. 25. *Chaetacme* large young leaves: percentage of monthly feeding observations.

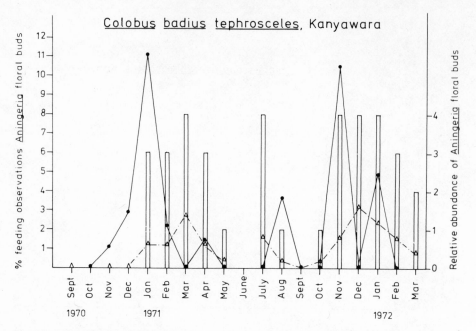

Fig. 26. *Aningeria altissima* floral buds: monthly feeding observations and relative abundance (5 trees). Symbols as in fig. 15.

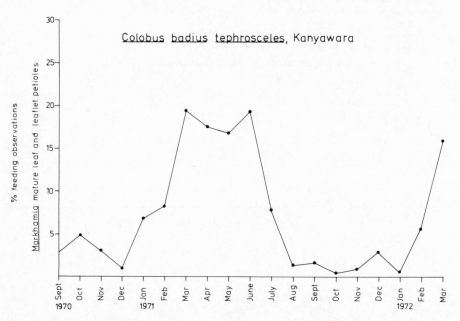

Fig. 27. Mature leaf and leaflet petioles of *Markhamia*: percentage of monthly feeding observations. The 10 trees evaluated monthly had a great abundance of mature leaves in all 18 months they were sampled. The monthly mean score of abundance for mature leaves ranged from 3.5 to 4, with an average of 3.9.

appear to be seasonal food for badius. This is a long season, lasting about 7 months and separated from the next by 6 months (fig. 27). A final variety of food usage is provided by *Fagaropsis*, the mature leaf petioles of which comprised an important food item in only 2 months of the entire study, February–March 1972.

Given a larger phenological sample, particularly for tree species that are asynchronous, many of the monthly food preferences of the red colobus might be found to have closer correspondence to their availability than the preceding examples indicate. Thus, although a species may be asynchronous in an absolute sense, its entire population in a particular area might show statistical trends toward more or less synchrony. The larger the sample of this population, the more precisely will one be able to demonstrate the degree of synchrony.

Table 36 gives a breakdown of the top 10 food species, regardless of item, for each month. Again, monthly variation is striking. The pattern for *Newtonia* remains unchanged from that described for specific items because, with very few exceptions, only its leaves of undetermined age were eaten. A similar case holds for *Bosqueia*. For other species, however, the trends toward seasonality described above tend to break down or become less clear in this analysis, because these species provide more than one category of food item for the red colobus. When the different food items for a given tree species are out of phase with one another, they extend the number of months in which the species is important as a source of food. For example, the small young leaves of *Aningeria* eaten by red colobus represent a phytophase that generally precedes floral bud and flower production. Consequently, the number of months in which *Aningeria* occurs as a prominent food species for the monkeys increases. Similar patterns exist for several other species, e.g., *Markhamia, Teclea, Celtis africana,* and *Celtis durandii.*

Table 39 summarizes the monthly distribution of feeding observations among food items, regardless of species. Seasonality is even less prevalent here than in the preceding two analyses, and there is no obvious correlation with wet and dry seasons. At least one of the 5 categories of young leaves occurred among the top 5 food items in every month except February 1971 and August 1971. Even in these two months the combined percentage for all classes of young leaves eaten by the monkeys was very high, placing the general category of young leaves well among the top 5. The petioles of mature leaves were among the top 5 in every month except February 1971, when they ranked seventh, and in September and October 1970, when petioles of leaves of an undetermined age were among the top 5. Because these last two months were early ones in the study, my ability to determine leaf age was not so well developed, and the petioles may, in fact, have been from mature leaves. Flowers or floral buds were among the top-ranking 5 food items in both wet and dry seasons: October 1970; January, February, April, July, August, September, and November 1971; and January 1972. In 5 other months at least one of these items was among the top 10. However, they did not occur at all in 3 of the monthly samples, which occurred both in rainy and dry seasons, i.e., March, May, and December 1971. In June 1971 (dry) and March 1972 (wet) the combined percentages of flowers and floral buds were less than 1%. At least one of the 3 categories of fruit occurred among the top 5 food items in only 4 months: November 1970; May, August, and December 1971. These months included both wet and dry seasons. Mature leaves ranked among the top 5 items even less often. Of the 4 categories covering mature leaves (entire mature leaves, pieces, apical tips, and basal parts), at least one was in the top 5 only in September 1970 and December 1971. Perhaps of relevance is the fact that neither flowers nor floral buds occurred in the sample of December 1971, which was also one of the few months when fruits ranked high as a food. The tree phenological data for this month indicate that relatively few flowers were available then; perhaps as a consequence the monkeys turned to mature leaves and fruits.

Another way of examining the feeding data for seasonality is to compare food habits of a particular month in different years. In this study data were collected for 7 months, September–March, in two consecutive years. Considering only the top 5 specific food items (table 38), it is seen that in each pair of months for consecutive years there are at least 2 specific items which rank high in both years. As mentioned above, *Newtonia* leaves of undetermined age were popular in every month and consequently shed no light on seasonality. Comparing September 1970 and 1971 it is seen that *Celtis africana* leaf buds were high-ranking items in both years. *Celtis durandii* flowers constituted 5.1% of the September 1970 diet and just fell short of the top 5, whereas a similar phytophase (floral buds) constituted 16.2% of the diet in 1971, ranking second. In October of both years, *Bosqueia* young leaves ranked among the top 5. In October 1970, *Celtis durandii* young leaves of unspecified size made up 7.3% of the diet. In the same month of the following year, 3 size classes of *Celtis durandii* young leaves were fed upon. When these 3 are summed they yield 7.1% of the diet for October 1971. Young leaves of *Millettia* ranked high among foods eaten in November of both 1970 and 1971. Similarly, the large young leaves of *Bosqueia* were important items in December 1970 and 1971. The flowers and floral buds of *Aningeria* were among the top 5 foods in January of both years. The overlap for February 1971 and 1972 was provided by the petioles of *Markhamia* mature leaves and leaflets. In this same month *Celtis africana* flower buds comprised 3.4% (not high enough for top 5) of the diet in 1971 and 4.0% in 1972. Similarity in diet was somewhat more pronounced in March 1971 and 1972 with the petioles of *Markhamia* mature leaves and leaflets and *Bosqueia* large young leaves ranking among the top 5 in both years. Furthermore, the small young leaves of *Lovoa* comprised 8.1% of the diet in March 1971 and 5.7% (not high enough for top 5) in March 1972. These data lend support to the hypothesis that certain aspects of red colobus food habits are seasonal. Additional support is gained for this suggestion if one allows a month's leeway in the seasons. For example, *Celtis durandii* young leaves were among the top 5 food items in October 1970, but in 1971 they were important foods in September. *Millettia* young leaves ranked high in November 1970 and October 1971. In January 1971 *Markhamia* mature leaf and leaflet petioles, and *Celtis africana* leaf and floral buds were among the top 5. Although not important in January 1972, they did achieve high ranks in February.

Monthly overlap in diets can be expressed more precisely by computing overlap percentages. Each month was compared with every other month from November 1970 through March 1972 inclusive. The percentage overlap in diet for each pair of months was computed by addition of percent values shared for any and all food items in the 2 months concerned. For example, if in November 1970 the red colobus diet consisted of 50% small young leaves of *Markhamia*, 25% *Markhamia* mature leaf petioles, and 25% *Newtonia* leaves of unspecified age, and in January 1972 their diet consisted of 10% small young leaves of *Markhamia*, 50% *Markhamia* mature leaf petioles, 10% *Newtonia* leaves of unspecified age, and 30% large young leaves of *Bosqueia,* the percentage overlap in diet for these 2 months would be 45%. This is the same method used by Holmes and Pitelka (1968) in their study of diet overlap among sandpipers.

Analysis was restricted to the last 17 months of the study because of greater reliability in the data and methodological consistency. Diet overlap in the 136 monthly pair combinations was extremely variable (table 40). The range in diet overlap was 9.3–50%, and the mean was 24.3%. A trend toward seasonality is not particularly striking. Examining the percentage overlap in the diet of the same months in subsequent years reveals that overlaps were not, on average, much greater than the mean overlap of 24.3%. The 5 months considered are November–March, and the overlap between the same month in subsequent years ranges from 21.5 to 40.1% with a mean of 29.1%.

High percentages in overlap often occur between contiguous months, e.g., January–February 1971, 41.5%; May–June 1971, 50.0%; October–November 1971, 46.1%; and December 1971–January 1972, 48.0%. Each of these examples probably reflects one period of similar dietary preference by the red colobus. That is, the particular food preferences lasted for a period which encompassed at least 2 monthly samples.

The absence of a strong trend toward seasonality in diet may be partially attributable to the variability in rainfall. Although in terms of many years there are definite wet and dry seasons, when one compares the rainfall for any two years this pattern is not always apparent (table 41). In the 17 months considered in this analysis, some months which are typically dry were, in fact, wet, e.g., January 1972, and some normally wet were dry, e.g., November 1970. A simple hypothesis would relate similarity in monthly rainfall with similarity in monthly food availability and thus in the monthly diet of red colobus monkeys. There is a weak correlation between similarity in monthly rainfall and percentage overlap in the monthly diet of red colobus. The difference in rainfall for each pair of months was compared with the overlap in diet for the same pairs (tables 40 and 41). These data were plotted for all 136 pair combinations, resulting in a graph of 136 points comparing the difference in rainfall and the percentage overlap in diet for each pair of months. This distribution was tested with the Olmstead and Tukey Corner Test for Association as described in Sokal and Rohlf (1969, pp. 538-40). The absolute value of the quadrat sum was 9.5, with $0.10 > p > 0.05$. The lack of a stronger correlation may be due to several factors. Our ignorance of the relation between rainfall and the occurrence of the various phytophases of different forest trees may be of great importance. For example, how much rainfall and how many days or months in advance must it fall for a particular phytophase to occur? Certainly the apparent predominance of phytophasic asynchrony among many of the tree species counteracts seasonality in food availability and thus in diet. In habitats with more predictable and seasonal rainfall and with food species that are highly synchronous, one would predict a greater correlation between similarity in rainfall and in diet.

The diversity of the red colobus diet also shows considerable monthly variation. Computation of indices of diversity[3] for food species and for food items (regardless of species) resulted in nearly parallel patterns during the 19-month sample period (fig. 28). The food item indices are usually slightly lower because fewer categories were involved than in the computation of food species indices. There are 3 apparent exceptions to this general parallel pattern. In June 1971 the food species index dropped because *Markhamia* constituted 43.4% of the diet, but the food item index changed very little because 7 different parts of *Markhamia* were eaten, so that a relatively high diversity of food items was maintained. Although the species index rose in October 1971, the food parts index fell. This was because 45.3% of the diet was composed of only 2 categories of food — medium-sized and large young leaves. A similar situation exists for January 1972, where 2 items (large young leaves and leaves of unknown age) comprised 50.3% of the sample.

A causal explanation of the variation in these curves of food diversity seems beyond the scope of the available data. A reasonable hypothesis would relate variations in the diversity of diet to variations in food availability. However, the phenological data clearly indicate considerable intraspecific variation within the study area in the timing of various phytophases for many tree species. For example, *Celtis africana* leaf buds occur at very different times on different individual and neighboring trees. Consequently, the monthly correspondence between the availability of this item, as estimated from only 5 trees, and its proportional representation in the diet is not particularly great (fig. 18). To make reasonable comparisons between variations in diet diversity and food availability would require a much larger phenological sample.

In spite of these deficiencies, the data do suggest some trends. In April 1971 and February 1972, when food species diversity and item diversity were

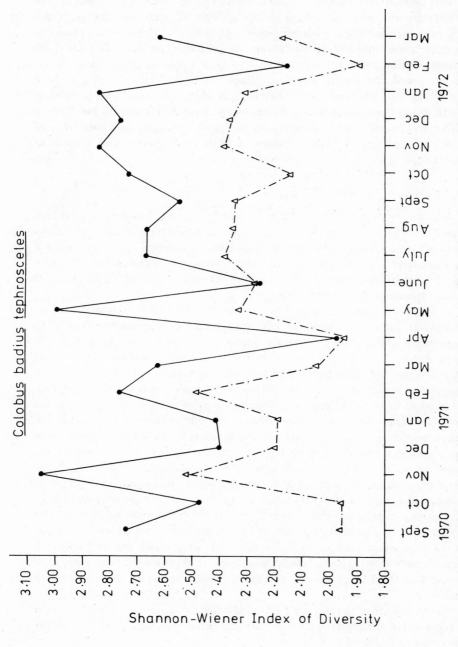

Fig. 28. Monthly variation in indices of diversity of diet. Food-species diversity is indicated by the solid line and food-item diversity by the broken line.

lowest, the monkeys fed very heavily on the *Markhamia* mature leaf and leaflet petioles and the leaf and floral buds of *Celtis africana.* In contrast, the 2 months with highest indices were not high in both these items. *Celtis africana* leaf and floral buds comprised only 8% and *Markhamia* petioles of mature leaves only 4% of the November 1970 diet. In May 1971 no *Celtis africana* buds were eaten, although 16.8% of the food consisted of *Markhamia* petioles of mature leaves. Although the food species diversity for June 1971 was low, the food item diversity was about average. The petioles of mature *Markhamia* leaves were an important food in this month, but buds of *Celtis africana* were not eaten. Apparently, only the combined popularity of these 2 items results in low indices of both food species and food item diversity.

Food Preferences — Selection Ratios

Although the data in tables 34 and 35 clearly indicate that the monkeys feed more on some tree species than others, in terms of food preferences or selectivity they make no allowance for several important parameters that may bias their relative values. The frequency with which a particular species is fed upon may be affected by its density (number of trees per unit area) and by the size of its crown (potential food-producing area).

Density estimates were based on enumerations of trees at least 10 m tall (table 32). Indices of crown size were computed for 10 specimens each of 13 common species. Only mature specimens were considered. Two measures were made: maximum crown depth and maximum crown diameter. A Blume-Leiss optical height finder was used to measure the maximum and minimum height of foliage, the difference of which was the crown depth. Maximum crown diameter was determined with a tape measure and was measured from the approximate edge on one side of the crown, through the trunk to the approximate edge on the other side. Precision in both these measures was probably not very great, but any source of error or bias was believed to be uniform for all species. Consequently, comparison of the relative measures is valid. Ideally, one would like to estimate the potential food-producing area of the tree, but because most trees have irregular crown shapes one cannot use crown depth and width to compute surface area or volume. I use, therefore, the sum of the maximum crown depth and maximum crown diameter as an index of crown size, which makes the fewest assumptions about crown shape. Average values of crown size were computed for each of the 13 species, i.e., the mean crown depth plus the mean crown width.[4]

An estimation of the relative food-producing area provided by each of these 13 species is the product of their density and mean crown size, which I call the *cover index* (table 42).

When the percentage of feeding observations which is comprised of a particular tree species is divided by this index of cover, a selection ratio is obtained which gives a better idea of the feeding preferences of the monkeys because it corrects for interspecific differences in density and crown size.

The 13 species for which cover indices were computed comprised 75.35% of the red colobus diet for the 12-month period of April 1971 through March 1972. The selection ratios clearly changed the apparent preferences for several species when compared to the percentage composition of these species in the annual diet (table 42). For example, 15.11% of the annual feeding observations were from *Markhamia*, placing it second in rank of importance. Its selection ratio, however, places it in eighth position among these 13 species, indicating that much of its popularity as food may be a function only of its relatively large cover area. A similar case exists for *Celtis durandii*. *Celtis africana* still remained the top-ranking species, because, although it provided little cover, it was still heavily fed upon, indicating that it was a highly preferred food species. *Lovoa* was fed upon relatively little, but part of this is due to its low density, resulting in a low index of cover in the study area. Its selection index places it in fifth position among these 13 species, indicating that the monkeys preferred it more than is reflected by its proportion in the annual diet. In contrast, the selection ratio of *Diospyros* is very low and consistent with its low proportion of the total feeding observations. Clearly the red colobus were selecting against this the most common species in the area. *Parinari* seems to present a similar case, because, although relatively low in numbers, it has a huge crown area and yet is still unimportant as red colobus food.

Sample bias

There was one unavoidable parameter which may have had an important effect on the results of the feeding observations. Different tree species offer different conditions for observation. Obviously, the monkeys were more readily observed in leafless trees, and foliage density and the distribution of branches largely determined how much could be seen. This almost certainly had some effect on the feeding data. In an attempt to give some indication of the importance of this variable, the 13 species discussed in the previous section were given a score based on my impression of how readily monkeys could be observed in them. Four classes of visibility were recognized, ranging from excellent (I) to poor (IV) (table 42). All appraisals were made in compartment 30. Deciduous trees, such as *Celtis africana* and *Celtis durandii*, were given a classification intermediate to that when they were fully foliated and when they were virtually leafless. Thus, *Celtis africana* was placed in category II, but when it was in the leaf-bud phytophase it afforded excellent observation conditions (category I). This probably contributed greatly to the high percentage of *Celtis africana* leaf buds in the red colobus diet. In contrast, the excellent visibility afforded by *Parinari* had little effect on its low proportion in the diet. Interspecific comparisons of food preferences of red colobus are most meaningful between those species of the same visibility class or similar classes.

Food Habits of Other Red Colobus Groups in
Other Parts of Compartment 30

In conjunction with a study by J. F. Oates on *C. guereza*, a few feeding
observations were made of other red colobus groups in the northwest portion
of compartment 30, where his main study group of guereza lived. The main
objective of these observations was to verify that the food habits of other red
colobus in compartment 30 were not substantially different from those of the
main study group (CW), and that they were consistently different from those
of *C. guereza* living in precisely the same section of forest.

Fifty-one feeding observations were made on 5 mornings in October and
December 1971. No new food species or items were added to the list compiled
from observations on the CW group. However, pieces of mature leaves of
Chrysophyllum gorungosanum constituted a larger portion of the sample
than for any monthly sample of the CW group. Also, large young leaves of all
species in this sample of 51 comprised a greater percentage (43.13%) than in
any sample for the CW group. These differences could be attributed to the
small sample size or to slight differences in tree species composition in
different parts of compartment 30. Percentage composition of other items
corresponded very well with the data for the CW group: floral buds 7.84%,
leaf buds 7.84%, mature leaf petioles 21.56%, young leaf petioles 5.88%,
and small young leaves 5.88% (compare with table 35). The 13 species and
their proportions of this sample of 51 are as follows: *Bosqueia* 25.49%;
Aningeria 23.52%; *Chrysophyllum g.* 9.80%; *Ficus dawei* 7.84%; *Mimusops* 5.88%; strangler fig sp. 5.88%; *Strombosia* 5.88%; *Funtumia* 3.92%;
Markhamia 3.92%; *Olea w., Lovoa, Celtis africana,* and *Teclea* each 1.96%.

In 25 feeding observations of red colobus in the area occupied by his main
study group, J. F. Oates (pers. comm.) recorded only one new food species
between December 1971 and February 1972. On one occasion 2 red colobus
ate the mature leaves of the liana *Hippocratea graciliflora*. All of his other
observations conform with mine throughout compartment 30. It seems that
the food habits of red colobus in this part of compartment 30 were not
substantially different from those of the CW group.

Nutritional Analysis of Red Colobus Foods

In an attempt to better understand the selectivity in the diet of *b. tephrosceles,* nutritional analyses were performed of young and mature leaves of
Celtis durandii and *Markhamia* (table 43).[5] The young leaves of *Celtis
durandii* were clearly preferred by the red colobus over mature leaves of this
species (tables 34 and 35). Table 43 suggests that the higher crude protein of
the young leaves may be the basis of this preference. The nutritional basis of
their dietary preference of leaf petioles over the actual leaves of *Markhamia*
cannot be answered with these data, because at the time the nutritional
analysis was performed I mistakenly believed that the badius preferred only
the basal part of the leaf petiole. I was later to learn that they prefer not only

leaf petioles but also leaflet petioles (petiolule) of this compound leaf. Since the leaflet petioles were unfortunately included in the specimens analyzed as leaves, the contrast between leaf and petiole composition is reduced and any nutritional basis for dietary preferences is masked.

There are consistent differences between young and old leaves which may explain some of the nutritional basis of the red colobus preference for young growth. The young leaves are higher in crude protein, fat (ether extract), and phosphorus, whereas mature leaves are higher in crude fiber, lignin, ash, and calcium. The crude protein content of these foods is extremely high, particularly for the young leaves of *Celtis durandii,* even when compared to the cultivated foods of man, but the implications of this cannot be discussed until more is known about other foods and the efficiency with which the red colobus digest them.

Miscellaneous Feeding Behaviors of *C. b. tephrosceles*

Intraspecific niche separation. Within this subspecies a peculiar foraging pattern was observed, which is best described as foraging upside down. The foraging monkey hangs by all four limbs beneath a large tree branch, its body approximately horizontal to the branch (pl. 28). The monkey then climbs forwards and/or backwards while suspended beneath the tree branch, covering up to 5 m in this manner. Occasionally the monkey will place its mouth to the undersurface of the branch and appear to eat something from it. Sometimes it suspends itself by 3 limbs only as it removes something from under the branch with one hand and then places it in its mouth. The tree branch which it forages under is usually, if not invariably, covered with mosses and lichens. This type of foraging has been observed to occur in many tree species including *Strombosia, Parinari, Newtonia, Aningeria, Fagaropsis,* and *Funtumia.* I suspect that the monkeys are searching for and occasionally eating certain types or parts of lichens, mosses, and relatively immobile insects (larvae and pupae). This foraging pattern is not performed by adult males and is primarily an activity of juvenile badius. Forty-eight cases were described from at least 4 different badius groups in compartment 30 of the Kibale Forest between 11 May 1970 and 25 December 1970 (table 44). Each case was separated from other similar cases by at least one hour and usually by a much longer period. Of these 48 cases, 64.6% were performed by juveniles, much more than could be expected if this pattern were uniformly performed by all age classes. Monkeys that were approximately the size of adult females but not sexually mature, and those that were slightly smaller but larger than juveniles, also foraged in this pattern much more than could be expected by chance. Adult females foraged upside down much less than expected. Adult males and infants never foraged in this manner. I suggest that adult males and most adult females are too heavy in body weight to perform this type of foraging efficiently and, perhaps, safely. Infants who still spend large portions of their time clinging to adult females may not have the strength to climb beneath branches in this manner. Subadults and juveniles seem best adapted to this foraging pattern and are thus best able to exploit the food niche of the undersurface of large moss and lichen-covered horizontal branches. This ability effects a type of intraspecific food niche separation within *C. b. tephrosceles.*

Foraging behavior and possible consumption of invertebrates. Observations of red colobus eating insects are few (tables 34 and 35), but this is more likely due to the difficulty of clearly observing and identifying the insect as it is being eaten than to the rarity of the event. There were innumerable cases in which the badius were suspected of catching and eating insects or other invertebrates, but because of the small size of the object and the rapid movement of the monkey the food object was not identified.

Foraging behaviors which suggested to me that the monkeys were looking for and occasionally finding and eating invertebrates include: unrolling mature leaves that have been curled, as when a cocoon is formed inside; closely examining and handling dead and dry leaves and occasionally eating something from them (once a silky, cocoon-like object was eaten); handling and examining many mature leaves in succession and then suddenly and rapidly grabbing something with one hand and eating it; and the same pattern as preceding but with handfuls of moss and lichens instead of mature leaves. Sometimes the actual catching of the item was slow, as when an animal carefully picked an object off a mature leaf or from among moss with the fingers after having sorted through the vegetation, suggesting that it was catching larvae or pupae. Once I observed a badius violently scratching itself and then eating something off the back of its hand and then from the tree branch, indicating that it was eating and being bitten by ants.

The low frequency with which galls were eaten (tables 34 and 35) is also dependent upon poor observation conditions. Most of the scores attributed to the category of pieces of mature leaves probably included insect galls.

In a very common foraging pattern, the monkey sits in one place and then carefully and methodically turns over and examines the undersurface of many mature leaves within its reach. Occasionally the monkey places its mouth to the undersurface of one of these leaves and appears to eat something off it. *Strombosia* is the tree species in which this type of foraging is most commonly performed, but it has also been observed with the mature leaves of *Aningeria, Chrysophyllum g., Markhamia, Funtumia, Lovoa, Symphonia, Celtis durandii, Diospyros, Blighia u., Monodora, Parinari, Mimusops, Premna, Dombeya,* and *Balanites.* Examination of many fallen and mature *Strombosia* leaves revealed that a few of them had lepidoteran pupae and the remains of such pupae on the undersurface; so it may be these that the red colobus are searching for and eating.

C. b. tephrosceles also feed on dead wood. This has been observed on several occasions among at least 3 groups. Most members of the group will feed on a dead, but still erect tree trunk, often clustering together in physical contact while feeding on a particular part of the trunk. The high incidence of agonistic encounters at such times indicates a choice food item at stake. I was once threatened by 2 adult males as I came close to them while they fed on a dead tree trunk. The bark of the tree is removed, if it is not already off, and then the mouth is usually applied directly to the wood. When first removing the bark of a dead tree, the monkeys often carefully examine it and the wood beneath it and then sometimes eat something. Certainly some wood is eaten, but the extensive insect tunnels throughout these dead trunks suggest that the monkeys may also be eating larvae or pupae (pl. 29). They sometimes probe the wood with their fingers, which also suggests they are after insects. Infrequently, I have found insect pupae in these dead trunks. Pieces of dead wood are sometimes carried several meters away from the actual trunk, where they are examined and fed upon. During one particular all-day watch, the CW group remained feeding at one particular dead tree trunk for 5 consecutive hours. The effect of this activity after several months is to greatly reduce the size of the dead trunk. One also finds that the ground at the base of the tree is compact and free of vegetation as a result of the monkeys' coming to the ground to feed on the base of the trunk. I have also seen monkeys gnawing on dead branches in live trees, but this is not common.

Other badius subspecies.   Observations of possible invertebrate feeding in the other subspecies studied are limited to 3 cases. Once with both *b. temminckii* and *b. badius* a monkey ate either the dirt and/or termites from an arboreal termite tunnel. A *b. preussi* once ate something out of a piece of dead wood and then dropped the wood.

Drinking.   During the entire study, *b. tephrosceles* were observed to drink water on only two occasions. Drinking was not observed in the other subspecies. Both cases occurred during the rainy season. On 1 October 1970, 4 badius drank in succession from the same tree hole. The duration of each drink was about 5 sec., but one lasted at least 12 sec. Two of the 4 monkeys drank several times. On 4 October 1971 an adult male dipped his head into a hole in the trunk of a *Celtis africana* 3 times in succession; each time he lifted his head, water dripped off his muzzle. One hour and 43 minutes later another adult male and a juvenile drank in succession from the same hole. Drinking is obviously an infrequent event among red colobus, and it seems paradoxical that it was observed at a time of year when there was an abundance of surface water on leaves and branches and when they were feeding heavily on young growth with a high water content. Perhaps it is only during the rainy season that the tree holes contain enough water for them to drink. In any event, drinking standing water does not appear important to their physiological well being.

Consumption of Food from Trees Treated with Arboricide

As part of its forest management practice the Uganda Forestry Department often follows the felling of a compartment by treating undesirable tree species with an arboricide designed to kill them. The arboricide used is called Finopal. Twenty-five ounces of Finopal are mixed with 5 gallons of heavy diesel to give a 3% solution. Finopal is described as 2:1 mixture of 2,4-D and 2,4,5-T with a minimum of 80% weight-for-weight total active ingredient. The tree to be killed is first girdled completely around its circumference and the arboricide is then sprayed onto and above the cut. In view of the considerable literature and confusion on the possible teratogenic (fetopathic) effects of arboricides containing 2,4-D and 2,4,5-T and the dioxin impurities (Orians and Pfeiffer 1970, Epstein 1970), it seemed worthwhile to determine whether, in fact, monkeys fed on trees that had been treated with arboricide. A large portion of compartment 13, located about 1.5 km from compartment 30, was completely poisoned with arboricide even though it had not been clear-felled or otherwise exploited for timber and contained large areas of swamp forest, which will never become valuable forest whatever forest management is applied. The reasons for this action were not clear, but the situation did provide a convenient opportunity to study the response of the monkeys to it. Nine hours and 20 minutes were spent observing red colobus in this compartment on 28, 30, and 31 March and 1 April 1971. In this time they fed on the following items of arboricide-treated trees: *Parinari* small young leaves and leaf buds (9 observations each) and fruit (once); *Newtonia* leaves (3 times); and an unidentified item of *Newtonia* (twice); leaves of *Pygeum africanum* (6 times); *Macaranga s.* young leaves (6 times) and flowers (once); an unidentified part of *Symphonia* (once); and young leaves of an unidentified tree (once). Some of these trees had been treated only a few days before the observations were made, but most had been sprayed 6-8 weeks earlier. An analysis of a sample of the arboricide by Dr. E. A. Woolson of the U.S. Department of Agriculture indicated that the total dioxin content was less than 0.1 ppm, which he considered a harmless amount. Whether the 2,4-D or 2,4,5-T had any physiological effect on the monkeys is not known, but laboratory evidence implicating them as harmful compounds are not unequivocal (Epstein 1970). The most apparent effect of the arboricide treatment on the monkeys is its immense alteration of their habitat.

Food Habits of Other Subspecies Studied

The red colobus of Senegal (*b. temminckii*) which were surveyed lived in habitats classified as undifferentiated savanna woodland, forest-savanna mosaic, and coastal forest-savanna mosaic (Keay 1959) (pls. 30–32). Those of the Ivory Coast (*b. badius*) and Cameroun (*b. preussi*) were living in moist forest at low altitudes (Keay's classification), i.e., lowland rain forest. Relatively few data were collected on food habits during the brief surveys of these three subspecies, and no attempt was made to quantify their food preferences. However, table 45 clearly shows that they share many of the dietary peculiarities of *b. tephrosceles* in the Kibale Forest. For example, entire mature leaves were not eaten, although their petioles and apical tips were. Both floral and leaf buds commonly appear in their diets, as for *b. tephrosceles. Parinari excelsa* was the only species which was also fed upon by *b. tephrosceles;* the same parts were eaten as in the Kibale Forest. Young leaves appear to be less important in the diet of the West African subspecies, although this apparent difference could be an artifact of the time of year the sample was taken or due to my failure consistently to distinguish mature from young leaves. Seeds and fruits seem to be more important foods for *b. badius* and *b. preussi* than for *b. tephrosceles.* In view of the short period during which the observations were made and the lack of quantification, these comparisons should not be overemphasized.

Comparison with Other Studies on Colobinae

In 1969–70 Clutton-Brock (1972) made an intensive 9-month study of feeding and ranging behavior of the red colobus (*b. tephrosceles*) in the Gombe National Park, Tanzania. For 2 months in 1970 he made a similar study of 2 groups of red colobus in the Kibale Forest. Many of the differences in the results of Clutton-Brock's study in the Gombe and mine at Kanyawara are attributable to differences in the floristic composition between the two areas. For example, although *Celtis africana* was the most important single species in the annual diet at Kanyawara, it was absent from the Gombe forest. Although Clutton-Brock's methods of sampling food habits was somewhat different from mine, some of our results are directly comparable. Including unidentified species, at least 60 species made up the diet of red colobus at Gombe, comparable to the 68 species or more eaten by them at Kibale. Only 5 species were common to the diets of these two populations. A comparison of the percentages with which these 5 common species occurred in Clutton-Brock's 9-month study with their occurrence in the last 12 months of my study reveals the following, with Gombe percentages presented first: *Newtonia b.* 15.4% versus 9.9%; *Albizia gummifera* 3.6% versus 0.2%; *Pseudospondias m.* 1.5% versus 0.1%; *Markhamia p.* 0.5% versus 15.1%; and *Spathodea n.* <0.1% versus 0.07%. Although Clutton-Brock's estimates of "cover" for the various tree species are not equivalent to density estimates, they do give some idea of the relative abundance. *Newtonia* was quite abundant in his Gombe study area; thus its high occurrence in the diet of the red colobus population both there and at Kanyawara indicates that it was selected with approximately equal preference in the two areas. In contrast, *Markhamia* apparently occurred in much lower densities at Gombe than at Kanyawara, which may account for the pronounced difference with which it was eaten in the 2 areas. *Albizia gummifera* appeared to be more common at Gombe than at Kanyawara, which probably accounts for its higher incidence in the diet there.

The single most popular food species in the 2 areas constituted identical percentages in the diet. *Newtonia* made up 15.4% of the diet at Gombe, and *Celtis africana* 15.4% at Kanyawara.

Considering only the parts of plants eaten, regardless of species, one also finds interesting similarities and differences between the 2 studies. In the Gombe study, mature leaves made up 44.1% of the diet, in sharp contrast to 2.3% in the Kanyawara diet. Some of this difference can be explained by differences in the way we categorized food parts. Clutton-Brock did not distinguish between entire mature leaves and apical tip, basal part, pieces, and petioles, as I did. If these later 4 categories are added to the 2.3% mature leaf category for Kanyawara, a total of about 23% is reached, which compares more favorably to the 44.1% of Gombe but still leaves 21% to be explained. Clutton-Brock explains that he often had difficulty determining whether the red colobus at Gombe were eating young leaves or mature leaves of *Newtonia*, and so he generally classified these undetermined items as mature leaves, which would inflate his estimate of the mature leaf percentage in their diet. In similar cases of uncertainty, I classified the items as "leaves of unknown age." For the last 12 months of the Kanyawara study this item constituted about 10% of the diet, which, when added to the 23% of entire leaves, apical or basal leaves, and petioles of mature leaves, gives 33%. This, however, still indicates that mature leaves are a more important element in the diet of red colobus at Gombe than at Kanyawara.

Young leaves appeared to be of similar importance in the diets of both populations, with large young leaves ("new leaves") comprising 14.9% at Gombe and 14.7% (including the category of large young leaf petioles) at Kanyawara. Medium, small, and very small young leaves (including petioles) accounted for 23.8% of the Kanyawara diet and 17.1% at Gombe.

Leaf and floral buds ("closed shoots") were much more important (20.1%) at Kanyawara than at Gombe (2.8%). In contrast, fruit was twice as prevalent in the Gombe diet (11.4%) as at Kanyawara (5.7% including seeds). Woody stems were 2.9% of the Gombe foods, whereas what I consider to be a similar category, young twigs, were only 0.5% of the Kanyawara diet. There are no obvious methodological differences to account for these differences in food preferences, which are, apparently, real. One generalization does, however, seem applicable to the two populations: that young growth is the most important dietary constituent for red colobus.

Although Clutton-Brock did not compute indices of diversity for food species, I have done so, using the percentages for the 58 species listed in his table 20. Computing H as described earlier in this chapter gives an index of 2.651, which indicates a slightly less diverse diet than that reflected by the last 12 months of data for the badius at Kanyawara (table 35), where H equals 2.962. Certain other statistics also indicate that dietary diversity was similar between the two populations. Ranking the food species according to their proportional importance in the overall diet shows that the first 5 species constituted 55% of the diet at Gombe and 55.9% at Kanyawara. The first 10 food species constituted 78.4% of the Gombe diet and 75% at Kanyawara.

Selection ratios were computed somewhat differently for the Gombe data but also result in some dramatic changes in the apparent preferences for some food species. Perhaps coincidentally, selection ratios for both the Gombe and Kanyawara data placed *Newtonia* third as a food preference of red colobus.

Food habits of the Gombe red colobus showed considerable monthly variation, as did those at Kanyawara. *Newtonia,* however, appeared to be important in all months at Gombe ($\overline{X} = 15.2\%$, range 4.1%–33.6%), as at Kanyawara ($\overline{X} = 10.5\%$, range 5.1–21.2%).[6] There seems to be greater monthly variation, even a suggestion of seasonality, in the types of parts eaten at Gombe than at Kanyawara, especially for "shoots" (medium, small, and very small young leaves) and mature leaves. The diversity of diet also showed monthly variation in the two areas. Although he did not support his observation by a correlation test, Clutton-Brock suggests that food species diversity was greatest at Gombe in those months when the red colobus fed most on shoots and flowers. This is in marked contrast to the Kanyawara situation, where no such relationship was found. A Spearman Rank Correlation Coefficient Test comparing the percentage of flowers, buds, and young leaves of all classes (including petioles) in the red colobus diet with the food species diversity of the diet at Kanyawara over 17 months was not significant ($r_S = 0.125$, $p > 0.05$).

Clutton-Brock's 2-month study in the Kibale Forest was conducted on 2 different groups of red colobus. One group was located near the Dura River bridge on the Kibale Road (referred to as Bigodi in his thesis) about 8 km south of Kanyawara, and the other was at Kanyawara in compartment 30, but different from my major study group. The forest in the Dura River study area is sufficiently different floristically that one would expect corresponding dietary differences, which, combined with the short duration of Clutton-Brock's study there, makes it seem most worthwhile to restrict the comparison to his one-month study at Kanyawara (September–October 1970). Because he sampled every 15 minutes, Clutton-Brock's sample of red colobus food habits (N = 563) for this period was larger than mine (N = 123), but the results are nonetheless comparable. Of 19 food species recorded by Clutton-Brock for this period, 13 (68.4%) were also recorded in my October 1970 sample. Some species were present in similar proportions in the two samples, such as *Newtonia* (15.6% in his and 16.3% in mine). For some others the correspondence was not close, e.g., *Strombosia* (14.0% in his and 1.6% in mine). The entire percentage overlap for the two October 1970 samples, computed by summing the shared percentages of all food species (and disregarding parts) was 53.5%. Some of this difference could be attributable to differences in the sample size and to local differences within compartment 30 in the densities and phenology of various tree species. The range of Clutton-Brock's study group did not overlap with that of mine. A major point of agreement emerging from the two studies is that the Kanyawara red colobus appear to feed more heavily on young growth than do those of Gombe. The five most important food species in Clutton-Brock's sample at Kanyawara comprised 59.1% of the diet, and the first 10 species 79%, which correspond closely with the last 12 months of my study — 55.9% and 75% respectively.

Clutton-Brock's assistant, Janet Brooke, collected information on the food habits of 2 groups of black and white colobus (*C. guereza*) at the Dura River and Kanyawara during the same period. These results, presented by Clutton-Brock, clearly indicate pronounced differences from the red colobus. The black and white colobus have a less diverse diet, concentrating on *Celtis durandii,* which made up 70.8% of the diet at the Dura River and 88.3% at Kanyawara. The as yet unpublished report on a longer and more detailed study of this species at Kanyawara by J. F. Oates (1974) elaborates on this subject.

The only other study of red colobus monkeys in East Africa is the nine-day study of Nishida (1972) and his assistants on 2 groups of *b. tephrosceles* living in the Mahali Mountains of Tanzania. This study was conducted in 1969 near the Kansyana Camp, Kasoge, about 160 km south of the Gombe study area. It was secondary to Nishida's major study of chimpanzees. Fifty-seven food species were listed in this period, 21 of which are also listed by Clutton-Brock as present in the Gombe. Nishida states that the red colobus are "very fond of young leaves ... but eat a great amount of hard leaves. Flowers, fruits, and barks are also eaten, although supplementary." This is essentially consistent with the findings at Gombe and Kibale.

There is virtually no information available on the food habits of West African red colobus outside of my surveys there. Booth (1956) found only leaves in the stomachs of 33 *C. b. waldroni* which he collected in Ghana. Kuhn (1964) notes that *b. badius* in Liberia eat the leaves of *Triplochiton scleroxylon.* In the stomachs of collected specimens he found finely chewed hard fruits and once some very small leguminous fruits. He also suggests that *b. badius* eat the beans of the legume *Pentaclethra macrophylla.* These beans have a 30–36% oil content and are very high in protein but low in starch (Kuhn 1964).

The scant information published for other African colobinae (black and white colobus and olive colobus of Central and West Africa) indicates that they are primarily leaf eaters, occasionally taking flowers (Kuhn 1964, Booth 1956 and 1957, Jones 1972). The intensive field study of *C. guereza occidentalis* by J. F. Oates (1974) in the Kibale Forest demonstrates that the diet of this subspecies is more varied than these earlier studies indicate and includes leaf and floral buds, flowers, and fruits.

There have been many field studies of the langurs in India and Ceylon. Langurs, like colobus monkeys, belong to the subfamily Colobinae and are typified as leaf-eaters. The majority of these studies have concerned various populations of the Hanuman or gray langur, *Presbytis entellus.* Yoshiba (1967) found that their food in Dharwar District of Northern Mysore State "consisted mainly of leaves, shoots and soft stalks of trees, vines and grasses," but also included fruits and flowers or buds. Infrequently, they ate bark, insect galls on the leaves of *Terminalia tomentosa*, and lepidopteran larvae of one particular but unidentified species. He concludes that the plants most frequently utilized as foods are the more common species. The langurs drink standing water, but apparently this is not obligatory.

Poirier (1968a) in his study of Nilgiri langurs (*Presbytis johnii*) in the Ootacamund area of the Nilgiri Hills of Madras state, South India, mentions that they eat the leaves, flowers, and fruits of *Acacia melanoxyln* and *A. molissima*. He lists 12 other food species, but not the items fed upon.

The *Presbytis entellus thersites* of Polonnaruwa, Ceylon, have been studied intensively. Ripley (1970) lists 60 plant foods and includes mature leaves, young leaves (flush), fruits, seeds, flowers, and sap as food items consumed. Hladik and Hladik (1972) made a detailed study of the feeding ecology of *P. entellus* and *P. senex* at Polonnaruwa. Analyzing their diets on the basis of the estimated weight of food consumed as determined from direct observations (118.8 hours on *P. entellus* and 96.5 hours on *P. senex*), they conclude that *P. entellus* consumes 21% mature leaves, 27% shoots (young leaves and buds), 7% flowers, and 45% fruits. In contrast, *P. senex* was estimated to consume 40% mature leaves, 20% shoots, 12% flowers, and 28% fruits. *P. entellus* appears to eat more fruit and less mature leaves than does *P. senex*. The overall diet of *P. entellus* is also apparently more diverse than that of *P. senex*. A total of 42 species were fed upon by *P. entellus* in contrast to 28 species for *P. senex*. Ninety percent of the *P. entellus* food was obtained from the 23 top-ranking species, whereas approximately the first 12 food species contributed 90% of the *P. senex* diet. Even more impressive is the fact that the 2 most popular foods in the *P. senex* diet comprised 50% of their intake. The differences between the diets of these 2 langurs resemble the differences between those of the 2 colobus monkeys in the Kibale Forest, where badius feeds less on mature growth than guereza and has a more diversified diet than guereza.

RANGING PATTERNS OF *C. b. tephrosceles*

Introduction and Methods

Most studies of home range among mammals are based on capture-recapture data. In contrast, estimates and measures of primate home ranges are based on direct observations over relatively long periods of time.

Home range generally refers to the entire area occupied by an individual or a social group during an entire year. Some investigators would exclude areas infrequently used as being merely the object of apparently erratic forays. Operational definitions of such forays are usually vague, making replication difficult and imprecise.

Implicit in all measures of home range is the concept that one is actually measuring the area occupied in a physical sense. It has been suggested that the home range of an individual or social group be extended to include the area that can be "surveyed" by an animal in its search for and utilization of the various life essentials (P. Waser, pers. comm.). In such a concept the animal surveys a larger area than it occupies spatially, by employing its senses of hearing, smell, and vision. This concept is even more difficult to define and measure operationally, although it seems very reasonable from an intuitive point of view.

In most primate field studies the methods of mapping and measuring home range are described only in the briefest terms. It would seem, however, that the common method is to construct an aggregate map consisting of all the daily maps of the group's movements during the entire study. The area occupied on this aggregate map is then encircled and the surface area measured, either with a planimeter or by counting the number of squares such an area occupies when placed on graph paper. The inherent deficiency in this encircling method is the arbitration involved in determining where to draw the boundary around the area occupied, so that replication is difficult. Bias is introduced because lacunae not actually occupied by the group are often included, resulting in an overestimate of the home range. Although this method is the one most commonly used, its use has been far from uniform. The various estimates for different species are not more comparable because the same method has been used.

The "taut string line" method was used by Altmann and Altmann (1970). In this method a taut string is placed around the area occupied. The outermost points of the mapped movements are connected with one another by the taut string, producing what Altmann and Altmann call a convex hull. The area enclosed by the taut string constitutes the estimate of the area occupied. The taut string method has the advantage of being objectively replicable, but I feel this is far outweighed by the potential error introduced by including many lacunae. Furthermore, the estimate is greatly affected by whatever factors affect the meanderings of the group, regardless of the space occupied. For example, configurations of identical shape and area could be attained with this method both in an area of uniform habitat used equally by the monkeys and in one containing a huge area of unsuitable habitat not utilized by them at all.

Analysis of range utilization employing a grid has been used in at least 5 primate field studies (Rowell 1966, Struhsaker 1967b, Mason 1968, Altmann and Altmann 1970, and Clutton-Brock 1972). In this method a grid composed of quadrats of uniform size is superimposed on the map(s) to be analyzed. The amount of time spent or the number of entries into each specific quadrat is then tabulated. This method is useful for the analysis of the distribution of time in space. Its main advantage is its replicability. Although it has been employed in only one previous primate study (Clutton-Brock 1972), the grid system can also be used to estimate home range. The simplest and least critical way of estimating home range with this method is to count the number of quadrats occupied and multiply by the area of each quadrat. This method would be exact only in a case where the group occupied the entire area of each and every quadrat. Such is rarely, if ever, the case. Thus, although the grid method removes many lacunae and the arbitrary decision of where to draw the boundary, it does include small lacunae in most quadrats. One solution would be to use a grid system composed of quadrats equal in size to the animals studied. Pragmatics generally make this impossible. Obviously, the extent of this bias is dependent on the grid size selected. Determination of the appropriate grid size is another problem with this method. Factors guiding the determination of an appropriate grid size include: accuracy of the original mapping (quadrat size should be no smaller than warranted by the accuracy of the mapping); dispersion of the group mapped and the accuracy of this mapping; convenience for analysis; and the purpose of the analysis, i.e., how accurate must the home range estimates be? Of pertinence to this latter question is the concept of home range mentioned above, which includes the area "surveyed" by the study animals. How large an area can an individual or group survey in the particular habitat at any one moment? The accuracy of the estimate of home range using the grid system can be improved by measuring the actual proportional area occupied in a sample of quadrats, as it was mapped. Does the group use 100%, 50%, 5%, or whatever, of the average quadrat? The reliability of this estimate in turn depends on the accuracy of the original mapping. The more compact the group's dispersion, the more accurate will be the mapping of the group's movements.

I used the grid system in my analysis. Daily ranging maps were maintained for all observations. Because of the compact nature and small size of the CW group, it was usually possible to delineate the approximate area occupied by the group on the map. They were usually dispersed over less than 50 m in any one direction. However, this method of mapping involves the inclusion of some small lacunae, because never were all members of the group in physical contact (directly or indirectly, as in a chain) at any one time, which would be the only case where encircling the group's dispersion did not include lacunae. Consequently, some small lacunae are included even in areas which on the map are indicated as being completely occupied. Once the data have been collected in this manner there can be no correction for this kind of imprecision. Estimation of the average proportional use of the quadrats improved my estimation of home range, but did not correct for the small lacunae included in this method of mapping.

The maps for the CW group were segregated and analyzed in 3 groups: (1) systematic monthly samples for the period of November 1970 through October 1971; (2) systematic monthly samples for the period of November 1971 through March 1972; and (3) all other observations outside the systematic samples for the entire study of the CW group (August 1970 through March 1972). Because the BN and ST groups were never observed for a complete day (i.e., for at least 11½ hours) and the total contact hours with them were relatively few, the map data for them were not segregated according to periods or types of observation, but were analyzed together.

A grid system composed of 0.25 ha. (50 m x 50 m) quadrats was placed over the daily maps, and the time that the group remained in each particular quadrat was tabulated. Quadrats were identified by row and column numbers (fig. 12). The number of days and the number of months the monkeys used each quadrat were also tallied. Because some quadrats were occupied simultaneously, the number of minutes tallied exceeds the number of minutes the group was actually observed.

The actual area occupied by the CW group during these samples was less than that estimated from the number of quadrats they used. This was because very few, if any, quadrats were ever fully occupied, as estimated from an analysis of 65 quadrats in an aggregate map of the CW group's movements during a 7-month period (February 1971 through August 1971). In this analysis all movements and sightings of the CW group during the period were transferred from the daily range maps onto one aggregate map. All 65 quadrats occupied in the first sectional map were analyzed (the map of the entire study area was divided into 6 sectional maps for convenience in the field). This section was the one most heavily used by the CW group. From the aggregate map, measurements were made of the proportion of each of the 65 quadrats that was actually occupied by the CW group during the 7-month period. On an average, only 25.2% of each quadrat was occupied. The range of occupancy was 1.0–72.5%. If these measures are representative of all quadrats used by the CW group, then a more accurate estimate of their actual home range is much less than the estimate based on the number of quadrats used. The proportion of occupancy may, however, increase with more days and months added to the sample. Tables 46 and 47 show that in the 2 periods of systematic sampling the average number of days that the CW group occupied any particular quadrat was 2.5 and 2.3. Thus the average quadrat is unlikely to be occupied any more fully than the cumulative area occupied within 2–3 days. Curves of the cumulative area occupied in 5 quadrats were determined for the systematic samples from November 1970 through October 1971 (fig. 29). One of these was used more days (10) than any other quadrat in this 12-month sample period; 2 were used for more minutes than any other quadrat; and 2 were used for approximately the average number of minutes (mean duration in the 260 quadrats was 190.8 min.). The quadrat used for the greatest number of days (quadrat -2 -12) had only 68.5% of its total area (0.25 ha.) used in this 12-month sample. One of the 5 curves indicates a definite plateau after 7 days of occupation (-2 -13). Three curves indicate a decline in the slope and appear to be approaching a plateau after 5 days of use. Only one quadrat curve indicates an increase in proportional use after 5 days (-4 -10 in fig. 29). These curves support the suggestion that the 65 quadrats analyzed and discussed above give a fair indication of the proportion of the quadrat area used during the entire 20-month study. A more accurate estimate of the home range would seem to be somewhere between 25% and 75% of the area indicated when one considers only the number of quadrats occupied.

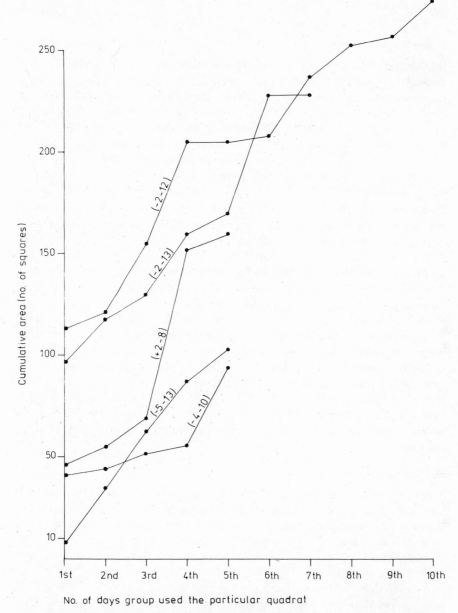

Fig. 29. Cumulative area occupied by CW group in selected quadrats during the systematic samples of the period November 1970 through October 1971. Each quadrat was divided into 400 equal-sized squares. The cumulative area represents the number of squares which the group occupied in the particular quadrat indicated. Quadrats are identified first by row number and then by column number as shown by the marginal numbers in fig. 12. For example, quadrat (–4 –10) is the quadrat forming the intersect between row –4 and column –10 in fig. 12. During the period considered, this quadrat was entered on 5 days by the CW group. On the first day, 41 of the possible 400 squares in the quadrat were occupied by the group. On the remaining 4 days on which the group used this quadrat, they occupied only an additional 53 squares, giving a total of 94 squares used in the quadrat.

Home Range

During the systematic sample days of the 12-month period November 1970 through October 1971, the CW group occupied 260 quadrats (table 46). In the systematic samples of the subsequent 5 months they used a total of 97 quadrats, only 17 of which were not used in the previous 12 months (table 47). Thus, of the total of 277 quadrats used by them in the systematic samples of all 17 months, only 6% were added in the last 5 months. Only 6 more quadrats were added to this number when all of the other observations from August 1970 through March 1972 were also considered. The total of 283 quadrats gives a first estimate of 70.7 ha. as the home range of the CW group during the 20-month study. As discussed above and in the previous section on methodology, a more realistic estimate lies between 25% and 75% of this area, or about 17.7 to 53.0 ha., i.e., 35.3 ha.

The map data for the ST group during the entire 20 months show that the group occupied 160 quadrats (40 ha., table 48). Using the same correction factor as for the CW group gives an estimate of about 20 ha. The actual home range of the ST group is certainly larger than this, because the monkeys were never studied for a complete day, nor were they studied on a systematic basis. Most of the map data for this group are opportunistic, i.e., the animals' position was mapped whenever they were seen, but they were rarely sought out.

The BN group was mapped in 66 quadrats (16.5 ha.) during the period of 24 November 1970 through March 1972 (table 49). Applying the above correction factor yields an estimate of 8.25 ha. However, the same qualifications apply here as with the ST group. Their position was mapped on only 27 days, none of which were full days and all of which were opportunistic. This estimate is surely too small.

Distribution of Time in Space

Tables 46-49 clearly demonstrate that the quadrats occupied were not used uniformly. More time was spent in some than in others. The range of utilization was great; for example, table 46 shows that the time spent in the 260 quadrats ranged from 2.5 to 1,060.5 minutes. Similarly, variation was great in the number of days and the number of months that the particular quadrats were used (tables 46-49).

An analysis of 53 days in which the CW group was observed for at least 11½ hours shows that they used 12 quadrats per day on average (fig. 30). The number of quadrats used per day varied from 4 to 24, with a coefficient of variation of 34.7%.

In the 17 months for which systematic samples were made of the CW group, the average number of quadrats used per month was 40.8 (range 26-65). This figure is based only on the days of systematic sampling for each month. However, the number of days and contact hours of systematic sampling per month was quite variable, particularly in the last 5 months, which probably accounts for much of the variation. Considering only the 7 months in which the

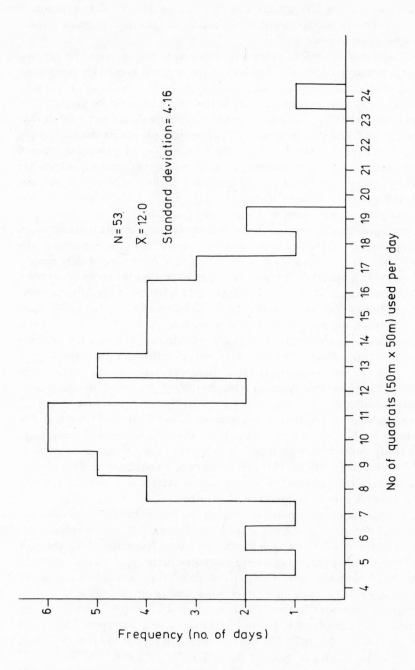

N = 53
X̄ = 12.0
Standard deviation = 4·16

No of quadrats (50m x 50m) used per day

Frequency (no. of days)

Fig. 30. Number of 0.25 ha. quadrats used per day by the CW group during systematic samples, including only complete days ⩾ 11½ hours of observation (N = 53 days).

systematic sample consisted of 5 full and consecutive days the mean number of quadrats used by the CW groups was 45.1 (range 27–65). Computation of indices of diversity in the monthly utilization of quadrats permits a more meaningful comparison.

The Shannon-Wiener information measure was computed for the ranging data for each month (table 50). This is the same formula as used for computing indices of food species diversity and is symbolized by H. The indices of ranging diversity or quadrat utilization diversity were computed using the proportional amount of time the CW group spent in the various quadrats each month. For example, if the CW group used only 4 quadrats in a particular month and the amount of time spent in each was equal (25%), the index of quadrat utilization diversity would be 1.39. If in another month they also used only 4 quadrats but spent 70% of their time in one and 10% in each of the remaining 3 quadrats, the index of quadrat utilization diversity would be 0.94, which represents a less diverse ranging pattern than in the previous example.

It seemed possible that some of the intermonthly variation in ranging pattern diversity (table 50) might be attributable to differences in the duration of the samples. For example, monthly samples which were comprised of only incomplete days ($< 11\frac{1}{2}$ hours) might be expected to have a higher index of diversity than samples comprised of complete days ($\geqslant 11\frac{1}{2}$ hours), because on incomplete days observations were usually not made between 1230 and 1530 hours. This was a relatively inactive time of day for badius. However, a Wilcoxon 2-sample test, comparing indices of quadrat utilization diversity for monthly samples consisting of only complete days with monthly samples consisting of only incomplete days, demonstrated that there was no significant difference in the indices for these two types of samples ($U_s = 26$, $p > 0.10$ one-tailed, $p > 0.20$ two-tailed).

There was some suggestion of seasonal variation in the utilization of the home range by the CW group. For example, the far north end of their range was used only in December 1970 and March 1972. The far southeast end was used only in June and August 1971. One means of an objective analysis of such tendencies is the computation of percentages of overlap in which the ranging patterns are compared one month with another.

In the analysis of intermonthly overlap in home range utilization the proportional distribution of time among the various 0.25 ha. quadrats was compared between all 17 months for the period of November 1970 through March 1972. Overlap percentages were computed in the same manner as in the comparison of monthly diets. The percentage overlap in quadrat utilization (ranging pattern) for each pair of months was computed by addition of percentage values shared by the 2 particular months for all 0.25 ha. quadrats concerned. For example, if in January 1971 the CW group spent 75% of its time in quadrat A, 20% in B, 5% in C, and zero in D, but in January 1972 it spent only 5% of its time in quadrat A, 10% in B, 30% in C, and 65% in D, then the percentage overlap in quadrat utilization between these 2 months would be 20% (the sum of 5% in A, 10% in B, and 5% in C). Examination of the 136 pair-combinations fails to reveal any striking seasonal trends in range utilization (table 51). The percentage overlap in quadrat utilization between like months in subsequent years is not particularly greater than the overlap between different months. In some cases the overlap was quite high between adjacent months, for example, January, February, and March 1972, but this was not a consistent trend, as exemplified by November and December 1970 and January 1971, in which there was little overlap. The absence of any outstanding pattern in these measures of overlap is similar to the result obtained when percentage overlap in diet was compared between months (table 40). It seemed reasonable, therefore, to ask whether there were not some relationship between percentage overlap in diet and percentage overlap in quadrat utilization. Do the monkeys have similar diets in those months in which they have similar ranging patterns? In other words, how well do the inter-monthly overlap percentages of diet and quadrat utilization correlate? In fact, they do not correlate very well. The data in tables 40 and 51 were compared using Olmstead and Tukey's corner test for association (Sokal and Rohlf 1969, pp. 538–40) and the absolute value of the quadrat sum was found to be 8.5, which indicates that the correlation is not significant ($p > 0.10$). Although I shall return to the subject of what affects ranging patterns, it is clear from this analysis that the manner in which the monkeys distribute their time in space is not strongly correlated with what they eat.

Group Movements

The distance traveled during any one day by the CW group of *b. tephrosceles* was extremely variable. The mean daily distance traveled in a sample of 54 days was 648.9 m, with a coefficient of variation of 97.1% (fig. 31). Distances traveled by the CW group were plotted each day on study-area maps having a scale of 1:2,500. Because the CW group was relatively small and rarely dispersed over an area exceeding 50 m in any dimension, it was generally possible to circumscribe on the map the approximate area occupied by the group at any one time. The times of all group movements were noted directly on the maps with the corresponding direction and distance of the move. This differs slightly from plotting the movement of the group's "center of mass" (Altmann and Altmann 1970), because I have also included the approximate edge of the group's dispersion. Only days in which the group was followed continuously for at least 11½ hours are included in this analysis. The 54 days are as follows: 2 October, 4–8 November, 1–5 December 1970, 3–7 January, 1–5 February, 1–5 March, 6–10 April, 12–15 May, 1–4 June, 2–6 August, 9, 10, 14 September, 4, 5, 31 October, 1 November and 1, 2, 3 December 1971.

The average distance traveled per day varied considerably from month to month (table 52). Included in this analysis are only those months for which the group was followed for 4 or 5 consecutive days. The mean daily distance traveled for each month was computed from these 4- or 5-day samples.

There was a positive correlation between the mean daily distance traveled per month and the monthly rainfall ($r_s = 0.511$, $p \approx 0.05$, one-tailed). In other words, the group traveled more each day in months with more rainfall than in months with less rainfall. The mean daily distance traveled for the 13 months shown in table 52 was compared with the rainfall for these months (table 41). This is consistent with a similar correlation between ranging pattern diversity and rainfall ($r_s = 0.417$, $p \approx 0.05$), but it should not be extrapolated from this that the CW group moved more on rainy days than on nonrainy days. In fact, there was no significant difference in the distance traveled on rainy versus nonrainy days (Wilcoxon 2-sample test, $p > 0.20$). It would appear that the increase in day range and ranging diversity are coincidentally correlated with increased rainfall, and are probably dependent on some other variable also correlated with rainfall but which remains undetermined.

The relation between the distance covered by the group and the time of day was also exceedingly variable (fig. 32). The distance traveled on each day was measured in terms of 30-minute periods, 15 minutes before the hour to 15 minutes after, and 15 after to 15 before. Data were measured and tabulated directly from the daily range maps. The coefficient of variation was high for every 30-minute period, ranging from 89.3 to 177.4% (fig. 32). The distribution of mean values does, however, indicate that they traveled more at certain times than others, with peaks occurring at 0730 and 1700–1730 hours. At 1230–1300 and 1900 hours the group usually moved least, as indicated by the low mean values at these times. The 1230–1300 hour low corresponds with a rest period (see section on activity cycles), and the 1900 hour low occurs at a time when the group is settling for the night.

There was no time period in which the group always moved; in other words, the lower range for all time periods was zero. Clearly, distance traveled by the group was not rigidly fixed to the time of day.

There are few data on daily ranging patterns for the 3 subspecies studied in West Africa, but they also indicate extreme variability in the distance covered per day. Estimates of the total distance traveled by two groups of *b. temminckii* on 2 different days were 1,116 m and 300 m. Another group of the same subspecies traveled 228 m between 0654 and 0800 hours, whereas a fourth group did not move at all for 3 hours and 42 minutes.

The following distances were estimated for groups of *b. badius*: 455 m in 3 hours and 40 minutes; 300 m in 45 minutes; 15 m in 5 hours; no move for 4 hours and 45 minutes.

The two estimates for *b. preussi* are consistent with this pattern of variability: 774 m in 10 hours and 55 m in 7 hours and 40 minutes.

Food as a Factor Influencing Differential Utilization of Range

Food is one of the more obvious parameters that might affect the group's movements and utilization of their home range. One way of testing this relationship is to compare the amount of time spent in a given quadrat with the

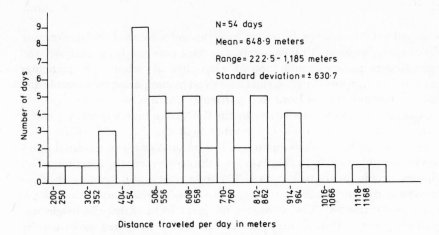

Fig. 31. Daily distance traveled by the CW group. Includes only complete days of ≥ 11½ hours of observation (N = 54 days).

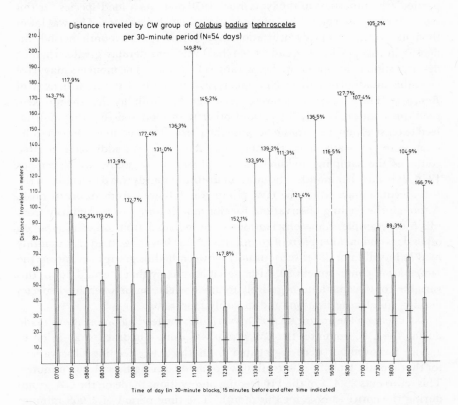

Fig. 32. Diurnal distribution of distance traveled by CW group (N = 54 days). The vertical line indicates the range in distance traveled; the horizontal line indicates the mean distance traveled; and the vertical column represents ± one standard deviation. The percentage at the head of each vertical line is the coefficient of variation for that time period.

density of various food species in that quadrat and with what the monkeys ate while in that quadrat. Unfortunately, the data have not been analyzed and segregated to permit examination of what they ate while in a particular quadrat, but an approximation can be attained by comparing the monthly diet with the amount of time spent in a particular quadrat for the same month.

Quadrat (+ 2 -8) was used more by the CW group than any other quadrat during the 12-month period of November 1970–October 1971. Of the total 1,060.5 minutes spent in this quadrat, 1,003 minutes were concentrated into 3 days in November 1970. In February 1971 the group spent the remaining 57.5 minutes there in the course of 2 days. The young leaves of *Bosqueia* comprised the most important item in their diet for November 1970, constituting 13.1% of the total (table 38). The density of *Bosqueia* 10 m or more in height was relatively great in this quadrat (12/ha., table 33) compared to its density estimate for the entire home range (2.8/ha., table 32). In contrast, *Strombosia*, the most abundant tree of this size class in quadrat (+ 2 -8), was not particularly important in the November 1970 diet. As a food species for this month it comprised only 2.8% of the total feeding observations. This was lower than its average monthly utilization (4.4%) over a 19-month period. Its density in this particular quadrat (44/ha.) was considerably greater than its density estimate for the entire home range (14.7/ha.). The monkeys may have used this quadrat heavily in November 1970 because of the high density of *Bosqueia*. The data do not, however, rule out the possibility that the monkeys used this quadrat heavily for reasons other than food and that they did not feed extensively on *Strombosia* because the particular food items preferred by them were not available on this species at that time. An additional variable concerns the sampling method at this phase in the study. In November 1970, as in all other months, the quadrat utilization study was based only on the 5 consecutive days when the CW group was followed from dawn to dusk. Although the feeding observations for this month were taken primarily from this 5-day sample, some observations were included from other groups, on other days and in other parts of compartment 30. It was assumed that monthly food habits did not differ substantially between badius groups occupying the same part of compartment 30. Although it is unlikely that it would do so, this variable could contribute negatively to any correlation between food-species density, quadrat utilization, and monthly food habits.

As a check on the importance of this variable, the feeding observations made during the 1,003 minutes the CW group was using quadrat (+ 2 -8) on 6, 7, and 8 November were segregated from the other feeding observations for that month. Only 28 observations were made during these 1,003 minutes. This represents 25.4% of the 110 feeding observations made on the CW group during the entire sample for this month. The time period of 1,003 minutes represents 26.4% of the 5-day sample (3,640 minutes). Thus, the proportion of feeding done by the CW group in this quadrat was, as expected, based on the amount of time they spent there. *Bosqueia* was fed on in 10.7% of these 28 observations, not very different from its proportion for the entire monthly sample. *Strombosia* increased to 7.1% in this smaller sample as compared to 2.8% for all of November 1970. In addition, badius foraged several times among mature *Strombosia* leaves while in quadrat (+2 -8). Certain other common species in this quadrat, such as *Trema, Chrysophyllum g.,* and *Teclea,* also increased in relative importance in the diet through this segregation of data. However, because the group occupied other quadrats at the same time they were in quadrat (+2 -8), certain food species that were rare or absent in quadrat (+2 -8) still appeared as proportions of the diet in this segregated analysis, e.g., *Markhamia* (21.4%), *Millettia* (25.0%), *Newtonia* (3.5%). In general, this restricted and segregated analysis indicates that the monkeys tend to feed most heavily on the most prevalent species present in the area they occupy. A notable exception for quadrat (+2 -8) was *Funtumia,* which, although common here (24/ha.), did not appear in the segregated analysis and was only 0.7% of the total November 1970 diet.

A similar analysis for quadrat (-5 -13), which was utilized more evenly over the study (1,590 min. on 15 days in 9 months out of a total of 80 days in 17 months), results in closer correspondence between monthly diet, monthly utilization of this quadrat, and food-species density. Examining only those 3 months (January–March 1972) in which the CW group used this quadrat for more than 200 minutes in the 5-day sample, it is seen that tree species with high densities in this quadrat ranked high in the monthly diet. In January 1972 the large young leaves of *Chaetacme* were the second most important item in the diet and *Newtonia* leaves the third (table 38). Considering only food species, *Chaetacme* was tied for second rank and *Newtonia* tied for fourth (table 36). Both of these species occur in much greater densities in quadrat (-5 -13) than throughout the entire home range (table 33). In February 1972, 4 of the top 5 specific food items were species which occurred in this quadrat in greater densities than estimated for all of the home range. *Markhamia* (mature leaf petioles) in third rank was the only exception in that its density in this quadrat was slightly less than that estimated for the whole home range, although it was still high, 48/ha. An examination of food species only for this month revealed a similar pattern (table 36). The results for March 1972 were consistent with the preceding 2 months. The second-, third-, and fifth-ranking specific food items were of species having higher densities in quadrat (-5 -13) than estimated for the entire home range. The first-ranking item (*Markhamia*) was of nearly typical density, and the fourth-ranked food (*Bosqueia*) was not present in this quadrat (table 33). Considering only food species, the first 3 ranks were occupied by *Markhamia, Newtonia,* and *Fagaropsis,* respectively. All three are represented by high densities here. A major advantage of this analysis of quadrat (-5 -13) for the months of January–March 1972 is that all of the feeding observations were taken from the CW group and virtually all during the 5-day sample when their movements were mapped, thus eliminating the possible source of negative correlation mentioned above.

A third analysis centers on quadrat (-2 -13) which was used for the greatest amount of time during the entire 17 months (1,706 min.) and was also a popular area with the CW group. They used it on 12 of the 80 sample days in 8 of the 17 months. In the systematic 5-day samples, in only 3 months did they spend more than 200 minutes in this quadrat: November 1970, March 1971, and February 1972. The density of *Bosqueia* was very great in the quadrat, and in November 1970 the young leaves of this species were the most important dietary item (13.1%; see table 38). *Newtonia* leaves were also important in this month (6%), and the one large specimen in this quadrat was frequently used by the CW group. When only food species for November are considered, *Bosqueia* and *Newtonia* still rank very high (table 36). These same specific food items were also important in the March 1971 diet. In addition, *Trichilia splendida* young leaf petioles comprised an important food (4.7%) for the red colobus in March. The single large specimen of this species in quadrat (-2 -13) is the only one in the area with the nearest other mature member of the same species being about 550 m away. Together these 3 species comprised 29.6% of the March 1971 diet. *Strombosia* and *Chaetacme*, 2 other abundant species in this quadrat, added another 8% to the foods of March. In February 1972, *Newtonia* comprised 21.2% and *Strombosia* 5.4% of the diet, and this may have been the main dietary reason that the CW group used quadrat (-2 -13) at that time. *Markhamia* was an important food source in February, but its density is not particularly great in this quadrat. Although only one feeding observation was made of *Trichilia s.*, the large specimen in this quadrat was used as a night sleeping tree on 2 days of the 5-day sample for February 1972 and may have been an additional reason for the monkeys' spending considerable time here.

The final analysis is of quadrats (-13 -15) and (-12 -15). About 80% of the time spent in these 2 adjacent quadrats by the CW group was in April 1971. Four of the top 5 food species comprising 78.3% of the April 1971 diet (table 36) were present in very high densities in these quadrats. *Celtis africana* (8/ha.), *Celtis durandii* (68/ha.), and *Millettia* (16/ha.) were all more abundant than estimated for the overall home range. The density of *Markhamia* was estimated at 52/ha., being about the same as the density estimate for the entire home range.[7]

These data, particularly for quadrats (-5 -13), (-2 -13), (-13 -15) and (-12 -15) clearly indicate that within their possible range of diet, red colobus feed primarily on those species of greatest density in the area where they spend most of their time. This does not answer the question whether they move to a particular place to feed on certain species or whether they move there for some other reason and feed on the most abundant foods available within their range of diet. For example, it is known that intergroup conflicts can greatly alter the direction and distance of group movements. Where the CW group's movements have been altered by such a conflict, the animals may then proceed to feed on the most common species available (see below). However, these species must be within their dietary range. Although *Diospyros* and *Uvariopsis* are clearly the most abundant species in quadrat (-5 -13), they were not fed upon by the CW group in the period analyzed (January–March 1972) and very rarely at any other time in the entire study (6 sightings for *Diospyros*, none for *Uvariopsis*).

When one considers the entire 17-month period for which reliable ranging data are available, it becomes apparent that the rather even and relatively heavy use of quadrats (-5 -13) and (-2 -13) is paralleled by the fact that they contain great densities of food species that ranked high in the total 19-month sample of food habits, such as *Celtis africana, Celtis durandii, Markhamia, Newtonia, Chaetacme, Strombosia, Bosqueia,* and *Teclea* (tables 34 and 35). In contrast, quadrat (+2 -8), which was used intensively but only in a very restricted period (3 consecutive days), lacked 2 of the top five food species (*Celtis africana* and *Newtonia*), had only one specimen of *Markhamia* (second-ranking food species), and had relatively few *Celtis durandii* (third rank). Clearly the abundance and distribution of food affects the manner in which red colobus distribute their time in space.

Ranging Diversity, Food, and Intergroup Relations

In an attempt better to understand the factors affecting the ranging patterns of the CW group, comparisons were made between the monthly indices of quadrat utilization diversity and various measures of food distribution and the frequency of intergroup encounters. This analysis has important implications for the more general problem of the relation between ecology and social organization (Struhsaker 1974). Basically, one would like to know the relation, if there is one, between food diversity and dispersion and ranging patterns. Does the monthly ranging pattern of a group of monkeys vary with its diet and food distribution? It has already been demonstrated that ranging pattern diversity of the CW group increases in months with heavier rainfall, but the ecological implications of this relationship are obscure. There is no obvious reason why rainfall per se should result in a more diverse ranging pattern among a group of monkeys. More likely the variation in rainfall correlates with some other ecological or behavioral factor, which in turn affects ranging pattern diversity. Diet is one of the most obvious possibilities.

Several parameters for each of the last 17 months of the study were compared and Spearman Rank Correlation Coefficients (r_S) computed and tested (one-tailed significance test) (also see Struhsaker 1974). There was no significant positive or negative correlation between the following:

1. Indices of food species diversity (fig. 28) and indices of quadrat utilization diversity (table 50) ($r_S = 0.078$, $p > 0.05$; fig. 33).
2. Average indices of dispersion for the top 4 food species of each month[8] (table 32) and indices of quadrat utilization diversity ($r_S = 0.129$, $p > 0.05$).
3. The combined indices of cover for the top 4 food species[9] (table 42) of each month and indices of quadrat utilization diversity ($r_S = 0.349$, $p > 0.05$, fig. 34).
4. The percentage of young growth (flowers, buds, and young leaves) in

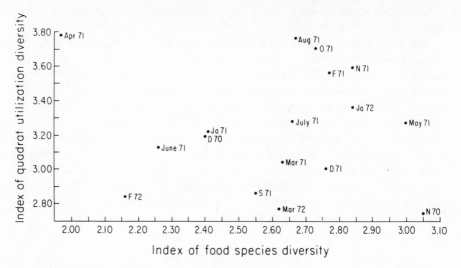

Fig. 33. Monthly plots of indices of quadrat utilization diversity versus food species diversity. Each point represents the results for a specific month, e.g., F72 is February 1972. $r_s = 0.078$.

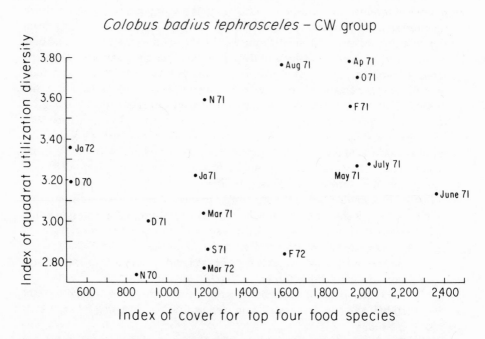

Fig. 34. Monthly plots of indices of quadrat utilization diversity versus the sum of the indices of cover for the top 4 food species in each month; abbreviations as in fig. 33. $r_s = 0.349$.

the monthly diet and indices of quadrat utilization diversity (r_s = 0.177, $p > 0.05$).

5. The mean daily distance traveled per monthly sample (table 52) and the indices of food species diversity (r_s = 0.209, $p > 0.05$).

To quote from the article of mine cited earlier, "These results are contrary to one's intuitive feeling that ranging patterns are related to and vary with food distribution. However, even if such a relationship exists, the nature of it is not readily predictable even on an intuitive basis. A simple correlation between dispersion, density or cover indices of food species and the ranging pattern of red colobus groups may not exist, because similar ranging patterns might result from very different trophic reasons. For example, restricted ranging patterns could result from either (1) feeding on a rare species with a clumped distribution, but having a high density of food per tree, or (2) feeding on a common and widely dispersed species also having a high density of food per tree."

In a previous section it has been shown that certain aspects of the badius monthly diet sometimes correspond to the relative availability of specific food items as estimated from the monthly phenological data. It seems plausible that the degree of phenological synchrony among food species may have an important bearing on the ranging patterns of red colobus. For example, if the animals are feeding on a common and widespread food species, they will not move far if only a few trees are bearing food in a restricted area, i.e., if it is an asynchronous food species. Similar complexities can also be expected when they feed heavily on a rare and widely spaced food species that lacks phenological synchrony.

In an attempt to evaluate this possible relationship, 2 separate indices of the monthly availability of young leaves for 8 different food species have been compared with the monthly indices of quadrat utilization diversity. The phytophase of young leaves was selected because it was an important item in the badius diet and because it was a phytophase whose relative abundance could be estimated with confidence for all food species represented in the monthly phenological sample. Only those 8 species of the 11 sampled each month whose young leaves made up at least one percent of the badius diet in either the 7-month or 12-month period were considered in this analysis. The same individual trees were evaluated at the end of each month and just prior to the monthly sample of the CW group. All were in the home range of the CW group. The trees considered in this analysis are: 5 *Aningeria,* 5 *Celtis africana,* 10 *Celtis durandii,* 5 *Lovoa,* 10 *Markhamia,* 10 *Parinari,* 10 *Strombosia,* and 5 *Teclea.* Each phytophase was evaluated on a 5-point scale (0–4) for each tree, 4 representing the maximum availability for a particular tree. Weighted mean scores were computed for young leaf availability for each month, combining the scores for all 60 trees evaluated. Weighted means were computed by first computing the mean score of young leaves of each species, multiplying this score by the number of trees in the sample of that species, summing this product for all 8 species, and dividing this sum by the total number of trees (60) represented by all 8 species. The monthly weighted means were rank-ordered and compared with the rank order of the corresponding monthly indices of quadrat utilization diversity for 16 months (November 1970 through March 1972, no phenological data for June). A Spearman Rank Correlation Coefficient demonstrated no significant positive or negative correlation between these two variables ($r_S = 0.015$, $p > 0.05$). In other words, there was no correlation between young leaf availability and diversity of ranging pattern.

The second index of young leaf availability is that of Koelmeyer (1959) and is referred to as a phenological index, expressed as: number of trees exhibiting a particular phytophase/total number of trees in the sample x number of species exhibiting a particular phytophase/total number of species in the sample x 100. This index was computed for the same 60 trees of the same 8 species for the 16 months and compared with the same monthly indices of quadrat utilization diversity. A Spearman Rank Correlation Coefficient again showed no significant positive or negative correlation between these two varibles ($r_S = 0.21$, $p > 0.05$).

In view of the pronounced degree of phytophasic asynchrony within the samples of several of the 8 species considered and the equally pronounced asynchrony between several of the 8 species, it is, perhaps, not surprising that there is no significant correlation between the above measures of young leaf availability and diversity of ranging pattern. In other words, the members of this sample of 60 trees are variable enough in their pattern of young leaf development to mask any relationship that may exist with badius ranging patterns. Neither of the two measures of young leaf availability even correspond, positively or negatively, with monthly rainfall ($r_S = 0.31$ and $r_S = 0.030$, $p > 0.05$). Perhaps larger samples of the various food species would yield a more realistic view of their degree of phenological synchrony and thus a better understanding of the relation between badius ranging patterns and food availability. On the other hand, there may not be a strong relation between food distribution and availability and the diversity of badius ranging patterns. The results indicate that the least diverse ranging pattern provides adequate food for the red colobus, at least on a short-term monthly basis. Any monthly ranging pattern more diverse than this is apparently in response to other variables.

The extensive and nearly complete overlap in home ranges of 2 other badius groups with that of the CW group, combined with the aggressive nature of intergroup relations associated with chasing, counterchasing, and the usual supplantation of one group by the other, has obvious effects on the movements of the groups involved in these encounters (see chapter 2).

It has been previously reported (Struhsaker 1974) that there is "a positive correlation between the number of days per total days in the monthly sample on which the study group had intergroup conflicts and the index of quadrat utilization diversity ($r_S = 0.520$, $0.05 > p > 0.01$; fig. 35). There is also a positive correlation between the number of days per monthly sample on which the CW group was proximal (within 50 m) to another red colobus group and the index of quadrat utilization diversity ($r_s = 0.564$, $p \approx 0.01$; fig. 36)." A more refined analysis further supported these results, showing a positive correlation between the number of days the CW group was proximal to another group per number of observation minutes in each monthly sample and the indices of quadrat utilization diversity for each month ($r_s = 0.451$, $0.05 > p > 0.01$). Clearly, in those months when the CW group interacted more frequently with other groups they had a more diverse ranging pattern.

Comparison with Other Studies of Red Colobus

Clutton-Brock's (1972) study of red colobus in the Gombe again provides the best comparative material. Unfortunately, differences in methodology make it difficult to compare data on home range and ranging patterns furnished by his study at Gombe and by mine at Kanyawara. Rather than plotting the daily movements of the monkeys on maps, Clutton-Brock tabulated at 15-minute intervals the specific "squares" (quadrats) occupied by one or more members of the group. Each square was delineated by footpaths and was about 100 x 100 yd. (circa 0.836 ha.) in the Gombe and Bigodi study areas, but in his Kanyawara study area each square was approximately one ha. If the groups of red colobus studied by Clutton-Brock behaved similarly to the CW group, they probably only occupied about 25–75% of any square.

In addition, because he used larger squares (0.836 ha. versus 0.25 ha.) than those on which I based my estimate of 25–75% occupancy, it is quite possible that his method tended further to overestimate the actual area occupied. Consequently, comparisons between the 2 studies of absolute values of area used per day or during an entire year are probably not very meaningful. Comparisons of orders of magnitude do seem worthwhile, however.

During the 9-month study at Gombe, Clutton-Brock found that a group of 82 badius occupied 137 squares or about 0.44 square miles (114 ha.) in 1450 observation hours. In his month-long studies in the Kibale Forest he found that a group of 64 badius in the Bigodi (Dura River Bridge) area used 111 squares (0.37 square miles or 95.8 ha.) during 210 hours of observation and a group of 58 at Kanyawara used 49 squares (0.19 square miles or 49.2 ha.) in 133.5 hours of observation. His studies at Gombe indicated that badius groups occupy about 65–74% of their annual range (number of squares) after 225 hours of systematic observation. From these figures he estimated that the annual home range of the group at Bigodi would be 0.51 square miles (132.1 ha.) and that of Kanyawara group would be 0.32 square miles (82.9 ha.). All of these estimates are greater than mine for the CW group at Kanyawara, which, based on the occupancy of 283 quadrats (70.7 ha.), with a correction reduction of 25–75%, was 35.3 ha. Part of the difference between Clutton-Brock's estimates and mine could be attributable to the fact that I was working with a smaller group (19–22 monkeys during the 17 months considered). However, my impression is that, among red colobus at Kanyawara, home range size is about the same for groups of all sizes. It is

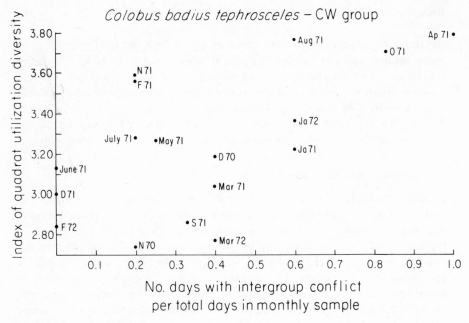

Fig. 35. Monthly plots of indices of quadrat utilization diversity versus the number of days on which the CW group had conflicts with other red colobus groups per total number of days in the monthly sample; abbreviations as in fig. 33. $r_s = 0.520$.

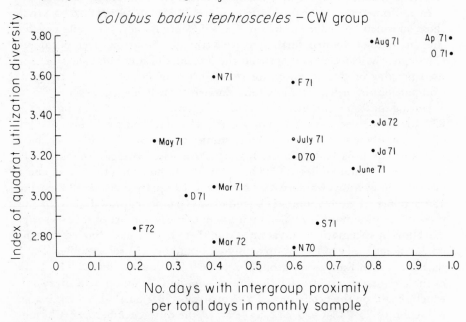

Fig. 36. Monthly plots of indices of quadrat utilization diversity versus the number of days on which the CW group was proximal to (within 50 m of another red colobus group per total number of days in the monthly sample; abbreviations as in fig. 33. $r_s = 0.564$.

remarkable how close Clutton-Brock's estimate for a group of 58 at Kanyawara comes to my estimate for a group of 20 at Kanyawara based only on the
number of 0.25 ha. quadrats occupied (82.9 ha. versus 70.7 ha.). Furthermore, if one applies a correction of minus 25-75% to his estimates and
considers the possibility of even greater overestimation because of his larger
quadrat size, then the home range of his Kanyawara group becomes very
similar to that of the CW group. In spite of correction factors, however, it
appears that the home ranges of his groups at Gombe and Bigodi are larger
than those at Kanyawara.

In his 3 study areas, Clutton-Brock found that the red colobus groups used
their range differentially. Certain areas were much preferred to others, as in
my own study at Kanyawara. He also suggests that there was "seasonal" variation in range utilization by his Gombe group, but this would be impossible to
establish in any study of less than 2 years' duration. His suggestion that the
number of squares (quadrats) used per month by the Gombe badius is directly
related to the proportion of shoots, flowers, and fruits in their monthly diet is
not substantiated by statistical testing. Using the data in his figure 53, I
computed a Spearman Rank Correlation Coefficient and found that there
was no significant correlation ($r_s = 0.40$, $p > 0.05$). This too is consistent
with my findings at Kanyawara.

Of particular interest are Clutton-Brock's findings at Gombe which show
that the use of specific quadrats is directly related to (1) the frequency with
which badius feed in them, (2) the estimated vegetative cover in them, and (3)
the density within the quadrat of the top 3 food species in the "annual" diet.

Clutton-Brock expressed daily ranging patterns as the number of quadrats
occupied per day. This is not comparable to estimating the distance traveled,
as was done in my study at Kanyawara. Data comparable to his have been
calculated from my Kanyawara study, but the difference in quadrat size
between our studies may again be an important variable, as discussed above.
In order that the data from our studies can be compared, the number of
quadrats occupied per day must be converted to surface area units (hectares).
At Gombe his group used 3-29 quadrats (2.5-24.2 ha.) per day with a median
of 12.5 (10.5 ha.). The results of his shorter studies at Kibale are similar; at
Bigodi the group used 8-15 quadrats (6.7-12.5 ha.) per day with a median of
12 (10.0 ha.) and at Kanyawara his study group used 5-11 one-hectare
quadrats per day with a median of 9 ha. My CW group at Kanyawara
apparently had smaller day ranges (1-6 ha., mean of 3 ha.). Some of these
differences could be due to differences in group size and the greater
overestimation inherent in Clutton-Brock's larger grid system (0.836 and
one-ha. quadrats versus 0.25 ha. quadrats used in my study).

There were no apparent correlations between the number of quadrats used
per month or per day and the monthly rainfall at Gombe, in contrast to my
findings at Kanyawara.

Clutton-Brock found a positive correlation at Gombe between day-range size for a particular month and the total area used by the group for that month. I reached similar results for my CW group at Kanyawara. When the mean daily distance traveled by the CW group per month was compared with the monthly indices of quadrat utilization diversity for the 9 months (November 1970–June 1971 and August 1971) in which the systematic sample consisted of 4 or 5 complete and consecutive days, a significant positive correlation (r_S = 0.77, 0.05 > p > 0.01) was arrived at.

The badius group studied at Gombe moved back and forth between 2 valleys during each month. Although the intervals between these moves varied between months, this oscillating pattern of ranging persisted throughout Clutton-Brock's study. No similar pattern of alternate use of particular parts of the home range has yet been detected for the CW group at Kanyawara.

Comparison with Studies on Other Colobinae

Data for black and white colobus (*C. guereza*) come from 4 different studies. At Kanyawara, J. F. Oates (pers. comm.) has estimated the home range for a group of 11 guereza to be 15 ha. (encircling method) during his 14-month study. This same group used 128 quadrats of a 0.25 ha. area, which, without any correction factor, yields an area of 32 ha. The same reduction factor of 25–75% as was used for the CW group of red colobus yields an estimated home range of 16 ha. These estimates for home range are approximately half that for the CW group of badius. A consideration of differences in group size and biomass between the 2 species results in equivalent areas per group biomass. The estimated biomass of Oates's main study group of 11 guereza was 60 kg (J. F. Oates, pers. comm.) and that of the CW group, using a figure of 20 monkeys and extrapolating from the data given in the section on censuses (table 56), was 118.3 kg, i.e., approximately twice that of guereza. In other words, within a group the biomass density is approximately the same for guereza and badius. One major difference between the 2 species is that the home ranges of guereza overlap much less with the neighboring conspecific social groups than do those of red colobus groups; so in effect the population biomass density is much greater for badius than guereza (see section on censuses).

Marler's (1969) estimate of home range for 5 groups of guereza in the Budongo Forest, Uganda is 13.7 ha. (mean group size is 8) and Ullrich's (1961) for one group on Mount Meru, Tanzania, is 15 ha. Both are in agreement with Oates's estimate. An exception appears to be the 5 groups of guereza studied by Schenkel and Schenkel-Hulliger (1967) near Limuru, Kenya. Marler (1969) measured the range maps presented in their paper and concluded that they had considerably smaller ranges (1.0 to 3.4 ha.).

Home range estimates for langurs of India, Ceylon, and Malaysia show considerable variation. For *Presbytis entellus*, Jay (1965) found that groups ranging in number from 10 to 54 with an average between 18 and 30 had home ranges of 129.5 to 1,295.0 ha., with averages between 259 and 777 ha. Her minimal estimate is considerably greater than the maximum estimates of 3 other studies of this species. Sugiyama et al. (1965) found groups of about 15 entellus living near Dharwar to have home ranges averaging 16.8 ha. (range: 10.3–31.5 ha.). Vogel (1971) found that entellus living in high altitude forests in India had smaller home ranges (20 ha.) than those in the thorn forest at lower elevation (60 ha.); but population densities were approximately the same in the two areas, because smaller groups were found at high altitudes. Ripley (1967) writing of the entellus at Polonnaruwa, Ceylon, states: "One square mile of dry-zone forest can support about five to seven troops with an average of twenty-five members each. Range overlap may give each troop a range of ⅛ to ½ square miles," i.e., about 32.4 to 129.5 ha. However, measurements with a planimeter of the 4 home ranges depicted in her figure 1 show that the home ranges are much less than this, averaging about 18 ha. Differences in the home range estimates for different studies of entellus may be attributable to a number of factors, including differences in sampling time, in methods of estimating home range, and in food density.

Two studies have been made of *Presbytis johnii*. For 30 days Tanaka (1965) studied groups with an average of 15 members and estimated their home ranges at about 12.8 ha. (range 6–20 ha.). Poirier's (1968a) estimate of 64.8 to 259 ha. for groups averaging about 9 in number was based on 1,250 hours of observation. Differences between these 2 studies may be due to differences in the duration of the studies and to gross ecological differences between the 2 study areas.

In Malaysia Bernstein (1968) found that 5 troops of *Presbytis cristatus* had territories of about 20 ha. These groups averaged 32 in number.

Oates (pers. comm.) found that his group of 12 guereza at Kanyawara moved, on average, 535 m per day. This is remarkably similar to the average day-range of the CW group of badius — 648.9 m.

Day ranges or daily distance traveled by groups of *P. entellus* are also extremely variable. Jay (1965) gives figures of 1.6 to 3.2 km per day. Sugiyama et al. (1965) found that their groups moved 70–870 m per day, with an average of 360 m. A group of 23 entellus studied by Yoshiba (1967), also at Dharwar, moved rather more each day; typically 300–700 m (range: 60–1,300 m). The day ranges of the entellus at Dharwar seem much like those of the CW group of badius at Kanyawara.

Tanaka (1965) reports that groups of *P. johnii* move about 500 m per day. Bernstein (1968) found that groups of *P. cristatus* travel between 200 and 500 m per day, including backtracking, "with only a few days in which greater distances were recorded." These 2 species would appear to move somewhat less each day than the CW group of red colobus.

Few data are available for African forest cercopithecines. Aldrich-Blake (1970) estimates the home range of *Cercopithecus mitis stuhlmanni* groups, numbering about 14, to be 7.8 ha. DeVos and Omar (1971) present home range data for 4 groups of *Cercopithecus mitis kolbi* in Kenya. These groups ranged in size from 15 to 18 monkeys, and their home ranges were 13.2, 13.8, 14.2, and 16 ha. Chalmers (1968) estimates the home range of 2 groups of mangabeys (*Cercocebus albigena*) to be between 13 and 26 ha. These groups numbered about 17.6 and 25 and were studied in the extensively degraded forests of Bujuko and Mabira in eastern Uganda. Chalmers' estimates are considerably lower than those of a group of mangabeys living in the same forest with the CW group of red colobus (P. Waser, pers. comm.). In 20 months of fieldwork, Gautier-Hion (1971) found that talapoin (*Miopithecus talapoin*) groups in Gabon, which average 63 in number (range 59–80), have home ranges of 100 to 140 ha. and travel, on average, 2,323 m per day (range 1,500–2,950 m). The preceding data, in conjunction with the results for badius, clearly demonstrate the extreme interspecific variability in ranging patterns among the rain forest Cercopithecidae of Africa.

TEMPORAL DISTRIBUTION OF ACTIVITIES OF *C. b. tephrosceles*

An analysis of the distribution of various red colobus activities throughout the day is essential for determining the presence or absence of activity cycles and for understanding the extent to which their behavioral fluctuations are influenced by the time of day and the corresponding changes in temperature and sunlight. Such an analysis also permits an assessment of the proportional allotment of red colobus time among various behaviors. The monkeys' time budget is of particular interest when related to their dietary habits and for interspecific comparisons. Do leaf-eating monkey species spend more or less time feeding than insectivorous or frugivorous species?

Methods and Terms

Data were collected between August 1970 and August 1971, inclusive, with the exception of July 1971. The majority of data were collected from the CW group during the systematic monthly samples of 4 or 5 continuous days of observation. However, in some months, data were also collected on other days and from other badius groups in compartment 30 of the Kibale Forest. Each sample was collected within a 10-minute period centered on every hour and half-hour between 0700 and 1900 hours, inclusive.[10] For example, the sample for 0700 hours began at 0655 and ended at 0705 hours. Each month an attempt was made to collect 20 observations for each of these 25 half-hour sample periods (0700–1900 hours). Ideally, 4 observations would have been made each day of the 5-day systematic sample of the CW group for each of the 25 time periods, giving a monthly sample of 20 observations per time period and a total monthly sample of 500. This was rarely achieved, however. Often it was not possible to see clearly what 4 badius were doing in a particular sample, which meant that supplementary data had to be collected on additional days. Furthermore, during the first few months of the study this sampling method was still being developed and deviations were not uncommon. The total sample is comprised of 6,983 observations of activity.[11]

An activity observation was operationally defined as that activity first performed by the animal under observation for a duration of at least 5 continuous seconds. On any particular date no animal was scored more than once for a given sample period. For example, the 0900 hour sample is begun at 0855 hours. The first monkey seen is an adult female who is feeding, but, as I continue to watch her, she stops feeding before 5 seconds have elapsed. After feeding she sits, but before 5 seconds have elapsed she begins cleaning herself. This she does for 7 continuous seconds and, because this exceeds the 5 seconds arbitrarily decided upon as delimiting an observation, her score for the 0900 hour sample period on that date is "self-cleaning." The next visible monkey is then observed and scored accordingly. When 4 or 5 monkeys have thus been scored, the sample for that period is terminated. However, if the arbitrary sample period of 10 minutes (0855–0905 hours in this case) passes before 4 or 5 monkeys have been scored, then the sample is terminated anyway.

This sampling method emphasizes sustained or long-duration activities, such as feeding and resting. Short-duration and relatively infrequent behaviors, such as aggressive and sexual encounters, were scored opportunistically (see chapter 3), although they too were scored whenever they occurred for at least 5 seconds within one of the half-hour sample periods.

The activity observations were categorized as feeding, resting, locomoting, self-cleaning and grooming, clinging, and play. Other behavior patterns, such as copulation and fights, were scored as such, but occurred so infrequently (0.26% of the total 12-month sample) that they have not been analyzed in detail. Feeding is defined as the actual ingestion of food, chewing, or the manipulation of food prior to ingestion, which infrequently included foraging (see section on food habits). Resting comprises inactive states, when the monkey does not move for at least 5 seconds. Locomoting refers to climbing. Self-cleaning refers to self-examination of the fur and removal of particles. Grooming includes the same motor patterns, but is a social event in which one monkey cleans another. Both groomer and groomee are scored in the grooming category. Clinging is the pattern employed by infants and young juveniles as they clutch the ventral surface of adult females. Play covers several motor patterns, including gamboling, grappling, tugging, and mouthing and is usually a social event.

Monthly scores of some of these behaviors will of course be affected by changes in the age structure of the group. In months when the group has several young infants, the score for clinging will be high, and, as the infants mature, clinging will decrease and play will increase in score. If young animals make up a large proportion of the group, this will also have an effect on the scores of feeding and resting. Similarly, fluctuations in adult membership will affect the scores of these 2 categories. However, because most of the data were collected from the CW group, whose adult membership remained constant throughout the sample period and whose infant and young juvenile constituent was small and relatively constant, the maturation of individuals probably had little or no effect on the monthly scores of this 12-month sample.

Time Budget

The highest scores in the time budget sample comprised feeding and resting. Of the total 6,983 activity observations, 44.5% were feeding, 34.8% resting, 9.2% locomoting, 5.3% self-cleaning and grooming, 3.4% clinging, 2.6% playing, and 0.3% other activities. These percentages make no allowance for possible biases introduced by unequal sample sizes for the different time periods and for different months. To correct for this, mean percentages for each activity were computed for each month. The monthly percentage of an activity is the average percentage of that activity in all 25 time-period samples for that month. In other words, for each month the proportional distribution of the scores among the various activity categories was computed for each time period, and then the average percentage of each activity for all 25 time periods was computed to give the monthly mean. Mean percentages for the entire 12-month sample were calculated as the mean of these monthly means for each activity (table 53). These corrections made little difference in the proportional distribution of scores for the 12-month sample, with feeding and resting still comprising 79.4% of the sample.

Significantly more observations were made of feeding than of resting and, consequently, of all other activities ($\chi^2 = 82.87$, $df = 1$, $p < 0.001$). In this chi-square test the 3,104 observations of feeding were compared with the 2,427 of resting, and the expected value assumed was 2,765.5, i.e., equally distributed.

The interpretation of the high scores for feeding and resting may prove to be most fruitful when compared with similar activity samples for other species, particularly other species from the same habitat. The great amount of time spent in these two activities may be related to the large proportion of leaves in the diet, which, combined with the rumenlike digestive system of the badius, may necessitate frequent and/or long periods of feeding followed by similar periods of inactivity for purposes of digestion.

The scores for locomoting are more difficult to interpret because this category appears to relate to several different motivational states. Locomoting was tallied regardless of whether the monkey was moving from one feeding site to another in the same tree or whether it was participating in a group progression covering 100 m.

Scores for grooming, clinging, and play will depend on the age structure of the group and thus are also difficult to interpret with this kind of sampling.

Intermonthly Variation

There are no significant variations between months in the monthly percentage of scores for any of the 6 categories of activity (table 53). The G-test (Sokal and Rohlf 1969, p. 560) clearly shows high probabilities for the small

differences between monthly percentages for each activity as being attributable to chance (table 53). In this test the observed monthly percentages were
compared with the expected value, which was considered to be the mean of
the monthly means. For example, the expected value for feeding was 45.26%
(table 53). These results suggest that there was no seasonal variation in the
time budget of the CW group during this sample and that the slight change in
the age structure of the group had little effect on their activity.

Diurnal Variation

Both feeding and resting activities showed considerable variation throughout
the day and were not uniformly distributed among the 25 time periods (figs.
37 and 38). The G-test (Sokal and Rohlf 1969) demonstrated that the diurnal
variation in resting differed significantly from the expected variation
($G = 52.3$, $df = 24$, $p < 0.005$), but that the diurnal variation in feeding was
not significantly different from the expected values ($G = 33.6$, $0.10 > p >
0.05$). No significant differences were found in the diurnal variation of the
other categories: locomoting ($G = 21.7$, $0.9 > p > 0.5$); self-cleaning and
grooming ($G = 27.2$, $0.5 > p > 0.1$); clinging ($G = 6.0$, $p > 0.995$); and
playing ($G = 9.9$, $p > 0.995$) (table 54). In these tests the mean percentage
for a given activity at each time period was computed by summing the
monthly percentages for this activity at each time period and dividing by 12,
the total number of months in the sample. The mean percentage for each of
the 25 time periods was then compared with the expected percentage, which
was assumed to be the overall mean percentage of the particular activity
during the entire 12-month sample. These expected values are given in table
53. For example, the expected percentage of feeding was 45.26%. This was
compared with each of the 25 mean values given in fig. 37. Each of these 25
values was simply the average percentage of feeding at a particular time
during the 12-month sample.

Although the data suggest peaks of feeding in the early morning, mid-
afternoon, and late afternoon, the large standard deviations and the range of
variation for these measures clearly indicate that activity patterns are not
closely correlated with specific times of day (fig. 37). I suspect, although I
cannot prove with these data, that on any given day the badius alternate
peaks of feeding with peaks of resting, which more or less correspond with
specific hours of the day but which show enough variation to mask cyclicity
when data from several days are pooled. Periodicity within a day might be
demonstrated by collecting more data for each sample period of each day
from a given group and analyzing the intervals between feeding bouts on each
separate day.

In spite of the extreme diurnal variation in activity patterns, there is one
time period which does not seem to be affected by the deficiencies of the
sampling method. At 1830 hours there is a peak of feeding activity in which
the range of variation and standard deviation are small. Similarly, the

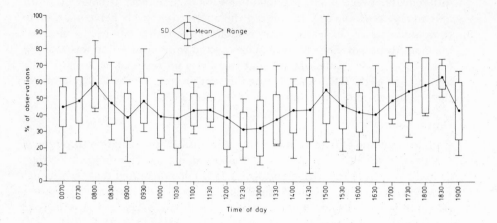

Fig. 37. Diurnal distribution of feeding observations (August 1970 through August 1971). SD is the standard deviation.

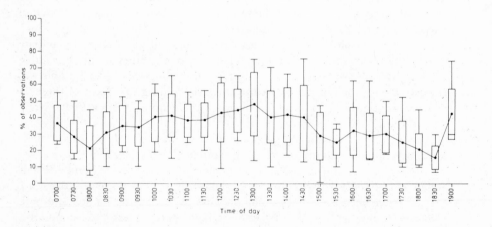

Fig. 38. Diurnal distribution of resting observations (August 1970 through August 1971). Symbols as in fig. 37.

amount of resting at this time is lowest and least variable. 1830 hours is also the only time period in which there is no overlap in the range of percentages of feeding and resting. Regardless of the many environmental parameters affecting activity cycles, there is a very high probability of badius feeding and a low probability of their resting at 1830 hours. The high predictability of this last feeding bout of the day suggests that the badius have an important physiological need to feed before nightfall.

In a further attempt to understand the presence or absence and nature of activity cycles, an analysis was performed which was designed to show the distribution of activity peaks during the daylight hours. For each month a mean percentage was calculated for each activity (table 53). A standard deviation was also calculated for each activity in each month, using the same half-hourly percentages. For example, feeding in August 1970 had a mean of 48.79% and a standard deviation of 23.16%. A peak for any particular behavior pattern was defined arbitrarily as any sample period in which the percentage of the particular activity equaled or exceeded the mean plus one standard deviation for that activity. In the example of feeding for August

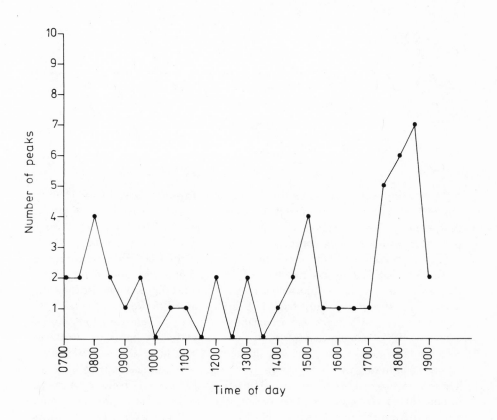

Fig. 39. Frequency distribution of peaks of feeding activity.

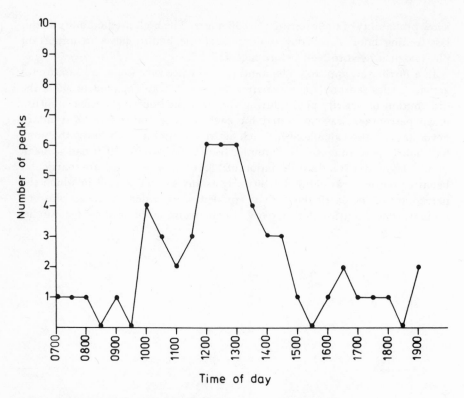

Fig. 40. Frequency distribution of peaks of resting.

1970, any time period in which the percentage of feeding was at least 71.95%
(48.79% plus 23.16%) was considered to be a time of peak feeding activity for
that month. For August 1970 there were 4 peaks of feeding, i.e., 4 time
periods in which feeding comprised at least 71.95% of the sample.

Similar calculations were made for all months for the feeding and resting
categories. All such peaks, as defined above, were then pooled for all 12
months of the sample and plotted as a frequency distribution against the time
of day (figs. 39 and 40). Most feeding peaks occurred between 1730 and 1830
hours inclusive, with peaks also occurring commonly at 0800 and 1500 hours.
These results correspond closely with those suggested by figure 37. Peaks of
resting were clearly most prevalent between 1200 and 1330 hours, inclusive,
and at 1000 hours (fig. 40), approximating an inverse relation with feeding.
These data support the hypothesis that badius do have relatively regular and
predictable activity cycles. The reason that activity cycles are not more
apparent from the raw data is probably that data from several different days
have been pooled in the monthly summaries. If activity cycles roughly
correspond to the time of day but are slightly modified by temporary local
conditions, such as rain storms, attacks by crowned-hawk eagles, and

intergroup conflicts, then the cycle of any given day might readily be shifted by 15–30 minutes, and when such data are pooled with other days the diurnal cyclicity of activity may be obscured.

Comparison with Other Studies of Colobinae

In his 8-month study of *C. b. tephrosceles* in the Gombe Park, Clutton-Brock (1972) gave particular attention to activity cycles and time budgets. At 15-minute intervals he recorded the first activity seen of all visible animals in the group. Not only is the sampling interval more frequent than in my study, but also the temporal criterion for scoring an activity is different. Clutton-Brock scored the first behavior seen, regardless of its duration. In my study, on the other hand, the criterion demanded that the behavior be sustained for at least 5 seconds before it was scored. The behavioral categories used by Clutton-Brock appear to be essentially identical to those used in my study.

The total scores for his 8-month study show that the category "inactive" (resting) was the most common, comprising 41–72% of the scores per day and a median of 54% per day. Feeding comprised 14–41% of the daily scores, with a median of 25% per day. These results are in sharp contrast to mine from Kanyawara, where feeding comprised 45.26% and resting 34.16% of the 12-month sample. That this difference is not a reflection of interpopulational differences is demonstrated by Clutton-Brock's data from his two study groups in the Kibale Forest. His Bigodi study group was scored as inactive for 50–65% of the day (median 60%) and as feeding for 17–36% of the day (median 25%) in the daily scores. Similar results were obtained in his study at Kanyawara, with inactive comprising 48–71% of the scores per day (median 60.5%) and feeding constituting 22–45% of the daily scores (median 29.5%). Clearly Clutton-Brock scored resting behavior about 20–26% more often and feeding about 16–20% less often than I did. We agree that the most likely explanation of this difference is attributable to differences in our sampling methods. By scoring the first activity seen, Clutton-Brock increased his chances of scoring "inactive" behavior. This happened because, as he has noted, when the badius are feeding they do not continuously place food in their mouth but often pause briefly (less than 5 seconds) between mouthfuls. Such pauses would be scored as inactive in Clutton-Brock's method. Stipulating that only behaviors sustained for at least 5 seconds are scored reduces the undue bias given to the category of "inactive." This is more than a semantic or an arbitrary problem. It is important to distinguish a brief pause between mouthfuls during a feeding bout from a 20-minute siesta. To classify the two events in the same category constitutes a serious misrepresentation of the badius time budget.

The proportional distribution of scores among the remaining activity categories in Clutton-Brock's study were similar to those in my study: moving 2–17% per day, median 8%; "mutual-grooming" 2–11% per day, median 5.5%; "self-grooming" 0.1% per day, median 0; and play 1–11% per day, median 3%; compare with table 53.

Clutton-Brock demonstrates that his badius at Gombe were more inactive in the wet season than in the dry and that they fed more in the dry than the wet season. He suggests that this difference, at least in part, may be attributable to seasonal differences in observation conditions. No inter-monthly differences in activity were found in my Kanyawara study.

Three distinct bouts of feeding were apparent on the majority of days in the Gombe study and also in Clutton-Brock's studies in the Kibale Forest. These peaks occurred at about 0800 hours, midday, and late in the afternoon. Our results agree regarding the early morning and late afternoon peaks of feeding, but differ in that the midday peak found by him was much more pronounced than in my study. Also the mid-afternoon (1500 hours) peak was less apparent in his study than mine.

The only other published data on colobine activity time budgets is that of Yoshiba (1967) for *Presbytis entellus* at Dharwar, India. The lack of any description of his methodology makes any comparison with or discussion of his results on diurnal variation virtually impossible. His longitudinal samples of 3 individual langurs for a total of about 44 hours indicate that they spend 44% of their time feeding, 33% resting, and 6% in locomotion. These results are remarkably similar to those of mine for the badius at Kanyawara (compare with table 53).

J. F. Oates's extensive study of *Colobus guereza* in the Kibale Forest (the results of which are not yet published) has yielded considerable information on the diurnal activity patterns and time budget of this species.

VERTICAL DISTRIBUTION OF ACTIVITIES OF *C. b. tephrosceles*

The spatial distribution of their time and activities among the forest strata may be of particular importance for understanding the ecological niche separation of badius from other primate species living in the Kibale Forest. It has, for example, been shown for several sympatric anthropoids in the rain forests of Cameroun that vertical stratification is prominent and may be an important means of effecting ecological separation among them (Gartlan and Struhsaker 1972).

Methods

The data on vertical distribution of activities were collected at the same time as those described in the preceding section. The methods and sample size and distribution are, therefore, the same. All heights were estimated by eye. My ability at estimating heights was checked against a Blume-Leiss Optical Height Finder. Simultaneous and independent estimates and measurements were made of 40 different trees along a road between compartments 30 and 14 in the Kibale Forest. Measurements with the height finder were made by Mr. Xavier Kururagire. Comparing my estimates with the measurements made with the height finder for these 40 trees revealed no significant difference (p = 0.098, T_S = 227.5, Wilcoxon Signed-ranks Test for 2 groups arranged as paired observations from Sokal and Rohlf 1969). The tallest tree measured 55 m, which seems of dubious reliability, but the next tallest tree was 44 m, which probably closely approximates the upper level of this part of the Kibale Forest.

Results

The badius at Kanyawara use a variety of heights in the forest, and very infrequently come to the ground (figs. 41 and 42). They have a definite preference for heights exceeding 9 m; 97% of all 6,983 observations occurred over this height. Combining all observations, regardless of activity, shows the most popular heights to be between 16.5 m and 27 m (fig. 41, steepest slope of cumulative percentage curve), with about 50% of the sightings being made in that layer. Only 25% of the observations occurred between the ground and 16.5 m and 25% at heights greater than 27 m. Because the predominant activity in this sample was feeding, it appears that the badius prefer the layer between 16.5 m and 25.5 m because their food is concentrated there. It is also possible that this is the only layer in which they are able efficiently or effectively to utilize food, even though the same food may occur at other heights. These height preferences will be most meaningful when compared to those of other primates living in the same section of forest.

Figures 41 and 42 indicate that self-cleaning, grooming, and resting occur at slightly greater heights than do feeding and locomoting. However, in the absence of an obvious statistical test which adequately compares these curves, the analysis has not been carried further. The chi-square or G tests are not adequate because they will only test whether there is a significant difference anywhere in either of the 2 curves compared. This makes the comparison meaningless because no single activity curve approaches a chance distribution (χ^2 distribution). A correlation coefficient test is not precise enough because it only compares ranks, and virtually all of the curves are positively correlated. The only pair of curves not positively correlated is locomoting versus self-cleaning and grooming (r_S = 0.30, $p > 0.05$).

Heights of activity were not systematically sampled for the other subspecies of badius studied. However, one outstanding point is worth mentioning. In the savanna of Senegal, *b. temminckii* often comes to the ground, in marked contrast to the other subspecies. This is particularly the case when the animals move from one grove of trees to another (pl. 31).

Comparison with the Gombe badius

The forest at Gombe was of lower stature than that at Kanyawara (Clutton-Brock 1972). Consequently, comparison of height preferences by the badius in the 2 areas is difficult. Clutton-Brock (1972) estimated the heights of 905 feeding observations and found that the majority of feeding occurred between

26 and 75 feet (7.8 and 22.5 m) above the ground, which is lower than the feeding heights preferred by the Kanyawara badius. He also found that the Gombe badius rarely fed below 7.5 m and rarely came to the ground. This is consistent with my findings at Kanyawara.

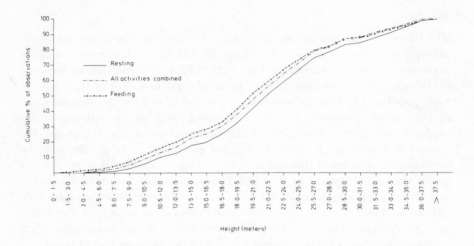

Fig. 41. Cumulative percentage distribution of heights of activity; feeding, resting, and all activities combined (August 1970 through August 1971).

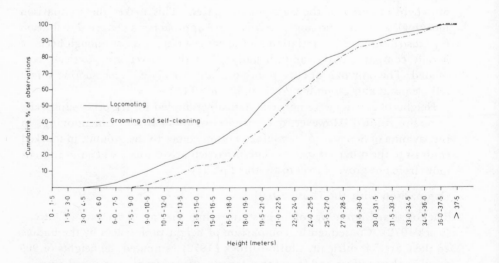

Fig. 42. Cumulative percentage distribution of heights of activity: locomoting, grooming and self-cleaning (August 1970 through August 1971).

CENSUSES OF ANTHROPOID PRIMATES IN THE KIBALE FOREST

Systematic censuses were made in the major study area at Kanyawara to permit estimates of the species diversity, relative abundance, and density of the various anthropoid species. Fewer censuses were made in other parts of the Kibale Forest having different floristic compositions and management histories. The results of these censuses allow a better understanding of the secondary productivity of the forest, the habitat preferences of the different primate species, the effects of forest management practices on primate populations, and, when compared with future censuses several years hence, an understanding of population dynamics.

Methods

All censuses were made at approximately the same time of the day and lasted about 5 hours (0730–1230 hours). Each time a group or an association of monkeys or chimpanzees was encountered,[12] I attempted to determine the number of species and individuals present. About 10 minutes was spent with each association before the search was continued. All observations were made within 10 m of the predetermined census route. About half of the censuses were conducted by myself alone, and in the other half I was accompanied by one field assistant. When the field assistant was present, no consultation was made until after we had left each association, in order to eliminate inter-observer influence and thus allow a check of observer reliability.

It usually requires about 60 hours of contact to obtain a complete count of a small group of red colobus and considerably longer for larger groups. Consequently, emphasis was placed on counting the number of social groups of each species rather than on counting the number of individuals in each association, because of the impracticality of the latter. Average group size for each species determined from intensive study of particular social groups permits extrapolation from the number of social groups counted in the censuses to the approximate number of individuals present. During each census, care was taken to avoid recounting any particular group. This source of error was only probable when the census route began and ended at the same place. In fact, the problem rarely arose.

Censuses of a particular area of the forest were always along the same route. The area most intensively censused was my major study area of red colobus monkeys, located in compartment 30 near the Kanyawara Forestry Station. The census route in this area was 4,020 m long, excluding 525 m of backtracking along one particular trail. The census route was roughly rectangular, covering both sides of the Nyakagera stream and encompassed by the hills of Nyakatojo, Nyamasika, and Butanzi (fig. 43). The vegetation along this route was predominately undisturbed high forest (*Parinari* forest, see description earlier in this chapter), but did include a small amount of swamp forest where the route crossed the Nyakagera twice and forest that was felled in 1969 (the felled area ran along one side of the census route for about

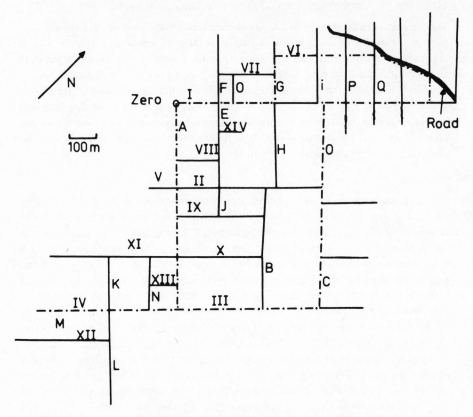

Fig. 43. Census route in compartment 30 of the Kibale Forest (see fig. 12), symbolized by dash-dot lines. Solid lines represent other trails in the study area.

115 m). Forty-four censuses of comparable nature were made along this route from 18 September 1970 through 16 March 1972. No censuses were made in June, July, or August 1971. Censuses were made in all the other 19 months of this period. The number of censuses per month was usually 2 or 4. One of the 2 censuses in January 1972 was excluded because it was accompanied by a heavy rain storm, which made it incomparable with the other censuses. In all other months, censuses were made on two consecutive days, starting in opposite directions of the circuit on alternate days. In those months with 4 censuses the pairs of census days were separated by one week.

The area of next most intensive censusing comprised compartments 13, 12, and 17, which are located north of Kanyawara and south of Sebitoli (fig. 44). Compartment 13 was felled in 1968 and subsequently selectively treated with arboricide to remove undesirable trees in 1968–69 (pl. 33). A small section of 13 remained untreated with arboricide in 1969, but treatment with arboricide was resumed in December 1970 and completed in April 1971. This latter area was about 1.0 to 1.5 km from the census route, and the actual arboricide operation had no immediate supplanting effect on the monkeys which would bias the censuses made in 1970–71. Compartment 12 was felled in 1967–68, and compartment 17 in 1967 (information from Fort Portal Forestry Department). The census route through these three compartments followed an old timber extraction road and was 6,100 m long. Twelve censuses were made along this route between 21 November 1970 and 12 November 1972, but one has been excluded from this analysis because it was interrupted by a heavy rain storm. The vegetation of this area is characterized by an extremely dense understory about 1–3 m high; scattered saplings and taller trees form a very open "forest" with no continuous canopy above the understory (pl. 33).

The third area is in the vicinity of the Dura River bridge on the Kibale Road, which runs between Fort Portal and Kamwenge (about 0° 28' N, 30° 23' E). This census route covered a circuit of trails 4,500 m long, which ran northeast of the road and then southeast across the Dura River, southwest back to the road and then northwest along the road and over the bridge back to the starting point. This section of the Kibale Forest is relatively undisturbed lowland rain forest and was also the site of Clutton-Brock's (1972) "Bigodi" study area. Only five censuses were made along this route between 28 October 1970 and 11 March 1971.

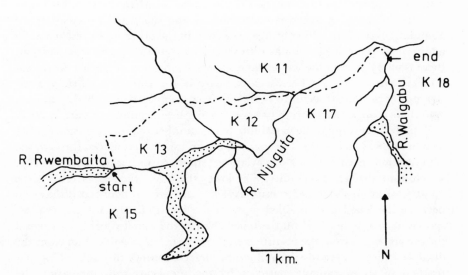

Fig. 44. Census route in compartments 13, 12, and 17 of the Kibale Forest, indicated by a dash-dot line; solid lines indicate streams and stippling indicates swamps. Numbers preceded by K refer to specific compartments, e.g. K 13 is Kibale compartment 13 (taken from Uganda Forest Department Gazetting Map).

Results

As mentioned earlier, seven species of monkeys and chimpanzees occur in the
Kibale Forest. The monkey species are: red colobus (*Colobus badius*); black
and white colobus (*Colobus guereza;* pl. 34), red tail monkey (*Cercopithecus
ascanius;* pl. 35), blue monkey (*Cercopithecus mitis;* pl. 36); lhoesti monkey (*Cercopithecus lhoesti*), sooty mangabey (*Cercocebus albigena;* pl. 37),
and baboon (*Papio anubis*).

In addition to the results summarized in table 55, monkeys that were
apparently solitary were seen in all three areas. In compartment 30 they
occurred in 15 (34%) of the 44 censuses, with a mean of 1.27 and a range of
1-3 per census (based on 15 censuses). Solitaries for the various species
occurred with the following frequencies: badius 3, guereza 3, ascanius 8,
mitis 3, and albigena 2. Solitaries occurred in 6 (54.5%) of the 11 censuses in
compartments 13, 12, and 17, with a mean of 1.67 and range of 1-3 per
census (based on 6 censuses). There were 3 solitary badius, 3 guereza, 3
ascanius, and one mitis. A solitary chimpanzee and a solitary albigena were
seen during the 5 censuses in the Dura River bridge area.

Although most of the species occur in all three areas that were systematically sampled, there are some pronounced differences in the abundance of
various species in the different areas (table 55). For example, the red colobus
and blue monkey are clearly more abundant in compartment 30 than in the
other two areas. The black and white colobus is more abundant in the
disturbed forest (compartments 13, 12, and 17) than in the other two
undisturbed areas. And the red tail monkey and sooty mangabey are more
abundant in the Dura River bridge area than in the other two forest types. In
addition to these proportional differences in primate species composition,
there is an apparent absence of some species from some of the areas. *Cercopithecus mitis* has never been seen or heard in the Dura River bridge area.
Cercopithecus lhoesti seems to be absent from compartments 13, 12, and 17.
And there is only one report of baboons having been seen in compartment 30.

Equally interesting is the variability in the number of groups of each species
seen per census (table 55). Comparison of variability is enhanced through the
computation of coefficients of variation (standard deviation/mean x 100),
which normalize the variability. The lower the C.V., the less variable is the
number of groups seen per census. Again there are pronounced differences
between the felled and unfelled parts of the forest. Except for guereza, all
species have a higher C.V. in the felled than in the unfelled area. In general
the census results from the undisturbed forest were less variable than from the
disturbed forest. Perhaps the reduction of food sources through felling has
resulted in wider ranging patterns by the monkeys, thus increasing the
variability with which they are seen in a given area, i.e., on the census route.

Relative abundance of the various species in the different areas is expressed
as the mean number of groups seen per census per km of transect (census

route). Thus, although nearly the same number of red colobus groups were seen on average in compartment 30 versus compartments 13, 12, and 17, when expressed as number per km of transect there were more in 30 (table 55). It is clear that with the exception of the black and white colobus and baboon, all species are less abundant in the disturbed forest (compartments 13, 12, and 17) than in the undisturbed forest (compartment 30).

If obtainable, measures of primate density would be preferable to measures of relative abundance. If one knew precisely the area censused, and the number of censuses were adequate, then the mean number of groups computed from all censuses would yield a good indication of the number of groups living in the area censused, i.e., the density. The difficulty lies in determining the size of the area censused. One method is to determine the average distance at which the groups of primates were first seen by the observer, i.e., the mean distance of detectability. In general, this tends to underestimate the area censused, because most groups are not first seen at the maximum distance which the observer is capable of seeing from the census route, but are usually seen at less than this maximum distance, because resting monkeys are often not detected until the observer is much closer than the maximum possible distance of detectability. Unfortunately, the importance of the distance of detectability was not appreciated until near the end of this segment of my study. An indication of detectability distance, however, can be gained from an analysis of the distances from the census route at which the monkey groups occurred. The location of all groups seen during censuses was either plotted directly on maps or described in the notes relative to the census route. This distance was often the same as the distance of detectability, but sometimes less. The distance from the census route was never greater than the detectability distance. Consequently, the distance from the route tends to underestimate the area censused, but this bias is probably constant for all habitats, thus permitting interhabitat comparisons. The mean distance from the route based on 281 sightings in compartment 30 was 22 m, which, if considered to be half the width of the census route, estimates the area sampled to be 17.6 ha. (44 m x 4,020 m). From these figures one computes that on average 3.8 groups of red colobus occupy 17.6 ha., or that there are about 22 groups per 100 ha. (1 km^2). This is a gross overestimate of the density, for it is known from my detailed studies of red colobus that three groups together occupy about 50 ha., as is discussed in detail later. It is of interest, however, to compare the mean distance of sighting from the census route with that computed for the monkeys in compartments 13, 12, and 17. Based on 58 sightings the mean distance is 53 m, or about 2.5 times greater than that for compartment 30. This supports the impression that a wider route was being censused through compartments 13, 12, and 17 than through 30. Consequently, the greater relative abundance of primates in the undisturbed area is not an artifact of better visibility but reflects a real difference from disturbed forest. I estimated that I could usually see about 50 m on

either side of the trail through compartment 30 and about 100 m on either
side in compartments 13, 12, and 17. This estimate gives a census area of
40.2 ha. for compartment 30 and 122 ha. for 13, 12, and 17. With these
estimates, densities were computed for the primates in the two areas, which
clearly reveal a much greater density in compartment 30 for all species except
the black and white colobus and baboons (table 55). This is not surprising in
view of the gross reduction of food resources for the primates resulting from
the felling and arboricide treatment in compartments 13, 12, and 17.

From the maximum number of red colobus groups seen during any one
census, and from individually recognizable groups, it was known that at least
nine groups of red colobus used the area censused in compartment 30.
However, this is not necessarily equivalent to density because some of the
groups surely used areas outside of the census area. Furthermore, it is not
consistent with the much more precise estimate of red colobus density, which
is based on detailed study of the movements of 3 specific groups of red colobus
during nearly 2 years of study. The area used by the CW, ST, and BN groups to
the virtual exclusion of other groups, where they spend the vast majority of
their time, was delineated on a map and then encircled with a planimeter.
The area was estimated to be 50.7 ha. Although these groups may not use
every bit of the area, any lacunae occurring therein are not used by any other
red colobus groups either. Group density based on these measures is 5.92
groups per 100 ha., or about 63% of the estimate in table 55.

If one assumes that an overestimate of similar magnitude is made for all
other species in compartment 30, as listed in table 55, a more realistic
estimate of density for them is 63% of the values listed there. Because of gross
differences in ranging patterns and social organization between the different
species this extrapolation may not be justified, but as a working hypothesis let
us assume it is. Estimates for all species in compartment 30 are taken to be
63% of those listed in table 55 (see table 56).

With these estimates on group densities it is possible to make some first
approximations of anthropoid primate biomass in compartment 30. The
major difficulty in estimating biomass is the lack of sufficient data on group
size, group composition, and weights for the different age-sex classes of each
species. I have attempted to calculate the average group biomass for typical
groups of each species. Estimates of mean group size are based on personal
observations in the Kibale Forest and information from J. F. Oates for
guereza and P. Waser for albigena. Weight estimations are taken from the
following sources for the species indicated: badius, Kingdon 1971 and
personal measurements of 4 specimens; ascanius, Haddow 1952 and Hill
1966; mitis, Hill 1966; lhoesti, estimated as approximately equivalent to
mitis; albigena, Malbrant and Maclatchy 1949 and Dorst and Dandelot
1970; guereza, J. F. Oates, pers. comm.; and chimps, Dorst and Dandelot
1970. Table 56 summarizes these approximations and indicates a biomass of
2,217.4 kg per 100 ha. Solitary monkeys of all species combined might add

another 100 kg per 100 ha. This will probably prove to be an underestimate, but an upper estimate using the group densities in table 55 would be about 3,578.5 kg per 100 ha. With more data on group size and composition and on body weights, particularly for the *Cercopithecus* species, a more realistic estimate of biomass could be computed, which would probably lie between these two figures. These estimates indicate a higher biomass density than previously expected for rain-forest primates. It is equivalent, for example, to about 55–89% of the ungulate biomass density in the Serengeti ecosystem (4,027 kg/km², Watson et al. 1969).

Species diversity for the three areas is expressed by the Shannon-Wiener Information measure (Wilson and Bossert 1971) and was computed using the mean number of groups per census (col. 2, table 55) for each species in each area. This measure is

$$H = -\sum_{i}^{N} p_i \log_e p_i$$

where H equals the amount of diversity in the species (N) occurring in each area and where p_i equals the proportion of the ith species. For example, in table 55 the mean number of red colobus groups seen per census is 3.8 in compartment 30. This represents 37.8% of the total mean number of groups per census for all species in this area. In this case $p_i = 0.378$. The indices of anthropoid primate species diversity for each of the three areas are as follows: compartment 30, 1.534; compartments 13, 12, and 17, 1.382; and the Dura River bridge area, 1.501. In terms of anthropoid primates, compartment 30 is the most diverse and compartments 13, 12, and 17 the least diverse. This diversity appears to correspond directly with tree species diversity in the two areas, with greater diversity occurring in the undisturbed areas.

Clearly the red colobus have the highest densities in both numbers and biomass among the primates of those parts of the Kibale Forest censused. Although their herbivorous diet and specialized digestive system may be related to their densities, they do not entirely explain them, for if they did, one would expect high densities to be achieved by guereza, which are also herbivorous and have similar digestive specializations. It is also evident that the red colobus are dependent on relatively mature rain forest, and that with extensive degradation of the forest their numbers decline.

Comparison with Other Studies

During his studies at Kanyawara, J. F. Oates conducted primate censuses similar to mine. Two of his censuses and part of a third were within the area of compartment 30 censused by myself. Oates's censuses did not follow exactly the same route as mine, but they do indicate similar results. Oates found in these 3 censuses an average of 2.67 groups of badius, 1.33 of guereza, 2.33 of ascanius, 1.00 of mitis, no lhoesti, 0.67 of albigena, and 0.33 of chimpanzees (compare with table 55).

Censuses comparable to those made in this have not been conducted elsewhere in Africa. The census results reported by Southwick and Cadigan (1972) from West Malaysia (formerly Malaya) are not directly comparable because each census was in a different area. However, of particular interest in their findings is that a colobine (*Presbytis melalophus*) was the most common monkey in all the rain forests which they censused. Similarly in Ceylon, it is the colobinae which have the highest densities. Eisenberg, Muckenhirn, and Rudran (1972) report the following biomass densities for Polonnaruwa: *Presbytis senex* 1,450 kg/km², *Presbytis entellus* 730 kg/km², and *Macaca sinica* 190 kg/km², giving a total anthropoid biomass density of 2,370 kg/km², which is slightly higher than that estimated for compartment 30 of the Kibale Forest. Their results for a higher altitude study area, Horton Plains, are lower: *Presbytis senex* 630 kg/km² and *Macaca sinica* 40 kg/km². Vogel's (1971) estimates of population density for *Presbytis entellus* in two study areas in India are lower than the red colobus densities at Kanyawara. In the upper monsoon forest of the Kumaon Hills he estimates that there are 97 langurs per km², and in the thorn forest of Sariska 104 per km². Yoshiba (1968) shows the considerable variation in population densities of *Presbytis entellus* throughout India, ranging from 2.7 to 134.7/km². While much of this variation is presumably a function of habitat differences, it is clear that even the highest densities are less than half those achieved by the Kanyawara red colobus.

The consistently higher densities achieved by the colobinae over the cerco-pithecinae in these studies are probably related to their digestive special-izations, which allow them more efficiently to utilize foliage, the most abun-dant food in a rain forest. Within the African colobinae, however, only the red colobus seem to achieve high densities in undisturbed and mature rain forest. It is unclear why, in the absence of red colobus, as in many rain forests of Cameroun, the other colobinae do not reach higher densities. In these areas the other colobinae (*C. satanas* and *C. guereza*) seem much less abun-dant than the cercopithecinae (Gartlan and Struhsaker 1972).

The correlation between primate species diversity and tree species diversity found in compartments 30, 13, 12, and 17 of the Kibale Forest is like that found in rain-forest habitats of different age and floristic composition near Idenau, Cameroun (Gartlan and Struhsaker 1972). In both areas the more recently disturbed and least diverse forest had the least diverse primate fauna.

INTERSPECIFIC RELATIONS

Anthropoid Polyspecific Associations in the Kibale Forest

Polyspecific associations among the various primate species were common-place in the Kibale Forest. In extreme cases I have observed as many as 4 species together simultaneously, including: (1) badius, ascanius, mitis, and lhoesti; (2) badius, ascanius, mitis, and albigena; (3) badius, guereza, mitis, and ascanius.

The first approach to understanding these interspecific relations was to determine which species tended to form associations with other species (polyspecific) and which tended to remain alone (monospecific). Using the census data described in the preceding section, the tendency for each species to form associations with other species as opposed to remaining alone was tested with the χ^2 one-sample test (Siegel 1956), which compares a particular species' frequency of occurrence in polyspecific groups with that in monospecific groups.[13] In cases where the χ^2 test gave marginal results or expected values were less than 5, the G-test with Yates' correction was applied (Sokal and Rohlf 1969). In this analysis consideration was given only to those associations in which it was confirmed either that the species concerned was alone or that at least one other species was associated with it. Species were considered to be in association when members of the different species groups were spatially intermingled.

In compartment 30, only 3 species tended to occur significantly more often in polyspecific associations than alone (table 57). For the remaining 3 species (guereza, lhoesti, and albigena) their occurrence in the 2 types of associations was as expected by chance. The number of chimpanzee groups seen was too small to warrant analysis. Three species in compartments 13, 12, and 17 occurred in both types of associations with equal probability (table 57). In contrast, guereza showed a significant tendency to be in monospecific as opposed to polyspecific associations. The samples for albigena and anubis were too small for analysis.

Clearly the tendency of primates to associate with other species is greater in compartment 30 (undisturbed, mature forest) than in compartments 13, 12, and 17 (disturbed forest). It is not apparent what ecological factors could result in these differences, although the gross differences in vegetation between the 2 areas are probably involved. Furthermore, as pointed out to me by J. F. Oates, the lower population density of most primates in the felled area (13, 12, and 17) could result in fewer polyspecific associations if these associations are typically chance phenomena and not "actively" formed by the species involved.

Affinities between Specific Pairs of Species

Fager's index of affinity (1957) allows us to consider what specific pairs of species associate with one another. Again, requirements of sample size (at least 5 groups of each of the 2 species compared) and statistical independence limit the use of this analysis for these data. Pooling data from all censuses for a given area violates the demand for statistical independence, but it does give samples large enough for analysis and at least permits a relative comparison between species and areas. Using the same census data we find that in compartment 30 the following pairs of species occurred together with significance ($p < 0.05$): badius-ascanius, badius-mitis, and ascanius-mitis. The only other pairs whose sample sizes were adequate for this analysis were badius-guereza and ascanius-guereza. Neither pair was significantly associated ($p > 0.05$). In compartments 13, 12, and 17 none of the pairs of species having adequate samples were significantly associated ($p > 0.05$). These included badius-ascanius, badius-mitis, badius-guereza, ascanius-mitis, ascanius-guereza, and mitis-guereza. Significant affinities between species occur only in the undisturbed forest. This is consistent with the results on the tendency of species to form polyspecific associations.

Having established that, within compartment 30, certain species tend to associate with one particular species rather than with others, we can now try to understand the nature of these interspecific relations through an analysis of the interspecific encounters of the CW group of red colobus.

Primate Associations and Interactions of the CW Group (*Colobus badius tephrosceles*) during Systematic Monthly Samples

The following account is based on the monthly systematic samples of the CW group taken between November 1970 and March 1972, inclusive. This sample consists of 53 complete days ($\geq 11\frac{1}{2}$ hours), totaling 645 hours of observation, and 30 incomplete days ($< 11\frac{1}{2}$ hours), totaling 229 hours. During the 53 complete days the CW group was associated (intermingled) with ascanius on 46 days (social groups on 44 and solitary adult and subadult males on 6), mitis on 28 days (no solitaries), lhoesti once (solitary adult male), albigena on 5 days (no solitaries), and guereza on 5 days (social groups on 2 and solitary adult males on 3). In the 30 incomplete days they were associated with ascanius on 24 days (no solitaries), mitis on 13 days (no solitaries), lhoesti once (a large solitary adult), and albigena on 3 days (no solitaries); they associated with guereza on no days.

The total number of interspecific associations of the CW group during all 83 systematic sample days was: ascanius 102 (6 of these with solitary males), mitis 57 (no solitaries), lhoesti 2 (both solitaries), albigena 8, and guereza 5 (3 of these with solitary adult males). The number and range of associations per complete sample day on which the CW group was associated with a particular species is as follows: ascanius 1.4 (1-3), mitis 1.3 (1-2), lhoesti 1, albigena 1, and guereza 1. That is, on 46 complete days in which ascanius associated with the CW group, they formed 65 associations with them (1.4 per day). The formation of more than one association in a given day was effected in several ways, including the formation of associations both with social groups and solitaries, the formation of associations with more than one social group of a given species, and the re-formation of an association with the same social group.

The duration of these associations was extremely variable. Considering only those associations in which the time of formation and disbandment were known with relative certainty, it is seen that the average duration of a given association between the CW group and ascanius was 130.4 minutes (range 1-666 min.; N = 51 associations) and with mitis 121.8 minutes (18-689 min.; N = 30). The 2 associations with lhoesti each lasted about one minute. In 5 associations with albigena the duration ranged from one to 91 minutes and in the 3 with guereza the duration was from one to 17 minutes.

The actual formation of associations between the CW group and other species involved no special behavior patterns, but rather they seemed to be formed as one species approached the other who was relatively stationary or as the 2 species converged in their simultaneous movements. Often, however, it was unclear who joined whom or whether the 2 species converged. In such cases, I would quite suddenly become aware that the 2 species were intermingled. In spite of these difficulties, there were a sufficient number of cases in which the active formation of the association was clearly seen. In 42 cases ascanius joined the CW group, and only twice was the association formed by the 2 species' converging. The CW group never joined stationary ascanius groups. In 26 associations with mitis, the CW group joined them twice, converged with them once, and were joined by them 23 times. At least once a solitary lhoesti joined them. On 6 occasions the albigena joined the CW group. The CW group joined a solitary guereza once. Clearly, most of the CW group's interspecific associations were formed by the active movement of the other species.

Similarly, the disbandment of associations involved no special behavior patterns. One species usually left the other. In all cases which were clearly observed, the other species left the CW group: ascanius, 40; mitis, 27; lhoesti, 1; albigena, 5; and guereza, 1. The fact that the majority of associations were both formed and disbanded by the other species rather than by the CW group of badius is probably a reflection of the more sedentary nature of the badius.

The majority of the CW group's interspecific associations were formed during the first 2 hours of the day. Considering only the data for ascanius and mitis and only for complete sample days, it can be seen that 50.8% of the CW group's associations with ascanius and 44.0% of those with mitis were formed between 0700 and 0900 hours (table 58). Most of the associations between the CW group and ascanius and mitis were disbanded between 0700 and 1130 hours; 60.5% for ascanius and 75.1% for mitis (table 58). The diurnal patterns are probably dependent on the diurnal movement patterns of ascanius and mitis, both of whom seem to move about more in the morning than at midday or in the early afternoon.

The specific social groups of species which associated with the CW group also formed associations with other badius groups. During the systematic samples of the CW group, social groups of ascanius were twice seen to leave the CW group and join another group of badius. A group of mitis once left the CW group and joined the BN group of badius and on another day left the CW group and joined the ST group of badius. On a third day a mitis group was simultaneously associated with both the CW and BN groups of badius. The group of albigena which associated with the CW group also associated with many other badius groups. Although specific examples are not available for lhoesti and guereza, it is most likely that they too associate with more than one specific group of badius. Furthermore, the CW group definitely associated with more than one social group of ascanius; probably as many as 4. On one particular day they associated with 2 ascanius groups in succession. On 4 of the 6 days on which they were associated with solitary ascanius, they were also associated with ascanius groups, suggesting that the solitary ascanius may have been peripheral to social groups of ascanius. The data for other species is less certain, but it seems likely that the CW group of badius associated with 2 groups of mitis and 2 groups of albigena. It is not certain how many groups of lhoesti and guereza they associated with.

Grooming was the most common type of social interaction between the badius of the CW group and the other species. Grooming occurred in 15 of the 102 associations between the CW group and ascanius. A total of 23 grooming bouts occurred between these 2 species, with badius being the groomer in 21 and ascanius the groomer in only 2 bouts. Ten (43.5%) of these grooming bouts were between badius and solitary adult male ascanius, even though the associations with solitary ascanius represented only 5.9% of all the associations the CW group had with this species. The solitary male ascanius were groomed by badius in 8 of the bouts, and they groomed badius in 2 bouts. Adult female badius groomed these solitary ascanius in 4 bouts. Juvenile and young juvenile badius each groomed these solitary ascanius twice. One particular male ascanius (TN) formed associations with the CW group on at least 3 occasions; once while he was with an ascanius group (14 September 1971) and twice while he was solitary (4 October 1971 and 1 November 1971). On all 3 occasions he was groomed by badius. His association with them on 4 October 1971 was particularly interesting because in this association he was groomed by an adult female and a juvenile, both of whom he groomed when they stopped grooming him, and then by a young juvenile. When the young juvenile badius ceased grooming TN, it briefly handled TN's penis before leaving him. On this date he was associated with the CW group for at least 74 minutes. In the remaining 13 cases the badius groomed members of ascanius social groups. The groomers were adult females 4 times, a subadult female once, an approximate adult once, juveniles 3 times, and young juveniles 4 times. The ascanius who received this grooming were adult males in 3 bouts, adult females in 3 bouts, an approximate adult once, juveniles in 5 bouts, and an ascanius of undetermined age once. Clearly, it is the solitary adult male ascanius who are most frequently groomed and the adult female and immature badius who do most of the grooming.

Grooming between the badius and other species was much less frequent. Only 2 grooming bouts were observed between badius and mitis during the systematic samples. A juvenile badius groomed an old infant mitis once and a juvenile badius female groomed an approximate adult mitis once. Similarly, only 2 bouts of grooming were seen between badius and albigena during this sample. A young juvenile badius groomed a juvenile albigena that was approximately twice its size, and a juvenile badius groomed around the perineal swelling of an adult female albigena after handling the swelling. Grooming was not observed between badius and lhoesti or guereza.

Grooming bouts between badius and ascanius and between badius and mitis were initiated by either of the 2 species concerned. However, in general, it appeared that the ascanius usually approached and presented for grooming to the badius. Some of the presentations were unsuccessful in evoking grooming from the badius. In such cases the ascanius usually left the proximity of the badius to whom it was presenting within one minute of the initial presentation. Infrequently, a badius was unsuccessful in its attempt to groom a member of another species. For example, a young juvenile badius approached an adult mitis and reached out with one hand as if to groom the mitis. The mitis did not look toward the badius, but climbed away, apparently not noticing this attempt at grooming.

Play was the next most common social interaction between badius and other species. During the systematic samples of the CW group, interspecific play occurred in 4 of the 102 associations with ascanius and in 3 of the 57 associations with mitis. The total number of play bouts with these 2 species was 4 and 3 respectively. The 4 play encounters between badius and ascanius involved the following partners: 2 young juvenile badius with a juvenile and a young juvenile ascanius; a young juvenile badius with a juvenile ascanius of approximately equal size; a young juvenile badius with 2 juvenile ascanius; and a young juvenile badius with an old infant and an adult ascanius. Physical contact between the species occurred in 3 of these 4 bouts. In one of these bouts the 2 species mouthed one another about the neck and shoulders and grappled with one another's arms. In another bout a young juvenile badius approached and then pulled on the tail of an old infant ascanius. This old infant moved away and clung to the ventral surface of an adult ascanius. The young juvenile badius then approached this pair and pulled on the tail of the adult with the clinging infant. The adult ascanius responded by staring with extended forequarters (a threat gesture) toward the young badius, which in turn gamboled around them. In still another encounter it was again a young juvenile badius which made the invitation to play. It did so by hopping up and down in front of 2 juvenile ascanius, which were about the same size as the badius, and then grabbed toward them with one hand. The ascanius jerked back from this grabbing motion and failed to engage the badius in physical play.

The 3 play bouts between badius and mitis observed during the systematic sample were also initiated by the badius. Once an old infant badius pulled and tugged on the head of a juvenile mitis that was about twice the size of the badius. The mitis remained seated and showed no response. In another bout a juvenile badius approached and reached out with one hand toward 2 adult mitis, one of which had a clinging infant. One of the adults stared and bobbed its head toward the badius, which was one meter away. The badius then bounced up and down (gamboled) in front of the mitis, which then moved another meter away, whereupon the badius fed. In the third case an old infant badius and an infant mitis of approximately equal size mouthed and grappled with one another and displayed the open-mouth play gesture to one another (the mouth is held approximately two-thirds open and the unclenched teeth are partially exposed).

One play bout seen between badius and albigena involved a juvenile badius of the CW group and a juvenile albigena of similar size. They both gave the open-mouth play gesture, grappled with one another's arms, mouthed one another on the nape and sides of the neck, and chased and counterchased one another. This bout lasted more than one minute.

Aggressive encounters between badius and other species were relatively infrequent during the systematic samples. There was only one aggression between badius and ascanius, aside from the ambiguity of some of the play encounters. In this case, adult male CW supplanted an adult ascanius that was seated near an adult female badius. Only once was aggression observed between badius and mitis, when an adult female badius spatially supplanted an adult female mitis with a clinging infant. Considering the relative infrequency with which the CW group was associated with albigena, aggression between them was relatively frequent. Once a juvenile albigena slapped toward a young juvenile badius that had been grooming it. This was apparently instigated when the albigena saw me, and, as a consequence, the grooming bout was disrupted. In the same association and only 9 minutes later an albigena slapped several times toward a juvenile badius. Each time the badius moved its head back and avoided the slap. The badius remained seated, looking at the albigena from about 0.5 m. The only aggression between badius and guereza occurred when a young juvenile badius spatially supplanted a solitary adult male guereza, which was considerably larger than the badius. During the systematic samples the badius were joined by chimpanzees only once. When the chimpanzees saw me they fled, evoking shrieks, screams, and chists from the badius, which were obviously directing these vocalizations toward the chimpanzees.

Four social interactions whose significance was not apparent occurred during the systematic samples. Once an old infant badius approached and made physical contact with an adult ascanius, without further developement. In another case a young juvenile badius groomed an ascanius of similar size and then appeared to mount the ascanius as if to copulate. A young juvenile badius once approached and slowly reached out with one hand and touched a juvenile mitis of similar size. The mitis showed no response. On one occasion a juvenile badius handled the large and crimson perineal swelling of an adult female albigena. It then groomed around the swelling until the female albigena climbed away. As she climbed away, the adult male badius CW followed her, chisting continuously (see chapter on vocalizations). An adult male albigena then followed male CW.

Additional Observations of Interspecific Relations of *C. b. tephrosceles*

Outside of the monthly systematic samples of the CW group, additional observations were made of interactions between badius and other primate species. All of these observations were made between 6 May 1970 and 17 March 1972, and all but one were in the Kibale Forest.

An additional 16 grooming bouts were observed between badius and ascanius. These occurred in at least 4 social groups of badius other than the CW group; 3 at Kanyawara and one in the Dura River bridge area. In addition to the grooming pairs observed during systematic samples, the following bouts were seen: an adult female badius groomed a juvenile ascanius; an adult female badius groomed an adult female ascanius; an adult male badius (POP of the ST group) groomed an adult female ascanius; a juvenile ascanius groomed an old juvenile badius that was ⅓ larger; an adult ascanius groomed a juvenile badius; and an adult ascanius and adult female badius groomed one another in succession. A large ascanius once presented for grooming to adult male badius HIF (ST group), but he failed to respond.

Four additional bouts of play were observed between badius and ascanius. In one of these bouts both the ascanius and the badius displayed the open-mouth play gesture. Gamboling, chasing, counterchasing, mouthing, and grappling with hands and feet were components of these play bouts. A subadult female badius once supplanted 2 juvenile ascanius as they approached her. She did so by merely looking rapidly toward them, and they jumped aside.

In addition to the grooming partners observed in the systematic samples between badius and mitis, the following bouts were seen outside these samples: an adult female badius groomed an adult mitis; once 2 juvenile badius simultaneously groomed an adult male mitis, and one of them groomed the mitis for a total of 9.5 minutes; in another case a young juvenile and a juvenile badius simultaneously groomed a solitary adult male mitis; and once a juvenile badius and a juvenile mitis of approximately equal size groomed one another in succession. Only one additional play bout was observed between these species, involving an old infant badius and a young juvenile mitis that was slightly larger than the badius. They gamboled together and grappled with one another's arms. Another case of aggression involved a juvenile badius and an old juvenile mitis of approximately equal size. They slapped toward one another and screeches were heard. This encounter terminated when an adult badius rushed toward the pair and the mitis fled. A juvenile badius once gave the species-specific present type I gesture twice in succession to an adult mitis. The mitis remained seated, showing no response. The final social interaction between these 2 species involved an infant badius which approached an old juvenile or subadult mitis. The mitis took the infant badius, which clung to the ventral surface of the mitis, and carried it 3 m into dense foliage and out of view.

In addition to the 2 grooming bouts between badius and albigena during the systematic samples, 3 other grooming encounters were seen. These involved 2 groups of badius other than the CW group; one at Kanyawara and the other about 22 km south of Kanyawara but also in the Kibale Forest. An adult badius once groomed an albigena of similar size; an adult female badius groomed a juvenile male albigena of similar size; and a juvenile badius groomed an old juvenile albigena about twice its size. Finally, in the Dura River bridge area a subadult male and an adult male badius were seen moving and intermingled with a group of at least 23 albigena. No other badius were heard or seen in the vicinity, and I conclude that these 2 "solitary" males were associated with the albigena group.

The only association between badius and baboons (*Papio anubis*) was seen outside the Kibale Forest along the riverine forest of the Mpanga River Gorge, just upstream of the Mpanga Falls and about 53 km south of Kanyawara. Aside from being intermingled, there was no social interaction between the 2 species. Baboons are relatively common in the Dura River bridge area of the Kibale Forest. I expect that with more observations there the 2 species will be found to associate more frequently than is indicated by the present study.

Outside the systematic samples, badius and guereza were seen in association only 5 additional times. Four of these involved solitaries, and none involved overt social interactions. A solitary adult male guereza was associated with the CW group of badius once. A solitary old juvenile badius remained with a guereza group for more than an hour. In another case an old juvenile or subadult male badius moved with J. F. Oates's main study group of guereza for at least 2 consecutive days. The final case concerned a solitary subadult badius that was with 2 approximate adult guereza. Ten days earlier a badius that greatly resembled this animal and was probably the same individual was seen moving alone.

Only one additional encounter was observed between chimpanzees and badius. In this case at least 2 chimpanzees made an apparent attack on the CW group of badius. There was much leaping about in the trees by both species, evoking many shrieks, wahls, and screams from the badius. This interaction was not clearly observed, but it seems likely that the chimpanzees were attempting to catch, kill, and eat a badius, as reported from the Gombe Stream of Tanzania (Goodall 1965).

In terms of social interactions, the relations of *b. tephrosceles* with ascanius and mitis are predominately "friendly." Those with albigena are also predominately friendly, but with a higher relative frequency of aggression than with the former 2 species. The relations with guereza and lhoesti are passive and infrequent. Those with chimpanzees are basically "unfriendly."

Why do these different primate species form associations with *b. tephrosceles?* The topic of ecological niche separation, with particular reference to food resources, has not been dealt with here and is the topic of current research and publications in preparation. Still, it can be said that the food habits of *b. tephrosceles* are sufficiently different from those of the other species to indicate that these interspecific associations are not formed because of common food habits (also see chapter 6). The grooming data strongly suggest that solitary ascanius form associations with the badius for the purpose of getting cleaned. However, this explains only a small portion of the interspecific associations. It is likely that one major advantage of such associations is to increase the efficiency of predator detection without increasing intraspecific competition for food. I suspect, however, that two major factors determining the formation of associations between badius and other primate species are the group densities of the other species and the overlap in home range of these other species with the badius. In other words, the element of chance may be extremely important in the formation of the badius interspecific associations. Although the factor of *chance* has been considered in the application of Fager's index of affinity to the census data, and it has been shown from these data and analysis that some species associate together more than can be expected by chance, when one considers the frequency with which the various species formed associations with the CW group of badius during the systematic monthly samples, one finds a positive correlation between the estimated group densities of a particular species (table 56) and the frequency with which it associated with the CW group (Spearman Rank Correlation Coefficient, $r_s = 0.991$, $N = 6$, $p = 0.01$). That is, the greater the group density of a particular species, the more likely it will associate with the CW group of badius. These data and analysis suggest, in contrast to the census data, that the factors of chance and interspecific overlap in home ranges play important roles in determining the frequency of interspecific associations.

Nonprimate Interspecific Relations of *C. b. tephrosceles*
The majority of nonprimate interspecific relations have already been mentioned in the chapter dealing with vocalizations. In general, they are alarm responses to the flight or alarm calls of other species such as squirrels, duikers, and pigs. Many of the other interactions might be termed "false" alarms, as when the badius responded vocally to nonpredatory birds that flew near them. Aside from these types of interactions, the badius had relatively few encounters with other nonprimates.

The most important of these interactions was with the crowned hawk-eagle (*Stephanoaëtus coronatus* L.). This large eagle is probably the only avian predator of red colobus in the Kibale Forest and, aside from infrequent predation by chimpanzees and man, is probably their only serious predator there. Only 2 encounters between this eagle and the CW group were observed during the entire 83 days of systematic sampling from November 1970 through March 1972. Once the eagle soared about 100 m overhead but was apparently unseen by the badius, for they did not respond until it had soared some distance away and then began giving its characteristic call (a 2-phase whistle). The badius responded to the whistling with uh! calls. In the other encounter a crowned hawk-eagle flew in silently among the CW group, at least 15 m below the upper tree canopy. This evoked chists, barks, uh!s, and chist-sqwacks from the badius. It was apparently an unsuccessful attack by the eagle. In a third case during the systematic samples, the CW group responded similarly to a large raptor that was probably a crowned hawk-eagle but which was not seen clearly enough for positive identification. It perched near the CW group but flew off as an adult male badius began moving toward it. As it flew away, more alarm calls were given by the monkeys, and adult male CW gave a leaping-about display, which ended with a branch-shake display. Outside the systematic samples and during the period of 6 May 1970 through 17 March 1972, only one other encounter was observed between a crowned hawk-eagle and badius. In this case the eagle flew in among an association of mitis, ascanius, and the ST group of badius. All 3 monkey species gave alarm calls. The eagle perched nearby, whereupon an adult male badius (male HIF) gave prolonged squeals (like those of old infant and young juvenile badius) and began climbing toward the eagle. When he was within 15 m of the eagle, the eagle departed, evoking more alarm calls from the monkeys. Apparently, male HIF was effective in supplanting the eagle.

Red colobus were seen in the proximity of elephants only once, in the Mpanga River gorge just upstream from the Mpanga Falls. A herd of 11 elephants foraged beneath a group of red colobus that were perched about 20 m above them. The badius looked toward the elephants but showed no other response. The rarity of associations between badius and elephants is due primarily to the fact that elephants rarely occurred in compartment 30 during this study.

One final interspecific relationship is worthy of note—the very common association between *b. tephrosceles* and the dung beetle *Gymnopleurus crenulatus* (Scarabaeidae). Wherever one finds red colobus in compartment 30, one is also certain to find these dung beetles. They forage beneath the red colobus, flying about 1-2 m above the ground, and then alight upon and roll into balls any available red colobus feces. It is possible that these beetles forage beneath all diurnal primates, but my impression is that they tend to be associated with red colobus more than with the other monkey species.

Interspecific Relations of Other Subspecies of Red Colobus

Although sympatric with many other primates, *b. temminckii* of Senegal and Gambia rarely formed associations with other monkey species. They were twice seen in association with *Cercopithecus c. campbelli* Waterhouse, 1838, and once with *Cercopithecus aethiops sabaeus* (L. 1766). Other monkey species found in the same forests with these red colobus, but with whom they were not observed in association, include: *Erythrocebus p. patas* (Schreber 1774), *Cercocebus a. atys* (Audebert 1797), and *Papio papio* (Desmarest 1820). The only other interaction observed between the red colobus of Senegal and another vertebrate species was near the village of Kambila. The red colobus were feeding in a small grove of trees proximal to the village, and domestic goats were feeding beneath them on the fruits knocked off the tree by the monkeys.

The *b. badius* in the Tai Reserve of Ivory Coast were frequently seen in association with *Cercopithecus d. diana* (L. 1758), *Cercopithecus petaurista büttikoferi* Jentink, 1886, and *Cercopithecus c. campbelli*. Less frequently they were associated with *Cercocebus a. atys, Colobus p. polykomos* (Zimmerman 1780), and *Procolobus verus* Van Beneden 1838. In a brief visit to another forest along the road between Béréby and Tabou, I saw *b. badius* in association with all of these species except petaurista and atys. Only once did a red colobus make physical contact with another species. This occurred when an old juvenile petaurista sat next to and in contact with an old juvenile badius. The white-crested hornbill, *Tropicranus albocristatus,* also associated with the red colobus and other monkey species of the Tai Reserve. Two crowned hawk-eagles once flew over a group of red colobus, with one of them giving 2-phase whistle. The badius responded with alarm calls (see chapter on vocalizations).

In the Korup Reserve of Cameroun, the *badius preussi* formed polyspecific associations with *Cercopithecus p. pogonias* Bennett 1883, *Cercopithecus mona* (Schreber 1775) *Cercopithecus nictitans martini* Waterhouse 1838, *Cercopithecus erythrotis camerunensis* Hayman 1940, *Cercocebus torquatus* (Kerr 1792) and *Mandrillus leucophaeus* (F. Cuvier 1807). Only one social interaction occurred between badius and another species. An adult female badius with a black infant clinging to her ventrum approached and spatially supplanted a juvenile nictitans. She then sat where the nictitans had been sitting.

In the very brief survey of *badius rufomitratus* on the lower Tana River in Kenya, animals of this subspecies were seen in association with *Cercopithecus mitis albotorquatus* Pousargues 1896. More extensive study will undoubtedly reveal that they also form associations with the other monkey species in these forest patches, including: *Cercocebus g. galeritus* Peters, 1879, *Papio cynocephalus,* and *Cercopithecus aethiops.*

Comparison with Other Studies

Polyspecific associations among African rain-forest monkeys have been observed in several other forests and countries by many other workers (Haddow 1952, Booth 1956, Jones and Sabater Pi 1968, Gautier and Gautier-Hion 1969, Aldrich-Blake 1970, and Gartlan and Struhsaker 1972). Similar observations have been made in Malaysia (Bernstein 1967). It is now well established that such polyspecific associations are common. Furthermore, that some monkey species tend to associate more with other species than to remain alone has also been found in Gabon (Gautier and Gautier-Hion 1969) and Cameroun (Gartlan and Struhsaker 1972). These studies have, like the census data in the present one, demonstrated that certain specific pairs of species tend to associate together much more than other pairs and more than could be expected by chance.

In Gabon most polyspecific associations were seen in the early morning and late afternoon (Gautier and Gautier-Hion 1969), as in my study. However, the Gautiers' finding that polyspecific associations among 3 *Cercopithecus* species at Belinga, Gabon, were stable throughout years has not been corroborated either by the studies in Cameroun (Gartlan and Struhsaker 1972) or by the present study.

These other studies have all been concerned primarily with cercopithecines, usually *Cercopithecus* species, and, in general, they all agree that "friendly," nonaggressive interactions between *Cercopithecus* species are rare and that most interspecific social interactions are of an aggressive nature (Gautier and Gautier-Hion 1969, Aldrich-Blake 1968, and Gartlan and Struhsaker 1972). This is in marked contrast to the interspecific social interactions of red colobus in this study, which were predominately "friendly."

None of these studies present convincing arguments regarding the adaptive significance of these polyspecific associations.

The crowned hawk-eagle has been implicated as a predator on African monkeys in several other studies, including Haddow 1952, Struhsaker 1967*b*, Brown 1970, and Gartlan and Struhsaker 1972. Most studies of African rain-forest monkeys conclude, however, that man is the major predator of monkeys (Haddow 1952, Booth 1956, and Gartlan and Struhsaker 1972).

MORTALITY

Little information is available on the causes of death among red colobus. Direct and indirect evidence collected during this study indicate 3 factors: disease, falls, and predators.

Disease

The indirect evidence implicating disease as a cause of death among red colobus is based on 4 cases. The most convincing case was that of an adult female *b. tephrosceles* that died in the Kibale Forest, compartment 30, on 26 July 1970. At 0812 hours she was first observed dangling by her hands from a tree branch about 28 m above the ground. At 0819 hours she fell 6 m into a lower tree. Here she lay on a branch and could barely lift her head. A young juvenile accompanying her squealed continuously, and when it touched her shoulder she lifted her head and forequarters, whereupon the juvenile briefly suckled. At 0904 hours the female climbed hand over hand along a branch in a feeble manner and several times tried to swing her hindquarters up onto the branch she was hanging from. She was unsuccessful at this and fell another 6 m at 0907 hours. This fall was stopped when she grabbed a branch with her feet. She dangled by her feet at this position for about 3 minutes before she was able to pull herself upright and then hung by her hands from the same branch. During this last fall, 2 adult male badius chisted and shrieked and approached to within 8 m of her. Several other badius also chisted and shrieked, one mitis in the association gave pyow calls, and one ascanius gave hacks. At 0911 hours she fell the remaining 16 m to the ground. At 0913 hours, 2 adult male badius climbed to a point 20 m above the female. Both of them shrieked, and then one of them descended to a point 9 m above her and looked down at her as she lay on the ground. At 0920 hours the group was quiet. At 0953 hours the female was still alive on the ground but was unable to lift herself up. Her tail swung into a vertical position a few times, and she rolled over onto her side once. Some of the badius chisted toward me as I approached the female to examine her. At 1038 hours an adult male badius climbed to a point 20 m above the female and chisted and branch-shook toward me. I was 15 m from the female, who was still alive. By 1318 hours she had shifted her position by 15 cm. At 1400 hours she was still alive, but large blue flies were settling on her. By 1507 hours she had moved 30 cm from her previous position and was still alive. At 1508 hours I left the female and her group. I returned and found her dead at 1645 hours. No other badius were in the vicinity. A cursory postmortem revealed no obvious internal injuries, parasitic infestation, disease, or broken bones. She weighed 5.75 kg and measured 530 mm from the tip of her nose to the base of her tail. Her tail length was 605 mm. The fact that this female was nursing a young juvenile indicates that she was not senile, and I suggest that she died from a virus infection.

Information on the other 3 cases is even more sparse; so I am even less certain that they died from disease. On 19 July 1970 at 0920 hours a young black infant male *b. tephrosceles,* estimated as less than 4 months old, was found lying on the ground in compartment 30 of the Kibale Forest (pl. 2.2). A group of badius was nearby but moving away from the infant. The infant gave low amplitude squeals (see chapter on vocalizations), but was obviously very weak. He wagged his tail slowly, but was unable to walk or crawl and could not stand on all fours. He could still clutch with his hands and occasionally lifted an arm, but he did not cling to a sapling or to my leg when placed there. At 0945 hours there was no indication of any badius retrieving the infant, which I then took back to my house. He died at 1300 hours. A postmortem revealed no broken bones, internal injuries, parasites, or obvious disease. His stomach contained only green vegetable matter. He weighed approximately 0.5 kg, had a body length of 220 mm, and a tail 278 mm long.

During the 15-day survey of the Forêt Classée des Narangs in Senegal I found 2 dead adult male *b. temminckii.* The first was found on 3 January 1970. His right side was extensively decomposed, but there were no external wounds on his left side, which, along with his incisors and canines, looked normal and healthy. Five days later he appeared much the same, with no sign of vertebrate scavengers having fed upon him. On the same date another adult male was also found lying dead on the ground, but he was in an advanced state of decay and it was not possible to conclude anything about possible causes of death.

Falls

Only 12 falls occurred during the entire study. All involved *b. tephrosceles,* and only one was fatal. Unfortunately, I did not observe the one fatal fall. On 30 November 1971 at 1700 hours I joined an assistant, Pius Kayenga, in the forest. He had been observing the CW group since early morning. He showed me a dead juvenile female hanging by one foot and lodged between 2 tree branches about 9 m above the ground. Kayenga said the female had fallen and hurt herself in the morning. She had not moved from this place since the fall. On the following day at 0653 hours, a large raptor-like bird, indistinctly seen, flew away from the vicinity of the carcass. At 0844 hours Kayenga retrieved the dead female. An autopsy revealed that her right leg and right shoulder were broken, apparently during the fall. She was about ⅘ the size of an adult female, and her nipples were not protruding beyond the body hair. She measured 420 mm from the tip of the nose to the base of the tail, and her tail length was 460 mm. The epiphyseal plates of her humerus, ulna, femur, tibia, and fibula were not yet closed, and molars number 2 and number 3, both uppers and lowers, were not yet erupted. I conclude that she died from injuries incurred during the fall.

The eleven nonfatal falls involved adult males twice, a juvenile male once, an old juvenile once, a juvenile once, young juveniles twice, and old infants 4 times. All but one occurred in the CW group. In only 5 of these falls did the monkey fall to the ground. In the other 6 the animal caught itself before reaching the ground. Those who fell to the ground were an adult male once, young juveniles twice, and old infants twice. All the old infants were retrieved by adult females. Young juveniles were not retrieved. The most severe case involved a young juvenile. The fall occurred during a group progression across a wide gap in the tree canopy along a stream course. Several other badius, including young juveniles, had already leapt across this gap, but the particular young juvenile concerned hesitated at the crossing point and, along with another young juvenile, began giving prolonged squeals. After several minutes hesitation the young juvenile leaped, but was far short of reaching the other trees and fell an estimated 26 m to the ground without interruption by any foliage. Contact with the ground was indicated by a loud thud. However, the young juvenile soon climbed into a tree 6 m above ground, where it sat for several minutes in an apparent dazed state.

Although relatively few falls were seen, I think that falling as a cause of mortality may be of particular importance in the population dynamics of red colobus living in high forest. The few data indicate that falling is especially important during the old infant and young juvenile ages, when the young monkeys are becoming independent of their mothers for carriage. I suggest that the highest mortality of red colobus occurs during the first year or two after the infant becomes independent in its locomotion, and that most of this mortality is caused by falls.

Predators

The role of crowned hawk-eagles as predators on red colobus has already been considered. Although they probably take an occasional red colobus, I doubt that they are very important in the broader aspects of population dynamics. Leopard are probably even less important as predators on red colobus, primarily because I doubt that a leopard would be a very effective hunter at heights greater than 10 m in the rain forest. Savanna populations of red colobus in Senegal may, on the other hand, be more severely affected by leopard, because in such habitats the monkeys often come to the ground.

In most of Central and West Africa, man is the major predator on red colobus, with Senegal an apparent exception. Some of the people I interviewed in Casamance Province of Senegal readily admitted to hunting and eating vervet monkeys (*Cercopithecus aethiops*), but denied hunting or eating red colobus. Outside of this province, most of the Senegalese are Moslems and, supposedly, do not eat any primate species.

In the Tai Reserve of the Ivory Coast people regularly hunt and eat all anthropoid primates. During my 25-day survey there, at least one red colobus was shot near the village of Troya, which was my base. On my last evening at Tai I dined with the Sous-Préfet, who served stewed red colobus. I found the meat to be tougher and less tasty than that of an adult male *Cercopithecus nictitans martini,* which was roasted for me in 1966 in the Mamfe Overside of Cameroun.

In the Korup Reserve, as for most of Cameroun, the people also hunt and eat all anthropoid primates. During my 17 days in the Korup, however, gunshots were heard on only 2 days and 3 nights. My impression was that hunting in the Korup was much less than in most of Cameroun's rain forests and that the people were concentrating on nocturnal duikers rather than diurnal primates. This may in part account for the very restricted distribution of *C. b. preussi,* which is found only in the Korup Reserve. The type specimen of this subspecies supposedly came from Kumba, where red colobus have not been seen for many years. Quite likely red colobus survive only in the Korup Reserve because of the relatively low human population density there and, consequently, the relatively low hunting pressure.

At present there is virtually no hunting of primates in the Kibale Forest of Uganda. The Batoro, who are the predominant people living around the Kibale Forest, do not, in general, hunt or eat primates. I encountered one exceptional Mutoro on 12 May 1970. He was carrying a dead red colobus, which he claimed he had speared in the Kibale Forest about 4 km from the Kanyawara Station. He planned on eating it, admitting that this was exceptional to his tribal ways. Until 1964 there were apparently a number of Bakonjo living near the Kibale Forest. Supposedly, they hunted diurnal primates quite extensively in the Kibale Forest, but precise details on this are lacking.

HABITAT DESTRUCTION AND CONSERVATION

In the section dealing with species' censuses in the Kibale Forest it was shown that extensive timber exploitation has an adverse effect on the populations of most anthropoid primates, but particularly so on the red colobus. Although quantitative data are not available for other subspecies, my impression in both the Ivory Coast and Cameroun was that red colobus there are also dependent on relatively undisturbed and mature rain forest for success. The situation in these 2 countries is, however, compounded by the fact that once an area of forest is felled, it becomes much more accessible to hunters who use the trails and roads put in for timber extraction. Thus, not only is the prime habitat of the red colobus degraded, but the monkeys become exposed to greater hunting pressure. Because red colobus are undoubtedly the easiest monkeys in Africa to hunt, they suffer from this effect of habitat destruction more than other species. As is suggested above, this may be precisely why the red colobus of Cameroun is so restricted in its present distribution.

The conservation of the majority of red colobus subspecies depends on the conservation of large areas of mature rain forest that are protected against timber exploitation and hunting. It is suggested that such protected areas could be turned to the economic advantage of the African countries concerned by promoting them as tourist attractions. The added advantage of these rain forest "parks" or nature reserves to applied and theoretical biological research is also obvious (Struhsaker 1972).

6 Concluding Remarks and Speculation

In this final chapter the life style of red colobus monkeys in the Kibale Forest is related to that of the other monkey species in the same forest and, where possible, to that of other cercopithecids elsewhere. The interspecific comparisons are stimulated by questions about ecological niche separation among sympatric species and about the interrelation of ecology, social organization, and phylogeny. How is competitive exclusion effected among rain-forest monkeys? What ecological parameters affect the grouping tendencies of the different species? In what way, if any, does the dispersion pattern of a monkey species' food affect its social organization? How do the similarities and differences in social organization between the various species of cercopithecids relate to their phylogenetic affinities, i.e., to their presumed evolutionary histories? The answers to these questions may be applicable not only to cercopithecids but to other animal groups as well. Several hypotheses have been offered in the past, but most such hypotheses have required considerable modification as more field studies were completed (Crook and Gartlan 1966, Struhsaker 1969, Crook 1970, Eisenberg et al. 1972). My own hypotheses will undoubtedly require modification as more data become available. Much of what I suggest here is based on preliminary observations and impressions — both my own and those of colleagues — and is highly speculative. The main objective is to stimulate further thought and research into these questions and to make such efforts more efficient by presenting what I consider to be the most reasonable hypotheses to be tested at the moment.

A comparison of the behavior and ecology of the two colobine monkeys in the Kibale Forest has recently been summarized by John Oates and myself (Struhsaker and Oates, in press). It is pointed out that these two folivorous monkeys are strikingly different in their behavior and ecology. The red colobus, with their large, multimale social groups, have extensive overlap in their home ranges with other groups of red colobus; their relations with these other groups is based on a dominance hierarchy, independent of spatial parameters. This kind of social structure, combined with their highly diversified diet of vegetable matter (especially young growth) results in a population and biomass density greater than that of any of the other anthropoid primates in the Kibale Forest. In contrast, the black and white colobus live in much smaller groups, usually having only one fully adult male per group. Although there is overlap in the home ranges of neighboring groups, they do defend territories against one another; that is, the intergroup dominance hierarchy is based on spatial parameters. Their diet is much more monotonous than that of the red colobus. At certain times of the year they rely heavily on the blades of mature leaves, something not yet observed in the red colobus. These differences in food habits and social organization may account for the lower population and biomass density of the black and white colobus as compared to the red colobus. Another pronounced difference between the 2 species lies in their intragroup spacing. The guereza (black and whites) are much more closely spaced than the badius and give the distinct impression of a clumped and spatially cohesive unit. The social life of the neonate is also radically different, with the guereza infant receiving much attention and handling by other females and immatures of its group, whereas the badius infant has no contact with members of its group, aside from its mother, until 1–3.5 months after birth. This, we believe, is related to the group territoriality of the guereza, with the neonate being completely integrated into the group at the earliest possible age. The lack of territoriality and the apparent mobility of juvenile badius between social groups may be related to the lack of attention of other group members to the newborn. Assortative mating occurs in both species but in different ways. In guereza, other males are excluded from the group; whereas, among the adult males of a badius group, differential reproduction results from a dominance hierarchy.

Although the biomass densities of the 2 main study groups of guereza and badius were similar, the total biomass densities of the 2 species were very different on account of the larger average group size of badius and the extensive overlap in home ranges of these groups. Differences in daily and monthly ranging patterns of the 2 groups of guereza and badius appeared to be related to differences in group biomass. Thus, although the badius group tended to travel farther each day and to distribute its time in space in a more diversified way than did the guereza group, these differences became relatively insignificant when the differences in group biomass were taken into account. Oates and I also found that guereza tend to feed at slightly lower heights in the forest than do the badius, but the major difference in their ecologies and the one which we believe effects niche separation between them lies in their food habits. Comparison of the monthly diets for the 2 species revealed that, on average, they overlapped in food habits by only 7%.

How can we account for these interspecific differences from an evolutionary standpoint? The following, extracted from our earlier report (Struhsaker and Oates 1975), summarizes our views.

"There are at least 3 obvious possibilities. The social structures may be specialized adaptations to the Kibale habitat or they may be adaptations which, although originally to a habitat different from the one in which they were studied, permitted a certain amount of success in Kibale. The third possibility concerns the phylogenetic proximity of the 2 species. Just how closely are they related? Obviously, these 3 possibilities are not mutually exclusive.

"*Adaptation to Kibale-like habitat.* In the first hypothesis we suggest that the small and cohesive social groups of guereza which defend fixed areas ("territories") against neighboring groups are adapted to the efficient exploitation of a monotonous diet, the chief component of which, *Celtis durandii,* has a high density, a high cover index (a function of crown size and tree density), and a uniform dispersion (Struhsaker, 1974). In addition, the phenology of this species is such that it provides an abundance of preferred guereza food at most times of the year. The concentration of guereza on a food species of this nature permits it to partition the habitat into territories and, in combination with their small group size, also permits relatively restricted ranging patterns on a daily and monthly basis. An important feature of guereza social behavior, which probably contributes to stability in group membership and to group defense of territories, is the attention given to newborn infants by other members of the group. Such attention, we suggest, is important to the rapid and complete integration of neonates into the social group. The close spatial cohesion betwen members of guereza groups is probably related to their emphasis on a monotonous diet and a high-density food source (both in terms of tree density and item density on a given tree) and not to their territoriality, because territoriality in other primate species is not necessarily accompanied by close interindividual spacing within the group, e.g. vervets (*Cercopithecus aethiops,* see Struhsaker, 1967). Differential mating in guereza is a result of the one-male social groups, from which other adult males are excluded. The ecological significance of this exclusion, if any, is not apparent, particularly because the extragroup males utilize the same food resources and in the same areas as do the social groups. The significance of this exclusion of all but one adult male from social groups is more probably involved with outbreeding, gene flow, and assortative mating, than with feeding ecology per se.

"The ecological adaptation of the badius social structure is less apparent. There is no obvious reason for having large, multimale groups in order to exploit a large home range and a great diversity of foods. Although it is tempting to suggest that a greater diet diversity dictates a greater diversity in ranging pattern the data for the main badius study group do not support this suggestion (Struhsaker, 1974). Neither was any correlation found between the pattern of food dispersion and ranging diversity. In fact, the badius ranging-pattern diversity was more closely related to the frequency of intergroup conflicts and intergroup proximity (Struhsaker, 1974). Greater interindividual distance between members of a badius group might, however, be related to their more diversified diet. By feeding on a greater variety of foods, the badius encounter a greater variety of food dispersion patterns, many of which are not densely spaced. The absence of group territoriality in badius may be related to their great dietary diversity, which in turn may dictate a home range larger than can be efficiently defended in a rain-forest habitat. The extensive overlap in the home ranges of 3 groups may permit each group to use a larger area than would be possible if, given the same group density, they defended territories. Their intergroup dominance relations effectively determine which group has priority of access to any food source, but the distribution of food sources are such that the supplanted group never appears to have difficulty in immediately locating another ample food source. The fact that these 3 groups appeared to defend and exclude other groups from an area of about 50 ha. may be viewed as a kind of territoriality and was probably related to food allocation.

"Several of our results and conclusions have been substantiated by Clutton-Brock (1972) during his 2-month study of badius and guereza in the Kibale Forest and his 9-month study of badius in the Gombe Park of Tanzania. He found, for example, that the guereza had a less diverse diet, moved less each day, and had a smaller monthly range than did the badius. He also speculated that guereza might be able to subsist primarily on mature leaves at certain times of the year — a speculation substantiated by the studies of Oates (in prep.). In an attempt to explain the adaptive significance of the badius social structure, Clutton-Brock (1972) suggested that a single large group having relatively little overlap in home range with other groups was advantageous '(1) because food is heavily clumped, so that each food source (usually a group of trees in shoot or flower) can usually support the greater part of the troop at one time, (2) because it permits an integrated use of the animals' food supply. If the whole area was used by several troops, troops would be likely to visit areas where the supply of shoots had been cropped immediately previously by another troop.' Although this hypothesis may be applicable to Gombe, which seems to have a much lower density of badius groups than Kibale, it clearly does not apply to the Kibale Forest, where there is extensive and nearly complete overlap in home ranges of badius groups. In our study it was found that groups supplanted one another, and on some occasions these supplantations were clearly over clumped food sources. In such cases it seemed that because of their dietary diversity and abundance of food, the supplanted group had no difficulty in readily locating another suitable food source. Furthermore, badius groups usually left a clumped source of food before they had depleted it, even without being supplanted by another group. Clutton-Brock (1972) suggests that the higher proportion of young growth in the diet of badius, combined with their larger group size, results in their having a larger day range than the guereza. However, it is clear from our studies not only that young growth usually constitutes a high proportion of the guereza diet, but that, when one considers the differences in biomass between groups of the 2 species, the differences in day ranges are virtually eliminated (see above).

"The lack of attention to newborn infants, or the lack of opportunity for other group members to express such attention, may result in a looser attachment of juvenile badius to their group. Although the data are sparse, there is some suggestion that the juvenile class is the most mobile element in the badius society and that they frequently shift between groups. This could be the way in which most outbreeding is achieved by badius.

"*Adaptation to habitat different from that of Kibale.* Surveys in many other parts of East Africa indicate that guereza are most successful in relatively young secondary forest and that they often succeed in narrow strips of riparian forest and exceedingly small patches of relict forest. They can successfully occupy old and mature rain forests but are much less numerous there than in younger forests. When viewed as a species adapted to secondary and riparian forests, the guereza is more clearly adaptive in an ecological sense, than if it is considered to be a species characteristic of mature rain forest. Many of the habitats in which guereza seem most successful are characterized by having few tree species and pronounced dry seasons when young succulent plant growth is usually absent. Such an environment would clearly select for a herbivore that could survive on a monotonous diet and, at certain times of the year, on mature leaves alone. The small area of some of these habitats would also select for small group size and close interindividual spacing because of the small and clumped nature of the food resources. Territoriality would be one means of allocating food resources among groups in such a situation. The suggestion that small group size is related to small or low-density food resources is indirectly supported by evidence from studies of other species. For example, 2 subspecies of red colobus, *b. temminckii* and *b. rufomitratus,* live in relatively dry woodlands where food is considered to be more seasonal and generally less abundant than in the rain forest. Both of these subspecies live in considerably smaller social groups than do the subspecies living in rain forests. Another example concerns the vervet monkeys (*Cercopithecus aethiops*) of Amboseli, Kenya, whose average group size declined in apparent response to a reduction in food supplies (Struhsaker, 1973). Clutton-Brock (1972) has also suggested that guereza may, in fact, be a species adapted more to a drier habitat than to the rain forest, again placing emphasis on their more monotonous diet and ability to subsist exclusively on mature leaves at certain times of the year.

"Considering the overall distribution of badius, we conclude that it is a species adapted primarily to the rain forest. Some populations and subspecies, however, are able to succeed in relatively dry and seasonal habitats, including savanna woodland, riparian and relict patches of forest. As noted above, these populations generally live in smaller social groups than do those living in rain forest. Presumably, this is a function of food availability. Clutton-Brock's (1972) suggestion that the dietary dependency of badius on young growth limits them to habitats which provide young growth throughout the year and, thus, restricts them to rain-forest habitats seems tenable, with the possibile exceptions provided by *b. temminckii* and *b. rufomitratus.*

"*Phylogeny.* As mentioned in the introduction [to Struhsaker and Oates], the taxonomy of the African Colobinae has not been worked out to the satisfaction of all. As far as this paper is concerned, the major question concerns whether or not badius and guereza are rightfully members of the same genus. The importance of this question lies in the assumption that, because species of different genera are phylogenetically more distant than species of the same genus, they have been separated over a relatively longer period of time and have had more different evolutionary histories than species of the same genus. The extent of these differences in the duration and nature of their evolutionary histories may be of considerable importance in explaining the current differences in the behavior and ecology of badius and guereza. Our data on the behavior and ecology of these 2 species demonstrates that they are very different. Marler's (1969, 1970) and our studies on the guereza and badius vocalizations also show pronounced differences. Whether the magnitude of the differences described by these studies warrants placing guereza and badius in different genera cannot be stated with certainty at this time. However, the osteological evidence which has been used to place them in the same genus is no more compelling than our data suggesting they should be placed in different genera. Clearly, more data of a different sort are needed before a satisfactory conclusion can be reached. Detailed and comparable data on karyology and serology could be extremely informative. Barnicot and Hewett-Emmett (1972) have examined red cell and serum proteins of badius, and Sarich (1970) presents information on the immunology of *Colobus polykomos,* but comparable data for badius and guereza are not available."

The other 5 monkey species living in the Kibale Forest belong to the subfamily Cercopithecinae, a different subfamily from the 2 colobus. Of these 5 species, only 3 have been sufficiently well studied to permit even preliminary descriptions of their ecology and social organization and speculation as to how they fit into the scheme of rain-forest ecology and the evolution of primate social systems. In the Kibale Forest the mangabey (*Cercocebus albigena*) has been studied in detail by Peter Waser; the blue monkey (*Cercopithecus mitis*) is currently under study by R. Rudran; and I am presently studying the red tail monkey (*Cercopithecus ascanius*). The impressions I have here are based on personal observations of all 3 of these species and lengthy discussions with Waser and Rudran.

The diets of all 3 species can be grossly classified as primarily frugivorous and insectivorous, although some buds and leaves are eaten by all. The 3 species live in relatively small social groups, ascanius numbering 15 to about 35, mitis about 10 to 28, and albigena 6 to 28. The ascanius and mitis typically have only one fully adult male per group, whereas the albigena usually have several. The ascanius groups are definitely territorial; the mitis groups may be territorial, but it is not yet clear because of the rarity of intergroup encounters; and the albigena groups overlap home ranges considerably but seem to practice nonaggressive avoidance of one another. In all 3 species, neonates are given considerable attention by other members of the group. This partially supports the hypothesis relating neonate popularity with territoriality.

The ascanius is the smallest monkey of the 3 but has the greatest biomass density, being nearly 4 times as dense as the mitis and 10 times the albigena. At this stage in our studies it would appear that ascanius and mitis are more insectivorous than albigena, but that ascanius exploits a greater variety of insects than does mitis. Although in biomass density the ascanius is much like the black and white colobus, it is about 8 times less dense than the red colobus. These 3 omnivorous monkeys are all more widely spaced within their groups than are either of the 2 leaf-eating monkeys, as expressed in terms of interindividual distance. Among the 3 omnivores it appears to me that within the van of each group the ascanius are more widely spaced (have a greater interindividual distance) than do either mitis or albigena. This spacing, I think, may be related to the greater importance of more mobile forms of insects in the diet of ascanius than in the diets of mitis and albigena. A greater interindividual distance is advantageous because it increases the effective search pattern of the group and also because mobile forms of insects are hunted by stealth and not by flushing. The latter point may also partially explain why the smallest of the monkeys feeds more on mobile insects; it is able to move most quickly and quietly. Although the interindividual distance may be greatest for ascanius, the overall spread of the group is probably most extensive among albigena. This may be due to fragmentation of the albigena groups into subgroups and also to the habit of adult and subadult males to move on the distant periphery of the group; a habit which may be associated with the location of super-abundant supplies of fruit. The mitis is probably intermediate to the ascanius and albigena in terms of overall group spread.

Following a similar trend are the data on home range size for the 3 species. Ascanius has the smallest home range (circa 25 ha.), mitis is intermediate (circa 100 ha.), and albigena has the largest (circa 400 ha.). The distance traveled each day is probably much the same for the 3 species. Daily ranging patterns are quite different, however. Ascanius tends to move back and forth over a rather limited area, whereas mitis and albigena groups tend to move more or less in a straight line, with the albigena usually covering a rather greater distance each day. I think these differences in group dispersion, annual home range, and daily ranging patterns are related to differences in food habits among the 3 species and the corresponding separation of feeding niches. I postulate that both mitis and albigena have larger home ranges, more diverse daily ranging patterns, and greater overall group spread as an adaptation to locating and exploiting superabundant sources of fruit. These superabundant sources of fruit are never a limiting factor to the more local and more insectivorous groups of ascanius. Consequently, the fact that the larger and more dominant mitis and albigena come into the range of an ascanius group to exploit a superabundant source of fruit is of little consequence to the ascanius. On a rather different scale, the same kind of relationship exists between the mitis and the larger albigena. Although the albigena can usually supplant the mitis from a superabundant source of fruit, the huge range of albigena (about 4 times that of mitis) means that it never competes very long or effectively with any one group of mitis.

Thus, one group of albigena moves around and exploits superabundant supplies of fruit in an area that encompasses the range of about 4 groups of mitis and about 16 groups of ascanius. And on a smaller scale, one group of mitis moves around and exploits similar supplies of fruit in the ranges of 4 groups of ascanius. The ascanius groups can be considered the "local" primates, which also use these superabundant supplies of fruit, but to which such supplies are less critical than to either the mitis or the albigena because of their greater efficiency at catching insects.

Preliminary results also indicate differences in the kinds of insects exploited by these 3 monkeys, as inferred from quantitative data on motor patterns used to catch prey and the kinds of microhabitats exploited. Although both mitis and ascanius tend to have similar proportions of unidentifiable objects (presumed insects) in their diets, ascanius uses a rapid grab with the hands to catch these items consistently more often than does mitis. We interpret this to mean that the ascanius is catching the relatively more mobile forms of insects and that these play a more important part in the diet of ascanius than in that of mitis. In contrast, mitis takes the majority of its unidentifiable food objects using a slow pick with the hands. We assume objects taken in this manner are relatively immobile forms of insects. Although ascanius also takes food objects in this slow manner, it does so relatively less often than mitis. Whether the factors involved in this difference are related to body size and/or neurophysiology, it is clear that a greater variety of insect forms are available to ascanius than to mitis. In essence then, the ascanius, by using the rapid grab as well as the slow pick, can exploit a larger proportion of the total insect population (mobile and immobile forms) than can mitis. Consequently, to obtain an equivalent amount of insect material in its diet, ascanius requires a smaller area than does mitis. Because it rarely uses the rapid grab and therefore has a smaller proportion of the total insect population available to it, the mitis must either cover a larger area or hit upon a superabundant source of insects to ingest as much insect material as does the ascanius. The mitis rarely hits upon a concentrated and superabundant source of insects, and it seems most reasonable to assume that its acquisition of immobile insects is enhanced by covering a large area. I would also like to suggest than within a particular area the exploitation of immobile insects by ascanius and mitis may be equivalent. With respect to its total diet, a specific group of ascanius appears not to eat as much immobile insect life as does a specific group of mitis. But the mitis group exploits more of this resource by covering a larger area. By way of further explanation, consider the hypothetical case where one group of mitis consumes 100 immobile insects in one day, but does so by moving through the home ranges of 2 groups of ascanius, each of which consumes 50 immobile insects on the same day. The total number of insects consumed by the ascanius is increased through the ascanius' exploitation of mobile insects by use of the rapid grab and is, thereby, equated to the mitis daily intake of insects.

The most distinctive feature of the albigena insectivorous diet is its exploitation of the insects living inside dead twigs. The twigs are torn open and the insects picked or licked out. This habit has not so far been observed in any of the other monkey species in the Kibale Forest.

Although the data from my studies in Cameroun are less detailed and of generally poorer quality than those from Kibale, there was a clear indication that *Cercopithecus p. pogonias* was ecologically separated from the other 3 common species of *Cercopithecus* as a consequence of its much larger annual home range and exploitation of superabundant food sources in a manner similar to that of the mitis and albigena groups in Kibale (Gartlan and Struhsaker 1972).

What features of social organization consistently correlate with the rain-forest habitat? Our studies in the Kibale Forest and elsewhere show that there are very few correlations. There is considerable variation in group size among the different cercopithecid species living in rain forests, such as the large multimale groups of *Miopithecus talapoin, Mandrillus leucophaeus*, and *Colobus badius*; the smaller multimale groups of mangabeys; and one-male groups of most *Cercopithecus* species; and the even smaller one-male groups of *Cercopithecus lhoesti preussi* and *Colobus guereza*. In view of the radical differences in the ecologies of rain-forest monkeys, these differences in social organization are not surprising. One feature, however, does seem common to most, if not all, species of rain-forest cercopithecids: the occurrence of solitary or extremely peripheral individuals (usually subadult and adult males). The rare appearance of solitary monkeys in open-country species is presumably a response to strong predator selection, which is less pronounced in the rain forest as I have suggested earlier (Struhsaker 1969).

Consideration of the finer features of ecology and social organization reveals a good correlation between the intragroup spacing (expressed as interindividual distance) and the food habits of a species. It appears, for example, that monkeys which feed on an abundant, dense, and inert supply of food, such as mature leaves, are more closely spaced than monkeys which feed on a sparse or widely spaced and mobile food, such as adult forms of insects. Extreme examples of this correlation are provided by the guereza on the one hand, whose short interindividual distance seems related to their monotonous diet of an abundant, dense, and inert food (leaves of *Celtis durandii*), and ascanius on the other hand, whose relatively great interindividual distance corresponds to their heavy reliance on mobile forms of insects which appear to be widely spaced and relatively sparse. The overall spread of the group may not be reflected by measures of interindividual distance but may, in fact, be related to the degree of dependence on locating superabundant supplies of fruit. Thus, one would expect to find the greatest overall group spread in the species which is the most opportunistic—that is, which relies most heavily on superabundant food sources. In the case of the 5 species studied in the Kibale Forest, albigena appear to be the greatest opportunists, and correspondingly they have the greatest overall group spread. Mitis appear to rank second in both of these features and ascanius third. Furthermore, I believe that detailed study would show the chimpanzees of the Kibale Forest to be even greater opportunists than the albigena and to have an even greater overall group spread. The positions of badius and guereza with regard to these 2 factors is unclear and possibly irrelevant because of their low reliance on superabundant sources of fruit.

Of relevance to this concept of the opportunistic omnivores having a large overall group spread as a means of locating superabundant sources of fruit, is the fact that the largest omnivorous monkeys have the largest home ranges and are generally able to supplant the smaller species from abundant sources of food such as fruiting fig trees. In a sense then, the whole scheme can be viewed as an outcome of body size. While the larger species can gain priority of access to fruiting trees, their larger body size limits the kinds of foods available to them. Larger monkeys cannot efficiently prey upon mobile insects, which smaller monkey species hunt effectively by stealth since their small body size permits quiet movement in the trees. As a consequence, the larger species are forced to cover larger areas and to have a greater overall group spread than the smaller species in their search for superabundant food supplies.

As mentioned above, within a given species the numerical size of social groups seems to be a function of food density. The example has been given of the vervet monkeys in the Amboseli Reserve of Kenya, which underwent a major population decline in response to a natural decline in food sources. This population decline resulted in a reduction in group size and not in the number of groups (Struhsaker 1973). Studies on semiwild Japanese macaques and rhesus macaques have clearly demonstrated that food provisioning can greatly increase population and group size. The observations reported earlier in this monograph for *Colobus badius temminckii* and *C. b. rufomitratus* indicate that group size in red colobus monkeys is also dependent upon food density. These 2 subspecies live in the smallest social groups and in the most marginal habitats of any of the red colobus subspecies studied.[1] However, attempts to correlate gross ecological parameters with interspecific differences in group size have so far been unsuccessful. One can, for example, find both large and small social groups in the savanna, e.g. baboons and patas, and likewise large and small groups in the rain forest, e.g., red colobus and black and white colobus. These interspecific differences within a given biome may, however, also be related to the food densities of each species. Comparison of the annual diet of guereza and badius reveals that about half of the feeding observations for guereza comes from only one species, *Celtis durandii*, whereas the 5 top-ranking food species of badius account for about half of its diet. The density of *Celtis durandii* in Oates's study area was 49.4 per hectare, whereas the combined densities of the 5 top badius food species totaled 103.4 trees per hectare.[2] Thus, the density of food accounting for 50% of the badius diet was slightly more than twice the density of food accounting for 50% of the guereza diet. Although the corresponding differences in group size and overall biomass density of the two species differed by much more than this, the relationship is the same; that is, badius live in larger groups and have a greater biomass density than guereza.

A consideration of phylogenetic affinities provides additional insight into the interspecific differences in grouping tendencies. Jolly (1966) has proposed a subdivision of the subfamily Cercopithecinae into three tribes: Cercopithecini, including the *Cercopithecus* spp. and *Erythrocebus patas*; Cercocebini, including the *Cercocebus, Papio, Mandrillus,* and *Macaca;* and Theropithecini, including *Theropithecus* and several fossil genera. The tribal affinity of *Miopithecus talapoin* is uncertain. With the exception of vervet monkeys, most, if not all, members of the Cercopithecini tend to live in small social groups of about 20 members, with only one fully adult male per group. Exceptions can be found within a given species, but the generalization seems to apply to the average group of most *Cercopithecus* species studied to date. Group size in the Cercocebini is extremely variable, with large groups occurring in both rain forest and savanna habitat, although the smallest groups are found in the rain forest. All species of this tribe do, however, live in multimale social groups, regardless of gross differences in their habitats and general ecologies.

No one has yet suggested tribal subdivisions within the colobinae subfamily. This omission, however, is an indication of ignorance rather than of close phylogenetic affinity of species within the subfamily. The majority of colobinae species live in small, one-male groups, but most red colobus subspecies and some populations of *Presbytis entellus* are exceptional in having large, multimale groups. More detailed study of the colobinae may lead to their subdivision into tribes, correlating with some of their differences in social organization.

Tables

TABLE 1
Weights and measurements of *Colobus badius tephrosceles* collected in Kibale Forest, Uganda

Age and sex	Date of collection	Means of collection	Weight (kg)	Head and body length (cm)1	Tail length (cm)	Entire forelimb length (cm)2	Hand length (cm)3	Entire hindlimb length (cm)4	Foot length (cm)5
dark infant male < 4 months	19 July 1970	"natural death"	0.5	22.0	27.8	17.0	5.7	21.7	8.7
adult female	26 July 1970	"natural death"	5.75	53.0	60.5	41.0	12.5	47.5	16.2
*subadult male	1 April 1970	shot	9.4	58.0	66.5	44.0	13.0	55.0	18.5
**adult male	1 April 1970	shot	10.5	63.0	70.0	47.0	13.0	55.0	19.5
juvenile female	1 December 1970	death from a fall		42.0	46.0	28.0		36.5	

1: tip of nose to base of tail   2: shoulder to tip of longest digit   3: wrist to tip of longest digit

4: hip to tip of longest digit   5: heel to tip of longest digit

* Accession number BM(NH) 71.2064 of skin and skull in the British Museum of Natural History, London

** Accession number BM(NH) 71.2065 of skin and skull.

TABLE 2

Hours of observation of *C. b. tephrosceles,* May 1970 through March 1972

Month	CW group	ST group	BN group	Other groups (usually uniden- tified)	Total
May 1970				36.9	36.9
June 1970				21.1	21.1
July 1970				32.2	32.2
Aug. 1970	27.5	3.8		11.6	46.9
Sept. 1970	19.6	29.1		53.7	102.4
Oct. 1970	63.1	39.1		27.3	129.5
Nov. 1970	73.3	50.6	1.8	12.7	138.4
Dec. 1970	61.8	31.1	0.2	4.7	97.8
Jan. 1971	66.1	10.2	0.2	17.0	93.5
Feb. 1971	70.7	3.7	0.1	7.3	81.8
Mar. 1971	62.5	2.5		10.9	75.9
Apr. 1971	74.7	1.2		3.4	79.3
May 1971	68.2	1.1		2.2	71.5
June 1971	53.2	0.1		0.3	53.6
July 1971	31.7	0.3		0.2	32.2
Aug. 1971	66.1	1.8	0.95		68.9
Sept. 1971	73.6	4.7		12.4	90.7
Oct. 1971	58.3	4.6		9.0	71.9
Nov. 1971	49.2				49.2
Dec. 1971	58.2	4.7	2.5	10.0	75.4
Jan. 1972	47.2	1.0	0.8	5.2	54.2
Feb. 1972	44.6	0.8		1.6	47.0
Mar. 1972	42.9	2.1	0.1	2.3	47.4
Total	1,112.5	192.5	6.7	282.0	1,593.7

TABLE 3

Counts and estimates of 9 groups of *C. b. temminckii* in Senegal and Gambia

Accurate counts ($\bar{X}$ = 24.5)

Total	Adult males	Adult females	Approx. adults	Old juveniles	Young juveniles	Clinging infants	?
34	?				≥ 6	0	28
21	≥ 2				≥ 2	≥ 1	16
29	≥ 2				≥ 4	≥ 1	22
30	?				≥ 1	≥ 1	28
*21	4	10	2	1	2	2	
*12	2	3	3	1	3	0	

Reliable estimates

Total counted	Total estimate	No. young juveniles	No. clinging infants
20	≤ 25	?	≥ 1
24	30	≥ 6	≥ 2
26	≤ 35	≥ 3	≥ 3

Mean of accurate counts and reliable estimates is 26.3

* accurate compositional data, both from E. Gambia

TABLE 4

Counts and estimates of 10 groups of *C. b. badius* in Tai Reserve, Ivory Coast (mean estimate: 40.1)

Total counted	Total estimated	Adult males	Unswollen adult females	Swollen adult females	Approx. adults	Old juveniles	Juveniles	Young juveniles	Old infants	Young infants	?
*51	51	3		≥2					8		38
22	>30	3		3	13				2		1
21	<35									2	19
27	>27	≥1		6					5		15
33	>33	4	11	2	7	1	2		6		
47	<55	6	11	4					11		15
36	≤45	5	8	1					5		17
28	<40	≥4		1					3		20
34	≤40	1		2					3		28
37	≤45	4	10	≥3	3		2	1	6	2	6

* Accurate count

TABLE 5
Counts and estimates of 7 groups of *C. b. preussi* in Korup Reserve, Cameroun (mean estimate: 47.0)

Total counted	Total estimated	Adult males	Unswollen adult females	Swollen adult females	Approx. adults	Juveniles	Young juveniles	Old infants	Young infants	Clinging infants of unknown age	?
25	> 25	≥ 1	≥ 8	≥ 1				≥ 7	≥ 1		7
67	≥ 80	≥ 2	> 2	≥ 3	≥ 2				≥ 1		57
25	40	3	2	3	9	4	2	≥ 1	≥ 1		10
38	≥ 50	3	7	4	12	1				7	
23	≤ 40	3	2	3	11	1		2	1		
24	> 24	5	2	9	6			1	1		
48	≥ 70	3	5	1	34			1	1	3	

TABLE 6

Counts and estimates of 14 groups of *C. b. tephrosceles* in Kibale Forest, Uganda, excluding CW, ST, and BN groups (mean estimate: 43.9)

Total counted	Total estimated	Adult males	Adult females	Approx. adults	Subadult males	Juveniles	Young juveniles	Old infants	Young infants	Clinging infants of unknown age	?
35	>35	6		5						5	19
42	≤60	3		9						9	21
44	<70	4		4						4	32
33	>33	4	3	6					1	5	14
44	≤60	5	16	1		10			1	1	10
26	>26	4	3	12		2			3		2
43	>43	3	8		1	3	7			1	21
49	>49	2	16	7		6	6	4	2	2	3
*28	28	4	12	1		1	2	6	1		1
48	>48	3	9	4		5	3	2	2		20
32	>32	≥7			1						24
50	<65										50
24	≤30	4	7	1	1	7	1	3			3
35	>35	≥5	≥9	5		5	2	3			6

* Accurate count

TABLE 7

Changes in size and composition of CW group of *C. b. tephrosceles*

Date	Adult males	Adult females	Approx. adults	Subadult males	Old juv. males	Juveniles	Young juveniles	Old infs.	Young infs.	?	Total	Apparent change
7 Nov. '70	3	8				5	4		1	1	22	One infant born on 2 or 3 November 1970.
7 Jan. '71	3	7				4	4		1		19	One adult female, one juvenile and ? disappeared between 4 December 1970 and 5 January 1971
7 April '71	3	7			1	4	3	1	1		20	One juvenile re-classified as old juvenile male; one young juvenile became juvenile; one young infant became old infant in early March 1971; one infant born between 6 March and 5 April 1971
17 April '71	3	5	1	1		3	2	1	2	2	20	Old juvenile male reclassified as subadult male; young juvenile disappeared; one infant born between 10 and 17 April 1971

Date										Total	Remarks
14 July '71	3	7		1	3	3	2			19	Oldest infant disappeared between 21 June and 11 July; one juvenile apparently re-classified as young juvenile; two young infants became old infants
30 Oct. '71	3	7	1 sub-adult female	1	5	3	2			22	Addition of one sub-adult female and 2 juveniles probably through a combination of maturation and immigration
16 Jan. '72	4	6	3		7		2			22	Subadult male attained adulthood; 3 young juveniles became juveniles
1 Feb. '72	4	2	8		3		2			19	3 disappeared
3 March '72	4	7			6	1	1	1		19	One old infant became young juvenile; 2 juveniles and one subadult disappeared between 16 January and 1 February 1972
weighted $\bar{X}$ % *	16.0	33.5	2.3	2.5	22.1	13.1	5.7	2.9	1.9		

* = $\dfrac{\Sigma(\text{No. mos.} \times \text{% composition})}{\text{Total no. of mos.}}$

TABLE 8
Intergroup conflicts of CW group of *C. b. tephrosceles* in the course of 83 days

Group interacted with	Other group supplants CW group	CW group supplants other groups	Aggressive but no supplantation	Uncertain
		Outcome -- frequency and %		
?	4 (21.1%)	13 (68.4%)		2 (10.5%)
DL				1 (100%)
HT		1 (100%)		
ST	9 (75%)	2 (16.7%)	1 (8.3%)	
BN	4 (33.3%)	7 (58.3%)	1 (8.3%)	

TABLE 9

Supplantations in CW group of *C. b. tephrosceles,* November 1970 through March 1972

November 1970 - 1 August 1971

	Supplantee						
Supplanter	AM LB	SAM	AF KT	AF ?	J	YJ	Total
AM CW	1	3	1	4	4		13
AM LB				1			1
AM ND						2	2
J					1		1
Total	1	3	1	5	5	2	17

2 August 1971 - March 1972

	Supplantee								
Supplanter	AM CW	AM LB	AM ND	SAM	AF GCW	AF ?	J	JM	YJ	Total
AM CW		1	1				1			3
AM LB				1						1
AM ND								1	1	2
SAM	1					1	2			4
AF ?					1	1				2
Total	1	1	1	1	1	2	3	1	1	12

AM = adult male, AF = adult female, SAM = subadult male (one particular individual), J = juvenile, JM = juvenile male, YJ = young juvenile, other letters indicate specific name of monkey.

TABLE 10, PART 1

Present type I in CW group of *C. b. tephrosceles,* September 1970 to 1 August 1971

	Recipient																
Sender	AM CW	AM LB	AM ND	AM ?	SAM	AF BT	AF GCW	AF KT	AF PGCW	AF ?	A	SA	?	J	YJ	OI	Total
AM CW																	
AM LB																	
AM ND																	
AM ?																	
SAM	4																4
AF BT	1	1															2
AF GCW																	
AF KT	1																1
AF PGCW																	
AF ?	3																3
A	1																1
SA	2	1															3
?			1														1
J	25	2	3	2						1				1	1		35
YJ	2	3								4					1		10
OI	1																1
Total	40	7	4	2						5				1	2		61

Abbreviations as in Table 9, plus A = approximate adult, SA = subadult,
OI = old infant

TABLE 10, PART 2

Present type I in CW group of *C. b. tephrosceles,* 2 August 1971 to 31 March 1972

Recipient

Sender	AM CW	AM LB	AM ND	SAM	AF BT	AF GCW	AF KT	AF PGCW	AF ?	A	J	YJ	OI	Total
AM CW														
AM LB														
AM ND				1										1
SAM		1												1
AF BT				3										3
AF GCW								1						1
AF KT														
AF PGCW				4										4
AF ?	1			1										2
A	1			1										2
J	5	4	3	8		1		1			1			23
YJ	1			1							1			3
OI	2	1												3
Total	10	6	3	19		1		2			2			43

TABLE 11

Present type II in CW group of *C. b. tephrosceles,* October 1970 through March 1972

| | \multicolumn{9}{c}{Recipient} | | | | | | | |
Sender	AM CW	AM LB	AM ND	AM ?	SAM	AF GCW	AF ?	J	Total
AM CW		4	15	3					22
AM LB			11						11
AM ND									0
AM ?			1						1
SAM								1	1
AF GCW									0
AF ?						1			1
J			1						1
Total	0	4	28	3	0	1	0	1	37

Abbreviations as in preceding tables

TABLE 12

Aggressive encounters with grabbing or slapping toward (no physical contact) in CW group of *C. b. tephrosceles,* August 1970 through March 1972

	Recipient											
	AM CW	AM ND	AM LB	SAM	AF ?	AF KT	A	J	YJ	OI	YI	Total
AM CW			1		1			2	2			6
AM LB		1						4	2	1		8
AM ND						1	1	4				6
SAM								1				1
AF ?	1							3	5			9
AF BT								1	1			2
J											1	1
Total	1	1	1	0	1	1	1	11	14	1	1	33

Sender

YI = young, dark infant; other abbreviations as in preceding tables.

TABLE 13

Lunges or chases in CW group of *C. b. tephrosceles,* August 1970 through March 1972

	Recipient												
	AM CW	AM LB	AM ND	SAM	AM ?	AF BT	AF GCW	AF ?	A	J	YJ	?	Total
AM CW		3	9		1	1		3		3	5	1	26
AM LB	1		3	1				3	1	5	2		16
AM ND	1						1			2			4
SAM	2		1							1	2		6
AM ?		1		1	1			1		1			5
AF ?										1	1		2
?			1									5	6
Total	4	4	14	2	2	1	1	7	1	13	10	6	65

Sender

Abbreviations as in preceding tables.

Note: Number of entries in this table exceeds the number of bouts (59) because in cases of multipartite encounters each participant is scored.

TABLE 14
Harassment during copulation in CW group of *C. b. tephrosceles*

Harassed male

Harasser	1 August 1970 - 1 August 1971			2 August 1971 - 31 March 1972					
	AM CW	AM ?	Total	AM CW	AM LB	AM ND	SAM	AM ?	Total
AM CW			0				1		1
AM LB	2		2						0
AM ND			0	6	1		1		8
SAM			0						0
J	5	1	6	4	4				8
YJ	4		4	8			2		10
?		1	1	1				1	2

Total copulation bouts 33   Total copulation bouts 75
Total copulation bouts with   Total copulation bouts with
harassment 10   harassment 24

Entries in this table are not mutually exclusive because some harassments involved more than one harasser.

Abbreviations as in preceding tables.

TABLE 15

Grooming roles in CW group of *C. b. tephrosceles* (first figure is percentage as groomer and second figure is percentage as groomee)

Month and (No. of bouts)	AM CW	AM LB	AM ND	SAM	AM ?	AF GCW	AF KT	AF PGCW	AF BT	AF ?	Approx. A	SA	J	YJ	OI	YI	?
Nov. 1970 (33)	3.0 / 15.2	3.0 / 6.0	21.0 / 15.2		0 / 6.0					55.0 / 15.2	3.0 / 6.0		12.0 / 15.2	3.0 / 12.0		0 / 9.0	
Dec. 1970 (17)	0 / 23.5	0 / 17.6	23.5 / 0		0 / 5.9					64.7 / 35.3	11.8 / 0		0 / 5.9	0 / 11.8			
Jan. 1971 (23)	0 / 13.0	0 / 0	4.3 / 13.0							78.3 / 13.0	4.3 / 8.7		4.3 / 17.4	8.7 / 21.7		0 / 13.0	
Feb. 1971 (18)	16.7 / 27.8	0 / 5.6	11.1 / 27.8							55.6 / 16.7	16.7 / 0			0 / 16.7		0 / 5.6	
Mar. 1971 (22)	4.5 / 18.2	0 / 13.6	13.6 / 9.1			0 / 0				63.6 / 40.9			4.5 / 0	13.6 / 18.2			
Apr. 1971 (58)	0 / 6.9	0 / 6.9	12.0 / 19.0	0 / 0		1.7 / 1.7			1.7 / 3.4	55.2 / 34.5	3.4 / 0	5.2 / 6.9	6.9 / 5.2	10.3 / 8.6	0 / 1.7	0 / 1.7	3.4 / 3.4
May 1971 (53)	1.9 / 15.1	0 / 20.8	5.7 / 7.5	0 / 5.7		9.4 / 3.8	3.4		9.4	49.1			9.4 / 5.7	11.3 / 11.3	0 / 1.9	0 / 7.5	0 / 1.9
June 1971 (39)	2.6 / 12.8	0 / 7.7	5.1 / 15.4	0 / 2.6		12.8 / 2.6	2.6 / 0		12.8 / 2.6	38.5 / 7.7	2.6 / 0		15.4 / 15.4	2.6 / 15.4	0 / 10.3	0 / 2.6	5.1 / 5.1
July 1971 (14)	0 / 7.1	0 / 21.4	0 / 7.1	0 / 0		14.3 / 0	14.3 / 0		14.3 / 7.1	21.4 / 7.1			14.3 / 7.1	21.4 / 28.6	0 / 7.1		
Aug. 1971 (60)	5.0 / 13.3	0 / 16.7	6.6 / 11.7	3.3 / 6.6		13.3 / 3.3	10.0 / 1.7		10.0 / 0	35.0 / 20.0			5.0 / 13.3	11.7 / 8.3	0 / 5.0		

Month (n)														
Sept. 1971 (49)	4.1	2.0	2.0		14.3	10.2	16.3	24.5	2.0			10.2	14.3	0
	6.1	14.3	26.5		0	4.1	4.1	18.4	0			8.2	6.1	2.0
Oct. 1971 (50)	4.0	6.0	0		18.0	10.0	12.0	14.0	12.0	0		12.0	12.0	0
	12.0	8.0	14.0		8.0	4.0	4.0	6.0	6.0	2.0		12.0	14.0	4.0
Nov. 1971 (49)	2.0	12.2	2.0		12.2	2.0	12.2	10.2	24.5			8.2	14.3	0
	12.2	12.2	16.3		2.0	0	6.1	6.1	12.2			10.2	14.3	4.1
Dec. 1971 (64)	0	7.8	0		10.9	12.5	9.4	21.9	12.5	4.7		7.9	12.5	0
	6.3	7.8	9.4		3.1	3.1	10.9	9.4	9.4	1.6		4.7	18.8	9.4
Jan. 1972 (47)	6.4	2.1	0		6.4	12.8	21.3	4.3	25.5	2.1		10.6	4.3	4.3
	6.4	14.9	8.5		2.1	0	4.3	2.1	6.4			12.7	17.0	19.1
Feb. 1972 (45)	2.2	15.6	0		0	4.4	11.1	20.0	17.8			28.8	0	0
	15.6	6.7	13.0		2.2	0	6.7	11.1	6.7			22.2	2.2	8.9
Mar. 1972 (44)	9.1	2.3	0	0	6.8	0	18.2	6.8	27.3		2.3	27.3	0	0
	9.1	6.8	13.6	2.3	2.3	0	2.3	6.8	11.4		0	25.0	0	13.6

Abbreviations as in preceding tables. SAM was first recognized in April 1971; GCW on 1 March 1971; BT on 17 April 1971; KT on 9 May 1971; PGCW on 30 September 1971.

TABLE 16

Grooming relations in CW group of *C. b. tephrosceles*, November 1970 through July 1971 (N = 277)

Groomer	Groomee																	
	AM CW	AM LB	AM ND	SAM	AM ?	AF GCW	AF KT	AF BT	AF ?	Approx. A	SA	J	YJ	OI	YI	?	Total	
AM CW			6						1								7	2.5%
AM LB									1								1	0.4%
AM ND	7	4			1	1		2	9	1			4				29	10.5%
SAM																		
AM ?																		
AF GCW	4	4	2	2							1						13	4.7%
AF KT		1											2	1	1		5	1.8%

Groomer																	Total	
AF/BT	1	3	3									6					13	4.7%
AF/?	23	13	15	2	2	1		1	23	2	3	17	28	6	10		147	53.1%
A	3	1	2						2			2					10	3.6%
SA			2						1								3	1.1%
J	4	4	1					2	6			5			1		23	8.3%
YJ	1	1	2	1	1			1	13	1			1		1		22	7.9%
OI																		
YI																		
?						4										4	4	1.4%
Total	39	31	36	4	4	4	3	4	55	4	4	28	37	7	13	4	277	
	14.1%	11.2%	13.0%	1.4%	1.4%	1.4%	1.1%	1.4%	19.9%	1.4%	1.4%	10.1%	13.4%	2.5%	4.7%	1.4%		

TABLE 17

Grooming relations in CW group of *C. b. tephrosceles*, August 1971 through March 1972 (N = 408)

Groomer	Groomee																Total	
	AM CW	AM LB	AM ND	SAM	AM ?	AF GCW	AF KT	AF PGCW	AF BT	AF ?	Approx. A.	SA	J	YJ	OI	YI		
AM CW		3	9	1	1	2											16	3.9%
AM LB																		
AM ND	6	6		1		1		1	5	5			3				28	6.9%
SAM			1							3							4	1.0%
AM ?																		
AF GCW	13	5	3	16		2	1			2				1			43	10.5%
AF KT		2	2	2						1			11	2	13		33	8.1%
AF PGCW	7	4		7		1	1						4	17			41	10.1%
AF BT	5	3	5	11		2	1			3		1	7	16			54	13.2%
AF ?	7	8	9	13		2		1	5	5			15	6	20		91	22.3%
Approx. A.										1	1						2	0.5%
SA			1	2			1										4	1.0%
J	3		9	1		1	4	4	10	12			8	1			53	13.0%
YJ		1	3					12	3	15			3				37	9.1%
OI													2				2	0.5%
YI																		
Total	41	32	42	54	1	11	8	18	23	47	1	1	53	45	33		408	
	10.1%	7.8%	10.3%	13.2%	0.2%	2.7%	2.0%	4.4%	5.6%	11.5%	0.2%	0.2%	13.0%	10.5%	8.1%			

Abbreviations as in preceding tables.

TABLE 18

Relation between perineal swelling and heterosexual mounting among 4 recognizable females in CW group of *C. b. tephrosceles*

Month	AF GCW		AF BT		AF PGCW		JF M	
	Mount-ing	Swell-ing	Mount-ing	Swell-ing	Mount-ing	Swell-ing	Mount-ing	Swell-ing
Mar. '71	X	?						
Apr. '71	-	?	-	?				
May '71	X	X	X_3	?				
June '71	-	?	-	?				
July '71	-*	X	-	?				
Aug. '71	X_1	$-_2$	-	?				
Sept. '71	-	X	-	?				
Oct. '71	X*	X	-	?	X	?	-	?
Nov. '71	X	?	-	?	X	?	-	?
Dec. '71	X*	X	-	?	-	X	-	X
Jan. '72	-	-	X	X	X	X	X	X
Feb. '72	-	-	X	X	X	X	X*	X
Mar. '72	-	?	X	X	X*	X	not seen	

X = definitely occurred

- = definitely did not occur

? = occurrence not determined

* = ejaculate seen on perineum

1 = perineum only noted on 2 August, but mounting occurred on 5 August

2 = perineum had a 2.5 cm deep pink knob, unlike typical swelling

3 = consisted of only 1 incomplete mount, which lacked pelvic thrusts

Other abbreviations as in preceding tables

TABLE 19

Monthly frequency of heterosexual mounts in CW group of *C. b. tephrosceles* (N = 160)

Month	Total observation hours	No. of mounts per hour of observation x 100			
		Complete (with pause)	Incomplete	Undetermined	Total
Aug. '70	27.5	0	10.9	0	10.9
Sept. '70	19.6	15.3	40.8	0	56.1
Oct. '70	63.1	0	0	1.6	1.6
Nov. '70	73.3	0	6.8	0	6.8
Dec. '70	61.8	3.2	11.3	3.2	17.7
Jan. '71	66.1	0	0	0	0
Feb. '71	70.7	0	0	1.4	1.4
Mar. '71	62.5	1.6	14.4	0	16.0
Apr. '71	74.7	1.3	1.3	0	2.6
May '71	68.2	0	4.4	0	4.4
June '71	53.2	0	3.8	0	3.8
July '71	31.7	0	0	0	0
Aug. '71	66.1	4.5	27.2	4.5	36.2
Sept. '71	73.6	2.7	17.7	1.4	21.8
Oct. '71	58.3	5.1	15.4	1.7	22.2
Nov. '71	49.2	4.1	14.2	0	18.3
Dec. '71	58.2	3.4	6.9	0	10.3
Jan. '72	47.2	6.4	19.1	2.1	27.6
Feb. '72	44.6	6.7	20.2	11.2	38.1
Mar. '72	42.9	4.7	21.0	4.7	30.4

TABLE 20
Heterosexual mounting relations in CW group of *C. b. tephrosceles*

22 August 1970 through 1 August 1971

Mounter	Mountee AF GCW	AF BT	AF ?	Approx. AF	J	?	Total
AM CW	8IC	1IC	5C 23IC 3?	1?		1C	6C 32IC 4?
AM LB			1IC	1IC			2IC
AM ND			1IC				1IC
SAM							0
AM ?				1C 2IC	1IC		1C 3IC
J			1IC				1IC
Total	8IC	1IC	5C 26IC 3?	1C 3IC 1?	1IC	1C	7C 39IC 4?

2 August 1971 through 17 March 1972

Mounter	Mountee AF GCW	AF BT	AF PGCW	AF ?	Approx. AF	SA F	JF M	SA	F	OJ	Total
AM CW	2IC	2C 2IC 1?	2C 3IC 2?	6C 17IC 2?	2IC		3IC 1?	2IC			10C 31IC 6?
AM LB	5IC	1?	1C 3IC	1C 3IC 1?		1IC	1C 1IC			1C 2IC	4C 15IC 2?
AM ND	1IC	1IC		1C 2IC		1?					1C 4IC 1?
SAM		5IC 2?	2IC	2C 10IC 1?	1IC	1C 6IC	1IC	2IC			3C 27IC 3?
AM ?		1?		1C	1C						2C 1?
Total	8IC	2C 8IC 5?	3C 8IC 2?	11C 32IC 4?	1C 3IC	1C 7IC 1?	1C 5IC 1?	2IC	2IC	1C 2IC	20C 77IC 13?

Abbreviations:
C = complete mount (with pause), IC = incomplete mount, ? = success of mount not determined, SAF = subadult female, JFM = juvenile female M, F = female, OJ = old juvenile, ? = undetermined identity and as in other tables.

TABLE 21
Interindividual distance sample summaries for 7 individuals of CW group of *C. b. tephrosceles*

Sample monkey	No. of samples	X̄ No. neighbors ≤ 2.5m per sample	X̄ No. neighbors > 2.5 & ≤ 5m per sample	No. samples with none ≤ 2.5m	No. samples with none > 2.5 & ≤ 5m	No. samples with none ≤ 10m	No. samples with physical contact	Distance of neighbors in m Mean	Range	Mode
AM CW	22	0.68	1.27	12	10	0	1	3.8	0-10	3.5
AM LB	24	0.63	0.79	14	12	3	0	3.3	0.3-10	3.0
AM ND	28	0.32	0.64	22	18	3	0	4.5	1-10	6.0
SAM	20	0.45	0.80	14	11	2	1	4.0	0-7	5.0
AF KT	20	1.45	1.00	1	9	0	16*	2.1	0-6	0
AF BT	22	1.45	0.86	4	12	1	10**	2.3	0-8	0
AF GCW	19	1.05	0.53	7	12	2	1	2.8	0-9	1.5

* all with a clinging infant

** all with a young juvenile

TABLE 22

Observed and (expected) frequencies of specific monkeys being ≤ 2.5 m from sample monkey in CW group of *C. b. tephrosceles*

Sampled animal	Neighbor										
	AM CW	AM LB	AM ND	AF BT	AF GCW	AF KT	SAM	4AF	3J	3YJ	2OI
AM CW		2 (0.83)	1 (0.83)	2 (0.83)	3 (0.83)	1 (0.83)	0 (0.83)	1 (3.33)	1 (2.50)	2 (2.50)	2 (1.67)
AM LB	3 (0.83)		2 (0.83)	0 (0.83)	3 (0.83)	0 (0.83)	0 (0.83)	0 (3.33)	6 (2.50)	1 (2.50)	0 (1.67)
AM ND	1 (0.50)	3 (0.50)		1 (0.50)	0 (0.50)	0 (0.50)	0 (0.50)	0 (2.00)	2 (1.50)	2 (1.50)	0 (1.00)
AF BT	2 (1.78)	0 (1.78)	1 (1.78)		1 (1.78)	1 (1.78)	1 (1.78)	4 (7.11)	1 (5.33)	17 (5.33)	4 (3.56)
AF GCW	3 (1.11)	4 (1.11)	0 (1.11)	1 (1.11)		1 (1.11)	3 (1.11)	2 (4.44)	1 (3.33)	4 (3.33)	1 (2.22)
AF KT	1 (1.61)	1 (1.61)	0 (1.61)	1 (1.61)	1 (1.61)		0 (1.61)	2 (6.44)	3 (4.83)	0 (4.83)	21 (3.22)
SAM	0 (0.50)	0 (0.50)	0 (0.50)	1 (0.50)	1 (0.50)	0 (0.50)		4 (2.00)	1 (1.50)	2 (1.50)	0 (1.00)

First number is observed frequency; number in parentheses is expected frequency.

TABLE 23
Measurements of *C. b. tephrosceles* vocalizations

Call type	Vocalizer	N	X̄ duration (sec.)	Range of duration (sec.)	Mean lower frequency	Range of lower frequency	Mean upper frequency	Range of upper frequency
						kHz		
Bark	2 unidentified males	2	0.065	0.05-0.08	0.90	0.50-1.30	4.40	3.00-5.80
	2 unidentified in CW group	8	0.10	0.07-0.16	0.85	0.70-1.00	3.83	3.75-3.90
	Male CW of CW group	3	0.08	0.06-0.09	0.83	0.60-1.00	6.66	5.50-7.50
	Male LB of CW group	19	0.10	0.08-0.13	0.58	0.50-0.80	4.53	3.20-6.00
	Male BN of BN group	2	0.12	0.09-0.15	0.85	0.80-0.90	4.55	3.60-5.50
Wheet	Male CW	30	0.73	0.08-2.09	2.26	1.00-4.80	8.32	2.25-13.00
	Unidentified male of CW group	1	0.58	-	2.10	-	8.00	-
	Male RF of ST group	4	0.60	0.30-0.76	1.89	1.75-2.00	6.37	6.00-8.00
	Unidentified male of BN group	3	0.77	0.65-0.96	2.17	1.90-2.40	7.96	7.90-8.00

TABLE 24
Measurements of *C. b. tephrosceles* chist calls

Vocalizer	N	X̄ duration (sec.)	Range of duration (sec.)	kHz			
				Mean lower frequency	Range of lower frequency	Mean upper frequency	Range of upper frequency
Male CW 16 Feb. '71	18	0.10	0.06–0.16	1.34	0.80–3.20	7.63	5.50–8.00
Male CW 20 June '71	12	0.06	0.04–0.08	1.46	0.90–2.40	6.46	3.60–8.00
Male CW 31 Dec. '71	5	0.06	0.05–0.07	0.08	0.08	7.34	6.50–8.00
Male LB 22 Dec. '70	7	0.06	0.05–0.07	0.91	0.50–1.40	4.50	3.20–4.80
Male LB 20 June '71	19	0.06	0.03–0.08	1.71	1.50–2.00	5.19	4.00–7.00
Male ND 22 Dec. '70	10	0.06	0.04–0.10	1.59	0.90–2.00	5.21	3.60–8.00
Male ND 31 Dec. '71	13	0.07	0.04–0.11	0.39	0.08–2.25	6.82	6.00–7.50
Unidentified male(s) of CW group	4	0.06	0.05–0.07	1.80	1.50–2.00	7.63	6.50–8.00
Male TTT of BN group	2	0.08	0.08	1.88	1.50–2.25	8.00	8.00
Unidentified male of BN group	4	0.095	0.07–0.12	1.23	1.00–1.50	6.78	6.00–7.50
Unidentified male(s) of ST group	22	0.09	0.04–0.18	0.97	0.40–2.00	5.48	4.00–7.00
Unidentified males of 1 or 2 groups	40	0.06	0.03–0.10	1.75	0.50–3.75	6.97	4.00–>8.00

TABLE 25

Percentage distribution of call types for males of CW group, October 1970 through March 1972

Vocalizer	Percentage of total bouts and (range of monthly percentages)						Total no. bouts in sample
	Wheet	Chist	Bark	Rapid quaver	Wah!	Shrill squeal	
Male CW	59.3% (0-100%)	34.9% (0-50%)	3.7% (0-28.6%)	1.6% (0.50%)	0.4% (0-12.5%)		246
Male LB	6.3% (0-25.0%)	53.0% (0-100%)	24.0% (0-87.5%)	7.3% (0-50%)	9.4% (0-33.3%)		96
Male ND	20.2% (0-50%)	46.7% (0-100%)	9.2% (0-50%)	17.4% (0-100%)	6.4% (0-21%)		109
Male SA*	32.4%** (0-60%)	27.0% (0-100%)			2.7% (0-7.7%)	37.8%*** (0-84.6%)	37

*   SA was not recognizable until May 1971; therefore his percentages are based on eleven months only.

**   Calculated from August 1971, when SA first gave this call, this figure becomes 48%.

*** Occurred only in May through July 1971.

TABLE 26

Frequency data on *C. b. tephrosceles* vocalizations

	Shrill squeal	Prolonged squeal	Wheet	Chist	Bark	Yelp	Nyow	Rapid quaver	Scream	Sqwack	Shriek	Wah	Uh!	Sneeze
Total no. of peaks & (no. of bouts) where per day freq. of bouts was>16	170 (858)		169 (4,840)	198 (3,511)	106 (1,333)								135 (645)	150 (2,818)
No. of days where frequency of bouts was>16	27	0	34	34	19	0	0	0	0	0	0	0	21	29*
Mean no. peaks & (no. bouts) per day in preceding sample	6.3 (31.8)		5.0 (142.4)	5.8 (103.3)	5.6 (70.2)								6.4 (30.7)	5.2 (97.2)
Range in no. peaks & (no. bouts) per day in preceding sample	3-11 (17-70)		2-8 (21-270)	4-10 (29-176)	4-8 (17-230)								4-10 (17-64)	3-8 (43-145)
Total no. of bouts in entire 34-day sample	939	88	4,840	3,511	1,411	45	63	15	42	39	75	146	765	2,818
Mean no. bouts per day based on 34-day sample	27.6	2.6	142.4	103.3	41.5	1.3	1.9	0.4	1.2	1.1	2.2	4.3	22.5	97.2**
Range in no. bouts per day based on 34-day sample	8-70	0-10	21-270	29-176	0-230	0-9	0-10	0-2	0-5	0-5	0-13	0-11	3-64	43-145

* During the first 5 days (4–8 November 1970) sneezes were not tabulated.

** Based on 29 days.

TABLE 27

Frequency distribution of interval length (no. of 30-minute sample periods) between peaks (≥ 1.5 x mean no. bouts) of calling, for *C. b. tephrosceles*

Note: 2 peaks occurring in contiguous periods are scored only once; no interval precedes 1st peak of the day and none follows the last peak.

Interval (no. of 30 min. periods)	Shrill squeal	Wheet	Chist	Bark	Uh!	Sneeze
Contiguous	54 (37.8%)	38 (28.4%)	70 (42.4%)	40 (46.0%)	41 (36.3%)	41 (33.9%)
1	24 (16.8%)	22 (16.4%)	16 (9.7%)	15 (17.2%)	19 (16.8%)	26 (21.5%)
2	18 (12.6%)	20 (14.9%)	15 (9.1%)	4 (4.6%)	11 (9.7%)	12 (9.9%)
3	10 (7.0%)	9 (6.7%)	23 (13.9%)	8 (9.2%)	14 (12.4%)	14 (11.6%)
4	8 (5.6%)	7 (5.2%)	9 (5.5%)	4 (4.6%)	1 (0.9%)	7 (5.8%)
5	4 (2.8%)	6 (4.5%)	6 (3.6%)	2 (2.3%)	7 (6.2%)	6 (5.0%)
6	7 (4.9%)	8 (6.0%)	8 (4.8%)	1 (1.1%)	2 (1.8%)	6 (5.0%)
7	3 (2.1%)	5 (3.7%)	6 (3.6%)	3 (3.4%)	4 (3.5%)	5 (4.1%)
8	5 (3.5%)	6 (4.5%)	2 (1.2%)	2 (2.3%)	4 (3.5%)	3 (2.5%)
9	1 (0.7%)		3 (1.8%)		4 (3.5%)	1 (0.8%)

	143	134	165	87	113	121
10	3 (2.1%)	3 (2.2%)	2 (1.2%)	1 (1.1%)	3 (2.7%)	
11	1 (0.7%)	3 (2.2%)	1 (0.6%)	1 (1.1%)	2 (1.8%)	
12	1 (0.7%)	3 (2.2%)	1 (0.6%)	1 (1.1%)	1 (0.9%)	
13		1 (0.8%)	1 (0.6%)	1 (1.1%)		
14		2 (1.5%)		2 (2.3%)		
15	2 (1.4%)	1 (0.8%)	1 (0.6%)			
16	1 (0.7%)		1 (0.6%)	1 (1.1%)		
17						
18	1 (0.7%)			1 (1.1%)		
19						
20				1 (1.1%)		
Totals	143	134	165	87	113	121

TABLE 28
Measurements of *C. b. temmincki* vocalizations

Call type	N	X̄ duration (sec.)	Range of duration (sec.)	kHz			
				Mean lower frequency	Range of lower frequency	Mean upper frequency	Range of upper frequency
Chirp	69	0.12	0.04-0.27	0.87	0.08-2.25	5.65	3.25-8.00
Chirp variation I	8	0.14	0.08-0.18	0.70	0.50-1.00	4.78	3.00-6.25
Chirp variation II	2	0.16	0.15-0.17	0.88	0.75-1.00	4.00	4.00
Nyow	40	0.16	0.10-0.24	0.66	0.25-1.00	3.66	1.50-8.00

TABLE 29
Measurements of *C. b. badius* vocalizations

Call type	N	X̄ duration (sec.)	Range of duration (sec.)	kHz			
				Mean lower frequency	Range of lower frequency	Mean upper frequency	Range of upper frequency
Chirp	18	0.10	0.08-0.16	0.41	0.08-1.25	7.26	5.00-8.00
Nyow	31	0.22	0.14-0.38	0.42	0.20-0.75	3.10	1.60-5.00
Exhalation of copulation quaver	142	0.58	0.06-1.89	0.44	0.08-1.00	2.45*	1.30-6.50*
Inhalation of copulation quaver	57	0.09	0.04-0.16	0.11	0.08-0.50	1.36**	0.80-3.00**

* Only 52 units were measured for this parameter.
** Only 19 units were measured for this parameter.

TABLE 30
Measurements of *C. b. preussi* vocalizations

Call type	N	X̄ duration (sec.)	Range of duration (sec.)	kHz Mean lower frequency	Range of lower frequency	Mean upper frequency	Range of upper frequency
Bark	32	0.16	0.08-0.28	0.61	0.08-0.90	3.18	1.75-6.40
Nyow	34	0.22	0.11-0.33	0.41	0.20-0.60	2.62	1.50-5.10
Short yelp	91	0.11	0.04-0.33	0.60	0.30-1.50	2.38	1.50-4.75
Long yelp	62	0.15	0.08-0.33	0.55	0.08-0.80	2.47	1.60-6.75
Prolonged sqwack	21	0.64	0.17-1.55	0.75	0.40-1.00	6.20	3.20-8.00
Yowl	37	0.55	0.16-1.52	0.59	0.30-1.00	2.89	1.70-6.60
Squeal and high squeal	12	0.62	0.23-2.00	0.68	0.08-1.30	6.17	3.0 -13.5
Exhalation of copulation quaver	315	0.14	0.01-1.56	0.49	0.08-1.50	2.58	1.25-6.00
Shriek, scream, and screech	33	0.30	0.12-0.60	0.75	0.40-1.00	3.57	2.00-6.00

TABLE 31

Subspecific comparison of spectrographic similarities in vocalizations for red colobus monkeys

Subspecies compared	Vocalizations concerned with intragroup cohesion						Other communicative vocalizations*	
	Chirp	Nyow	Bark	Long yelp	Chist	Wheet	No. similar	No. dissimilar or absent from one or both of pair
b. temminckii & b. badius	+	+	+	X			X	X
b. temminckii & b. preussi	-	-	-	X			X	X
b. temminckii & b. tephrosceles	+**	+	-	X	0	0	X	X
b. temminckii & b. rufomitratus	X	-	X	X	0	0	X	X
b. badius & b. preussi	-	-	-	-			Zero	6
b. badius & b. tephrosceles	+**	+	-	X	0	0	Zero	4
b. badius & b. rufomitratus	X	-	X	X	0	0	X	X
b. preussi & b. tephrosceles	-	-	-	0	0	0	Zero	8
b. preussi & b. rufomitratus	X	-	X	X	0	0	X	X
b. tephrosceles & b. rufomitratus	X	+	X	X	+	+	X	X

+ = similar; - = dissimilar; 0 = absent from one; blank = absent from both; X = not adequately sampled in one or both of pair

*Includes other vocalizations of social significance, excluding screams, squeals, shrieks, screeches and those inadequately sampled. They include: b. temminckii: none; b. badius: copulation quaver; b. preussi: short yelp, yowl, prolonged sqwack, copulation quaver, misc. tonal calls; b. tephrosceles: wah!, rapid quaver, uh!; b. rufomitratus: none.

**some barks of b. tephrosceles greatly resemble some chirps of b. temminckii and of b. badius.

TABLE 32

Results of strip enumeration of all trees $\geqslant$ 10 m tall in major study area of compartment 30, Kibale Forest

Total strip length 2,873m   58 sections of $\approx$ 50m length

Total area 1.43 ha.

469 trees, 51 species

Species	Number in sample	% of total	Density (no./ha.)	No. of 50m sections occurred in	Index of dispersion (variance/mean)
Diospyros abyssinica (Hiern) F. White	94	20.0	65.7	38	2.31
*Markhamia platycalyx (Bak.) Sprague	83	17.7	58.0	42	1.23
*Celtis durandii Engl.	49	10.4	34.3	33	0.91
Uvariopsis congensis Rabyns and Chesquiere	36	7.7	25.2	16	2.90
*Teclea nobilis Del.	30	6.4	21.0	17	2.39
Funtumia latifolia (Stapf) Stapf ex Schltr.	21	4.5	14.7	16	1.07
*Strombosia scheffleri Engl.	21	4.5	14.7	14	1.42
Parinari excelsa Sabine	15	3.2	10.5	10	1.84
*Chaetacme aristata Planch.	12	2.6	8.4	9	1.49
*Millettia dura Dunn	11	2.4	7.6	10	1.01
Pancovia turbinata Radlk.	9	1.9	6.3	6	
Dombeya mukole Sprague	6	1.3	4.2	5	
Cassipourea ruwenzorensis (Engl.) Alston	5	1.1	3.5	5	
Neoboutonia macrocalyx Pax	5	1.1	3.5	3	
*Bosqueia phoberos Baill.	4	0.9	2.8	4	0.95

TABLE 32—*Continued*

Species	Number in sample	% of total	Density (no./ha.)	No. of 50m sections occurred in	Index of dispersion (variance/mean)
Linociera johnsonii Baker	4	0.9	2.8	4	
Lovoa swynnertonii Bak. f.	4	0.9	2.8	3	
Olea welwitschii (Knobl.) Gilg. and Sch.	4	0.9	2.8	4	
Aphania senegalensis (Juss. ex Bernh.) Radlk.	3	0.6	2.1	2	
*Celtis africana Burm. f.	3	0.6	2.1	3	0.96
Chrysophyllum gorungosanum Engl.	3	0.6	2.1	3	
Fagaropsis angolensis (Engl.) Dale	3	0.6	2.1	3	
Leptonychia mildbraedii Engl.	3	0.6	2.1	2	
Mimusops bagshawei S. Moore	3	0.6	2.1	3	
Strychnos mitis S. Moore	3	0.6	2.1	3	
Balanites wilsoniana Dawe and Sprague	2	0.4	1.4		
Cordia millenii Bak.	2	0.4	1.4		
Ficus dawei Hutch.	2	0.4	1.4		
Monodora myristica (Gaertn.) Dunal	2	0.4	1.4		
*Newtonia buchanani (Baker) Gilb. and Bout.	2	0.4	1.4		0.98
Premna angolensis Guerke	2	0.4	1.4		
Pseudospondias microcarpa (A. Rich.) Engl.	2	0.4	1.4		
Rubiaceae sp. I	2	0.4	1.4		

Species			
Spathodea nilotica Seem.	2	0.4	1.4	
*Aningeria altissima (A. Chev.) Aubr. and Pellegr.	1	0.2	0.7	0.98
Alstonia boonei De Wild.	1	0.2	0.7	
Ehretia cymosa Thonn.	1	0.2	0.7	
Fagara angolensis Engl.	1	0.2	0.7	
Randia urcelliformis Hiern	1	0.2	0.7	
Rinorea cf. poggei Engl.	1	0.2	0.7	
Ritchiea albersii Gilg.	1	0.2	0.7	
Symphonia globulifera L.	1	0.2	0.7	
Trema guineensis Ficalho	1	0.2	0.7	
Turraea ?	1	0.2	0.7	
Xymalos monospora (Harv.) Baill.	1	0.2	0.7	
Rubiaceae sp. II	1	0.2	0.7	
Rubiaceae sp. III	1	0.2	0.7	
Sapindaceae sp. V	1	0.2	0.7	
Ficus sp. VI	1	0.2	0.7	
Sp. I	1	0.2	0.7	
Sp. IV	1	0.2	0.7	

Names and authorities taken from: Eggeling (1956), A. Hamilton (pers. comm.), Willis (1966), Wing and Buss (1970).

*Top 10 food species during April 1971 through March 1972.

TABLE 33

Results of quadrat enumeration of all trees ≥ 10 m tall in major study area of compartment 30, Kibale Forest

Each quadrat is 0.25 ha., total area 1.25 ha., 384 trees, 41 species

Species	Quadrat* and density (no./ha.)				
	(+2 -8)	(-2 -13)	(-5 -13)	½ of {(-1 -12) / (-2 -12)}	½ of {(-12 -15) / (-13 -15)}
Diospyros a.		8	268	16	52
Markhamia p.	4	16	48	8	52
Celtis durandii	20	40	60	12	68
Uvariopsis c.			76	8	4
Teclea n.	12	48	20	92	4
Funtumia l.	24			8	16
Strombosia s.	44	24			
Parinari e.					8
Chaetacme a.		12	20	20	
Millettia d.		4		8	16
Pancovia t.	12	24	4	24	
Dombeya m.		4	16	20	
Cassipourea r.		4	16		4
Neoboutonia m.		4			
Bosqueia p.	12	24		8	4

Species					
Olea w.		8	4		
Aphania s.	4				4
Celtis africana	8	8	8		
Chrysophyllum g.		4			8
Fagaropsis a.			4		
Mimusops b.		24		8	4
Strychnos m.					
Cordia m.			4		4
Monodora m.	4				
Newtonia b.			4	4	
Premna a.	4				4
Spathodea n.	4	8			
Alstonia b.	4				
Fagara a.					4
Symphonia g.					8
Trema g.		4			20
Ficus brachylepis Welw. ex Hiern					4

TABLE 33—*Continued*

Species	Quadrat* and density (no./ha.)				
	(+2 -8)	(-2 -13)	(-5 -13)	½ of(-1 -12) / ½ of(-2 -12)	½ of(-12 -15) / ½ of(-13 -15)
Ficus exasperata Vahl.		4			
Vangueria apiculata K. Schum.			4		
Dracaena steudneri Schweinf. ex Engl.		4			
Trichilia splendida A. Chev.		4			
Albizia grandibracteata Taub.				4	
Blighia unijugata Bak.				4	
Sp. I	4				
Sp. II	4				
Sp. III				4	

* quadrat designations refer to fig. 12.

TABLE 34

Percentages for each specific food item eaten by *C. b. tephrosceles,* September 1970 through March 1971, compartment 30, Kibale Forest (N = 1,164 feeding observations)

Key to categories (with total percentages)

a = fruits of unknown age (1.38)

b = ripe fruits, usually with some color other than green (0.94)

c = unripe fruits (2.41)

d = seeds (2.24)

e = flowers (6.73)

f = floral buds (7.58)

g = leaf buds (8.34)

h = young leaves:

    h? of unspecified size (14.06)
    h2 large (8.85)
    h3 medium-sized (2.10)
    h4 small (8.68)
    h5 very small (2.19)

i = petioles of young leaves:

    i? of unspecified size (1.90)
    i2 large (0.09)
    i3 medium-sized (0.52)
    i4 small (0.26)

j = mature leaves (2.04)

k = petioles of mature leaves and leaflets (9.20)

m = pieces of mature leaves (0.93)

n = apical tip of mature leaves (1.39)

o = basal part of mature leaves (0.09)

p = leaves of unknown age (13.98)

q = petioles of leaves of unknown age (3.75)

r = young stem tips (0.09)

s = small twigs, sometimes without bark, i.e., quite young (0.54)

t = galls (0.08)

u = unknown items (0.27)

Food species (and weighted $\bar{X}$ %)	Food item (and weighted $\bar{X}$ %)
Albizia gummifera (Gmel.) C.A. Sm. (0.88)	f(0.34) g(0.27) h5(0.10) k(0.09) q(0.08)
Aningeria altissima (9.92)	e(1.91) f(2.67) g(1.28) h?(0.94) h3(0.34) h4(1.99) h5(0.17) i3(0.26) k(0.19) q(0.17)
Balanites w. (0.35)	p(0.09) s(0.26)
Bosqueia (9.45)	h?(3.35) h2(5.32) h3(0.09) h4(0.09) p(0.60)
Celtis africana (12.43)	a(0.68) e(1.89) f(2.48) g(4.28) h?(1.04) h4(0.68) h5(1.38)
Celtis durandii (8.98)	a(0.35) b(0.17) c(2.33) e(1.37) f(1.13) g(1.38) h?(1.20) h3(0.17) h4(0.88)
Chaetacme (0.89)	h?(0.26) h2(0.44) p(0.10) s(0.09)
Chrysophyllum gorungosanum (0.63)	h?(0.10) j(0.18) m(0.08) n(0.27)
Cordia millenii (0.35)	h5(0.35)
Diospyros (0.18)	k(0.08) m(0.10)
Dombeya mukole (1.21)	h4(0.17) j(0.34) k(0.10) m(0.08) p(0.44) q(0.08)
Ehretia cymosa (0.85)	h2(0.17) i?(0.17) k(0.17) p(0.27)
Elaeodendron buchananii Oliv. (0.09)	k(0.09)

TABLE 34—*Continued*

Food species (and weighted X̄ %)	Food item (and weighted X̄ %)
Fagaropsis a. (0.27)	k(0.10) q(0.17)
Fagara a. (1.57)	h?(0.26) h2(0.18) h3(0.35) h4(0.26) i?(0.17) i3(0.09) i4(0.26)
Ficus spp. (Strangler fig ≥2 spp.) (includes Ficus natalensis Hochst.) (1.23)	a(0.09) h?(0.18) h2(0.35) h3(0.26) h4(0.17) p(0.18)
Ficus sp. (0.18)	h?(0.08) j(0.10)
Funtumia l. (2.04)	a(0.08) d(1.71) h2(0.08) j(0.08) p(0.09)
Lovoa s. (3.71)	h?(0.27) h2(0.09) h3(0.09) h4(2.24) i?(0.26) k(0.17) p(0.17) q(0.42)
Markhamia (8.47)	e(0.45) f(0.27) h?(0.26) h4(0.26) i?(0.61) k(5.67) q(0.95)
Millettia d. (3.69)	g(0.78) h?(1.55) h2(0.19) h3(0.28) h4(0.79) h5(0.10)
Mimusops b. (2.00)	b(0.09) g(0.09) h?(0.17) j(0.35) n(1.12) q(0.08) s(0.10)
Myrianthus arboreus Beav. (0.17)	h2(0.17)
Newtonia b. (11.54)	g(0.17) h?(0.78) h2(0.09) p(10.50)
Olea w. (0.25)	k(0.17) m(0.08)
Pancovia t. (0.26)	h2(0.26)
Parinari (2.60)	h?(1.82) h2(0.78)
Premna a. (2.01)	h4(0.10) k(0.54) q(1.37)
Pseudospondias (1.64)	e(0.68) f(0.51) h2(0.09) h3(0.09) h4(0.27)
Spathodea (0.43)	k(0.35) q(0.08)
Strombosia (3.40)	d(0.53) e(0.43) f(0.18) h?(0.17) h2(0.10) j(0.10) k(1.04) m(0.59) o(0.09) p(0.08) s(0.09)
Strychnos mitis (0.10)	h?(0.10)
Symphonia (0.17)	h3(0.08) r(0.09)
Teclea n.(1.68)	a(0.08) b(0.69) c(0.08) h?(0.44) h2(0.19) j(0.10)
Trema (1.13)	h3(0.09) h4(0.09) j(0.17) p(0.78)
Trichlia s. (1.12)	h?(0.08) h2(0.09) h4(0.09) i?(0.60) i2(0.09) i3(0.17)
Urera cameroonensis Wedd. (0.54)	h2(0.09) h3(0.09) h4(0.09) i?(0.09) k(0.09) p(0.09)
Epiphyte sp. (0.18)	t(0.08) u(0.10)

TABLE 34—*Continued*

Food species (and weighted $\overline{X}$ %)	Food item (and weighted $\overline{X}$ %)
Climber spp. (≥ 4 spp.) (0.09)	j(0.09)
Lichen sp. (0.17)	u(0.17)
Liana spp. (≥ 3 spp.) (1.29)	h?(0.51) h4(0.08) j(0.09) k(0.09) p(0.34) q(0.18)
Shrub sp. (0.09)	h4(0.09)
Tree sp. (1.12)	g(0.09) h?(0.08) h2(0.08) h4(0.17) h5(0.09) j(0.27) k(0.26) p(0.08)
Specimen #91 (0.17)	j(0.17)
≥ 2 spp. (1.19)	h?(0.42) h2(0.09) h3(0.17) h4(0.17) p(0.17) q(0.17)

TABLE 35

Percentages for each specific food item eaten by *C. b. tephrosceles*, April 1971 through March 1972, compartment 30, Kibale Forest (N = 2,898 feeding observations)

Key to categories (and total percentages)

a = fruits of unknown age (1.24)

b = ripe fruits, usually with some color other than green (1.38)

c = unripe fruits (2.22)

d = seeds (0.82)

e = flowers (2.98)

f = floral buds (8.88)

g = leaf buds (11.18)

h = young leaves:

   h? of unspecified size (0.80)
   h2 large (13.91)
   h3 medium-sized (6.45)
   h4 small (8.12)
   h5 very small (3.50)

i = petioles of young leaves:

   i? of unspecified size (0.03)
   i2 large (0.83)
   i3 medium-sized (2.14)
   i4 small (3.59)

j = mature leaves (2.32)

k = petioles of mature leaves and leaflets (18.53)

m = pieces of mature leaves (1.41)

n = apical tip of mature leaves (0.61)

o = basal part of mature leaves (0.21)

p = leaves of unknown age (10.22)

q = petioles of leaves of unknown age (0.21)

r = young stem tips (0.18)

s = small twigs, sometimes without bark, i.e., quite young (0.49)

t = galls

x = petioles of dead and dry leaves (0.03)

y = dead and dry leaves (0.04)

z = young leaves with galls or other insect damage (0.48)

u = unknown items

Food species (and weighted $\bar{X}$ %)	Food item (and weighted $\bar{X}$ %)
Albizia grandibracteata (0.38)	d(0.03) j(0.04) k(0.31)
Albizia gummifera (0.24)	e(0.04) h5(0.07) k(0.13)
Aningeria altissima (5.20)	e(0.52) f(1.55) g(0.72) h3(0.11) h4(0.45) h5(0.21) i2(0.07) i3(0.03) i4(0.03) k(1.51)
Aphania s. (1.40)	h4(0.03) i4(0.04) k(0.03) m(0.04)
Balanites w. (3.17)	h3(0.04) h4(0.03) h5(0.10) i?(0.03) j(0.11) k(2.21) n(0.03) o(0.04) s(0.13) z(0.45)
Bosqueia (4.06)	a(0.21) f(0.07) g(0.04) h2(3.24) h3(0.18) k(0.32)
Cassipourea r. (0.57)	h2(0.21) h3(0.10) h4(0.03) j(0.03) p(0.07) s(0.13)
Cardiospermum grandiflorum Sw. (0.23)	h?(0.04) h3(0.03) j(0.03) k(0.10) p(0.03)
Celtis africana (15.36)	e(0.45) f(1.93) g(8.39) h2(0.07) h3(0.10) h4(3.34) h5(1.04) j(0.04)
Celtis durandii (10.36)	a(0.90) b(0.35) c(1.08) e(1.03) f(3.40) g(1.04) h2(0.24) h3(0.41) h4(0.37) h5(1.31) k(0.16) x(0.03) y(0.04)
Chaetacme (4.45)	h2(3.34) h3(0.42) h4(0.22) k(0.41) p(0.03) s(0.03)
Chrysophyllum gorungosanum (0.67)	h2(0.07) j(0.11) m(0.46) n(0.03)

TABLE 35—*Continued*

Food species (and weighted X̄ %)	Food item (and weighted X̄ %)
Cleistanthus polystachyus Hook.f. (0.07)	c(0.07)
Cordia millenii (0.10)	h2(0.07) k(0.03)
Diospyros (0.14)	k(0.10) s(0.04)
Dombeya mukole (1.42)	g(0.04) h2(0.44) h3(0.56) h4(0.07) h5(0.04) j(0.07) k(0.04) p(0.13) z(0.03)
Ehretia cymosa (0.18)	h2(0.11) i2(0.04) p(0.03)
Elaeodendron (1.37)	c(0.04) g(0.04) h2(0.72) h3(0.14) h4(0.24) i2(0.04) j(0.04) n(0.04) q(0.07)
Fagaropsis a. (2.17)	h2(0.18) k(1.82) m(0.03) n(0.14)
Fagara a. (0.18)	h2(0.04) h3(0.07) h4(0.07)
Ficus spp. (Strangler fig ≥ 2 spp.) (includes *Ficus natalensis*) (1.07)	c(0.45) g(0.38) h2(0.03) h3(0.07) h5(0.07) i2(0.03) m(0.04)
Ficus brachylepis (1.11)	h2(0.21) h3(0.66) h4(0.24)
Ficus dawei (0.7)	h2(0.07)
Funtumia l. (1.96)	a(0.03) d(0.03) e(0.03) f(0.21) h2(0.20) j(0.65) k(0.18) m(0.11) o(0.14) p(0.34) t(0.04)
Lovoa s. (1.35)	h2(0.07) h3(0.17) h4(0.55) i2(0.07) i3(0.07) k(0.21) m(0.07) n(0.07) p(0.04)
Markhamia (15.11)	f(1.14) g(0.24) h2(0.03) h3(0.04) h4(0.73) h5(0.31) i2(0.21) i3(1.24) i4(3.03) k(8.00) m(0.14)
Millettia d. (4.73)	c(0.03) e(0.21) f(0.06) g(0.06) h2(0.96) h3(2.14) h4(1.06) h5(0.21)
Mimusops b. (2.50)	a(0.03) b(0.24) d(0.04) h2(0.03) j(0.55) k(1.08) m(0.20) n(0.23) s(0.10)
Myrianthus a. (0.33)	h2(0.07) h3(0.07) h4(0.04) i2(0.04) i3(0.07) i4(0.04)
Newtonia b. (9.85)	e(0.13) g(0.13) h?(0.69) p(8.83) s(0.04) t(0.03)
Olea w. (0.08)	i2(0.04) k(0.04)
Pancovia t. (0.12)	h3(0.06) h4(0.06)
Parinari (1.09)	e(0.03) f(0.03) g(0.10) h2(0.45) h3(0.03) h4(0.03) h5(0.10) k(0.17) n(0.04) s(0.11)
Premna a. (0.44)	h2(0.07) h3(0.03) h4(0.03) i2(0.14) i3(0.03) i4(0.04) k(0.07) q(0.03)
Pseudospondias (0.10)	e(0.03) h4(0.07)
Rhipsalis baccifera (J.S. Mill.)Stearn(0.11)	r(0.11)
Spathodea (0.07)	k(0.07)

TABLE 35—*Continued*

Food species (and weighted $\overline{X}$ %)	Food item (and weighted $\overline{X}$ %)
Strombosia (4.59)	d(0.72) f(0.07) h2(0.55) h3(0.49) h4(0.24) i2(0.15) i3(0.66) i4(0.28) j(0.07) k(1.04) m(0.18) p(0.14)
Symphonia (0.08)	h3(0.04) i3(0.04)
Teclea n. (3.78)	a(0.07) b(0.79) c(0.55) e(0.41) f(0.28) h2(1.34) h3(0.14) h4(0.07) i4(0.10) k(0.03)
Trema (0.69)	h2(0.45) h4(0.04) p(0.20)
Trichilia s. (0.72)	e(0.03) f(0.03) i4(0.03) j(0.10) k(0.28) m(0.11) n(0.03) o(0.03) s(0.04) u(0.04)
Urera c. (1.00)	e(0.07) f(0.11) h2(0.10) h3(0.21) h4(0.04) j(0.30) p(0.17)
Xymalos (0.10)	h2(0.10)
Epiphyte sp. (0.11)	r(0.07) u(0.04)
Climbers (≥ 4 spp.) (0.65)	h2(0.28) h3(0.04) h4(0.04) j(0.11) k(0.10) p(0.04) q(0.04)
Lichen sp. (0.17)	u(0.17)
Lianas (≥ 3 spp.) (0.49)	h?(0.04) h2(0.10) h4(0.03) h5(0.04) j(0.07) k(0.03) p(0.11) q(0.07)
Tree sp. (0.32)	h?(0.03) h2(0.07) h3(0.10) k(0.06) m(0.03) p(0.03)
Moss or lichen (0.04)	u(0.04)
pupae (0.03)	u(0.03)
Pamela (?tinctorum) a lichen (0.07)	u(0.07)

TABLE 37

Summary sheet for a selection of species eaten by *C. b. tephrosceles*, September 1970 through March 1972, compartment 30, Kibale Forest

	Celtis africana	Celtis durandii	Newtonia	Teclea	Bosqueia	Markhamia	Chaetacme	Aningeria	Millettia	Lovoa	Faga-ropsis	Strom-bosia
$\overline{X}$ (%) of observa-tions per month	9.6	8.7	10.5	3.4	6.8	13.2	3.4	7.1	4.2	2.3	2.0	4.4
Standard deviation	±11.71	±7.86	±4.07	±3.66	±7.71	±10.33	±4.50	±7.40	±5.10	±3.14	±2.90	±3.20
Coefficient of varia-tion	122.0%	90.3%	38.8%	107.6%	113.4%	78.3%	132.4%	104.2%	121.4%	136.5%	145.0%	72.7%
Monthly range in % of obser-vations	0-38.1	1.1-26.5	5.1-21.2	0-15.4	0-31.7	1.9-43.4	0-12.0	0-25.4	0-18.9	0-11.0	0-12.3	0.4-11.4

Note. Table 36 appears on the following pages.

TABLE 36

Rank order of 10 most common food species (regardless of item) eaten by *C. b. tephrosceles* and percentage of total feeding observations, September 1970 through March 1972, compartment 30, Kibale Forest

Month & N	I	II	III	IV	V	VI	VII	VIII	IX	X
Sept. '70 136	Celtis durandii 14.0%	Newtonia 12.5%	Celtis africana 11.7%	Parinari 11.0%	Mimusops 10.3%	Teclea n. 5.9%	Premna angolensis 5.9%	Millettia d. 4.4%	Aningeria 4.4%	Markhamia 4.4%
Oct. '70 123	Celtis durandii 22.8%	Newtonia 16.3%	Celtis africana 15.4%	Markhamia 13.8%	Bosqueia 9.8%	Liana spp. ≥2 4.1%	Parinari 4.1%	Teclea 2.4%	Strombosia 1.6%	Lovoa 1.6%; Premna 1.6%; spp.? tree ≥2 1.6%
Nov. '70 283	Celtis africana 15.9%	Bosqueia 13.1%	Millettia d. 9.2%	Aningeria 8.5%	Celtis durandii 7.8%	Newtonia 7.1%	Markhamia 4.2%	Lovoa 3.2%	Strombosia 2.8%; Premna 2.8%; Spp. ≥2 2.8%	Dombeya 2.8%; Albizia gummifera 2.8%
Dec. '70 104	Bosqueia 31.7%	Aningeria 20.2%	Newtonia 10.0%	Strombosia 4.8%	2 spp. 4.8%	Millettia 4.8%	Trema 3.9%	Funtumia 3.9%	Celtis d. 2.9%	Mimusops 2.9%
Jan. '71 189	Aningeria 25.4%	Celtis africana 22.1%	Newtonia 12.1%	Markhamia 8.4%	Strombosia 7.4%	Funtumia 4.7%	Lovoa 3.2%	Celtis durandii 2.1%	Balanites w. 1.6%	Chrysophyllum g. 1.6%; Olea w. 1.6%
Feb. '71 180	Newtonia 14.4%	Celtis africana 12.8%	Celtis durandii 12.8%	Pseudo-spondias 8.9%	Markhamia 8.3%	Aningeria 7.2%	Lovoa 5.0%	Funtumia 4.4%	Fagara angolensis 3.3%	Bosqueia phoberos 3.3%
Mar. '71 149	Markhamia 20.1%	Bosqueia 12.8%	Newtonia 12.1%	Lovoa 9.4%	Fagara a. 6.0%	Ficus natalensis 4.7%	Trichilia splendida 4.7%	Strombosia s. 4.0%	Ehretia c. 4.0%	Chaetacme a. 4.0%
Apr. '71 468	Celtis africana 32.7%	Celtis durandii 21.2%	Markhamia 18.2%	Newtonia 6.6%	Millettia d. 6.2%	Aningeria 4.3%	Balanites w. 2.6%	Funtumia 1.9%	Parinari 1.5%	Pancovia t. 0.9%
May '71 214	Markhamia 26.6%	Celtis durandii 13.6%	Millettia d. 7.9%	Newtonia 5.1%	Bosqueia p. 5.1%	Mimusops 5.1%	Chaetacme a. 4.7%	Funtumia l. 4.7%	Dombeya m. 4.2%	Balanites w. 4.2%

Month	n	1	2	3	4	5	6	7	8	9	10
June '71	166	Markhamia 43.4%	Newtonia 7.8%	Millettia d. 7.2%	Strombosia 6.0%	Celtis durandii 4.8%	Balanites w. 4.2% / Elaeodendron 4.2%	Urera camerunensis 3.0%	Teclea n. 2.4%	Lovoa 1.8% / Trichilia s. 1.8% / Premna 1.8% / Aningeria 1.8%	
July '71	130	Markhamia 20.0%	Teclea n. 15.4%	Bosqueia 7.7%	Strombosia 6.9%	Newtonia 6.2% / Balanites w. 6.2%	Trichilia s. 5.4% / Albizia grandibracteata 5.4% / Trema 5.4%	Elaeodendron 3.9%			
Aug. '71	220	Markhamia 18.2%	Aningeria 17.7%	Strombosia 11.4% / Strangler fig sp. 1 11.4%	Newtonia 6.8%	Celtis africana 5.9%	Teclea n. 5.5% / Urea camerunensis 5.5%	Elaeodendron 3.2% / Myrianthus 3.2%			
Sept. '71	351	Celtis durandii 26.5%	Celtis africana 18.5%	Teclea 7.1%	Ficus brachylepis 6.0%	Chaetacme 4.8%	Strombosia 4.6% / Markhamia 4.6%	Funtumia 4.0%	Aningeria 3.1%		
Oct. '71	212	Millettia 18.9%	Markhamia 11.3%	Chaetacme 10.8%	Newtonia 9.4%	Celtis durandii 8.0%	Bosqueia 7.1%	Celtis africana 6.1%	Dombeya 4.7%	Mimusops 4.2%	Fagaropsis 2.4% / Ficus brachylepis 2.4% / Aningeria 2.4% / Lovoa 2.4%
Nov. '71	200	Aningeria 15.0%	Millettia d. 12.0% / Chaetacme 12.0%	Newtonia 8.0%	Markhamia 7.5%	Mimusops 6.0%	Balanites w. 4.5%	Strombosia 4.0%	Teclea n. 3.5% / Fagaropsis 3.5%		
Dec. '71	230	Newtonia 11.7%	Chaetacme 11.3%	Celtis durandii 10.4%	Bosqueia 10.0%	Mimusops 8.7%	Markhamia 8.3% / Strombosia 8.3%	Teclea n. 6.1%	Aningeria 3.0%		
Jan. '72	183	Bosqueia 13.7%	Chaetacme 11.5% / Aningeria 11.5%	Strombosia 8.7% / Newtonia 8.7%	Balanites w. 8.2%	Teclea n. 4.9%	Fagaropsis 3.8%	Markhamia 3.3% / Trema 3.3%			
Feb. '72	297	Celtis africana 38.1%	Markhamia 6.1%	Strombosia 5.4%	Fagaropsis 4.0%	Celtis durandii 3.3%	Parinari 3.0% / Millettia d. 3.0%	Mimusops 2.4%	Teclea n. 2.0% / Elaeodendron 2.0%		
Mar. '72	227	Markhamia 21.6%	Newtonia 14.5%	Fagaropsis 12.3%	Lovoa 11.0%	Bosqueia 7.9%	Balanites w. 6.6%	Aningeria 4.4%	Dombeya 3.5% / Celtis durandii 3.5%	Elaeodendron 2.6%	

TABLE 38

Rank order of first 5 specific food items (regardless of species) eaten by *C. b. tephrosceles* (mainly CW group) and percentage of total feeding observations, September 1970 through March 1972, compartment 30, Kibale Forest

Month & N	I	II	III	IV	V
Sept. '70 136	Parinari young leaves 11.0%	Newtonia leaves of unknown age 9.6%	Mimusops apical tips of mature leaves 8.8%	Celtis africana leaf buds ---- 5.9% Premna petioles of leaves of unknown age ---- 5.9%	
Oct. '70 123	Newtonia leaves of unknown age 16.3%	Bosqueia young leaves 9.8%	Celtis durandii young leaves 7.3%	Celtis africana leaf buds ---- 5.7% Celtis durandii green fruits ---- 5.7% Celtis africana flower buds ---- 5.7%	
Nov. '70 283	Bosqueia young leaves 9.2%	Celtis africana leaf buds ---- 6.0% Newtonia leaves of unknown age ---- 6.0%		Celtis durandii unripe fruits 4.6%	Bosqueia large young leaves ---- 3.9% Celtis africana very small young leaves 3.9% Millettia young leaves of unspecified age 3.9%
Dec. '70 104	Bosqueia large young leaves 27.9%	Aningeria small young leaves 9.6%	Newtonia leaves of unknown age 5.8%	Millettia small young leaves ---- 3.8% Trema large leaves ---- 3.8% Funtumia seeds ---- 3.8% Newtonia young leaves ---- 3.8%	
Jan. '71 189	Newtonia leaves of unknown age 12.2%	Aningeria flower buds 11.1%	Aningeria flowers 10.6%	Markhamia mature leaf and leaflet petioles ---- 6.3% Celtis africana flower buds ---- 6.3% Celtis africana leaf buds ---- 6.3%	
Feb. '71 180	Newtonia leaves of unknown age 13.9%	Markhamia mature leaf and leaflet petioles 8.3%	Pseudospondias flowers 4.4%	Funtumia seeds ---- 3.9% Celtis durandii flowers ---- 3.9% Celtis durandii flower buds ---- 3.9% Aningeria leaf buds ---- 3.9%	
Mar. '71 149	Markhamia mature leaf and leaflet petioles 19.5%	Newtonia leaves of unknown age 12.1%	Bosqueia large young leaves 10.7%	Lovoa small young leaves 8.1%	Trichilia splendida basal petioles of young leaves 4.7%
Apr. '71 468	Markhamia mature leaf and leaflet petioles 17.5%	Celtis africana leaf buds 16.5%	Celtis durandii flower buds 9.0%	Celtis africana flower buds 8.8%	Celtis durandii flowers 6.0%
May '71 214	Markhamia mature leaf and leaflet petioles 16.8%	Celtis durandii unripe fruits 6.1%	Newtonia leaves of unknown age ---- 5.1% Bosqueia large young leaves ---- 5.1%		Celtis durandii fruits of unknown age ---- 4.7% Millettia medium-sized young leaves ---- 4.7%

Date					
June '71 166	Markhamia mature leaf and leaflet petioles 19.3%	Markhamia petioles of small young leaves 12.0%	Markhamia petioles of medium-sized young leaves 7.8%	Newtonia leaves of unknown age 7.2%	Elaeodendron large young leaves ----- 4.2% Millettia medium-sized young leaves ----- 4.2%
July '71 130	Teclea flowers 9.2%	Markhamia petioles of mature leaves and leaflets 7.7%	Markhamia flower buds ------- Newtonia leaves of unknown age ------- Teclea flower buds ------- Balanites petioles of mature leaves -------	6.2% 6.2% 6.2% 6.2%	
Aug. '71 220	Markhamia flower buds 10.0%	Aningeria petioles of mature leaves 7.7%	Newtonia leaves of unknown age 6.8%	Strangler fig. sp. unripe fruits 5.9%	Teclea small unripe fruits ------- 5.5% Strombosia petioles of mature leaves ------- 5.5% Celtis africana leaf buds ------- 5.5%
Sept. '71 351	Celtis africana leaf buds 16.5%	Celtis durandii flower buds 16.2%	Newtonia leaves of unknown age 9.1%	Celtis durandii very small young leaves 7.1%	Teclea yellow fruits 5.7%
Oct. '71 212	Millettia medium-sized young leaves 9.9%	Chaetacme large young leaves 9.4%	Newtonia leaves of unknown age ------ 8.0% Markhamia petioles of small young leaves ------ 8.0%		Bosqueia large young leaves ------- 4.7% Millettia small young leaves ------- 4.7%
Nov. '71 200	Aningeria flower buds 10.5%	Chaetacme large young leaves 10.0%	Newtonia leaves of unknown age 8.0%	Markhamia petioles of small young leaves ----- 5.0% Millettia medium-sized young leaves ----- 5.0%	
Dec. '71 230	Newtonia leaves of unknown age 11.7%	Bosqueia large young leaves 10.0%	Chaetacme large young leaves 9.6%	Mimusops entire mature leaves 6.1%	Teclea large young leaves 5.7%
Jan. '72 183	Bosqueia large young leaves 13.7%	Chaetacme large young leaves 10.9%	Newtonia leaves of unknown age 8.7%	Balanites petioles of mature leaves 6.0%	Aningeria flowers 4.9% Aningeria flower buds ------- 4.9%
Feb. '72 297	Celtis africana leaf buds 30.6%	Newtonia leaves of unknown age 20.9%	Markhamia mature leaf and leaflet petioles 5.7%	Celtis africana flower buds ------- 4.0% Fagaropsis petioles of mature leaves ------- 4.0%	
Mar. '72 227	Markhamia mature leaf and leaflet petioles 15.9%	Fagaropsis petioles of mature leaves 12.3%	Newtonia young leaves 8.8%	Bosqueia large young leaves 7.9%	Newtonia leaves of unknown age 5.7%

TABLE 39

Rank order of 10 most common food items eaten by *C. b. tephrosceles* (mainly CW group) and percentage of total feeding observations, September 1970 through March 1972, compartment 30, Kibale Forest

Month & N	I	II	III	IV	V	VI	VII	VIII	IX	X
Sept. '70 136	Young leaves of unspec. age 36.0%	Petioles of leaves of unknown age 15.4%	Leaves of unknown age 12.5%	Leaf buds 8.8%	Apical tips of mature leaves 8.8%	Flowers 6.6%	Ripe fruits 4.4%	Fruits of unknown age 2.2%	Unripe fruits 2.2%	Floral buds 1.5%
Oct. '70 123	Young leaves of unspec. age 33.3%	Leaves of unknown age 19.5%	Leaf buds 10.6%	Floral buds 9.8%	Petioles of leaves of unknown age 7.3%	Unripe fruits 5.7%	Flowers 4.9%	Petioles of young leaves of unspec. age 4.1%	Apical tips of mature leaves 2.4%	Ripe fruits 2.4%
Nov. '70 283	Young leaves of unspec. age 23.7%	Leaf buds 13.8%	Leaves of unknown age 9.2%	Unripe fruits 7.4%	Large young leaves 6.7%	Small young leaves 6.7%	Mature leaf & leaflet petioles 5.7%	Very small young leaves 4.6%	Mature leaves 4.6%	Petioles of leaves of unknown age 4.2% / Floral buds 4.2%
Dec. '70 104	Large young leaves 31.7%	Small young leaves 19.2%	Leaves of unknown age 12.5%	Medium-sized young leaves 5.8%	Mature leaf & leaflet petioles 4.8%	Seeds 3.8%	Floral buds 3.8%	Leaf buds 3.8%	Young leaves of unspec. age 3.8%	Petioles of medium-sized young leaves 2.9%
Jan. '71 189	Floral buds 18.5%	Flowers 18.0%	Small young leaves 15.9%	Mature leaf & leaflet petioles 14.3%	Leaves of unknown age 12.2%	Leaf buds 6.3%	Seeds 3.7%	Mature leaves 2.6%	Large young leaves 2.1%	Medium-sized young leaves 1.6% / Pieces of mature leaves 1.6% / Small twigs 1.6%
Feb. '71 180	Leaves of unknown age 17.8%	Floral buds 12.8%	Flowers 12.2%	Mature leaf & leaflet petioles 11.7%	Leaf buds 9.4%	Large young leaves 7.2%	Small young leaves 7.2%	Very small young leaves 4.4%	Seeds 3.9%	Unripe fruits 2.8%
Mar. '71 149	Mature leaf & leaflet petioles 25.5%	Large young leaves 22.1%	Leaves of unknown age 14.8%	Small young leaves 12.1%	Petioles of young leaves of unspec. age 7.4%	Medium-sized young leaves 6.0%	Seeds 3.4%	Fruits of unknown age 2.7%	Very small young leaves 2.7%	Small twigs 1.3%

Apr. '71 468
- Mature leaf & leaflet petioles 22.4%
- Floral buds 20.9%
- Leaf buds 20.3%
- Flowers 12.4%
- Small young leaves 7.7%
- Very small young leaves ---- 6.2% / Leaves of unknown age ---- 6.2%
- Medium-sized young leaves 2.1%
- Large young leaves 0.9%
- Petioles of small young leaves 0.4%

May '71 214
- Mature leaf & leaflet petioles 26.2%
- Large young leaves 22.4%
- Medium-sized young leaves 8.4%
- Unripe fruits 6.1%
- Leaves of unknown age 5.6%
- Fruits of unknown age ---- 4.7% / Petioles of small young leaves ---- 4.7%
- Petioles of medium-sized young leaves 4.2%
- Seeds 3.7%
- Small young leaves ---- 2.8% / Pieces of mature leaves ---- 2.8%

June '71 166
- Mature leaf & leaflet petioles 27.1%
- Petioles of small young leaves -- 12.7% / Leaves of unknown age --- 12.7%
- Large young leaves 12.0%
- Petioles of medium-sized young leaves 8.4%
- Unripe fruits 5.4%
- Medium-sized young leaves ---- 4.8% / Small young leaves ---- 4.8%
- Seeds 3.6%
- Pieces of mature leaves 2.4%

July '71 130
- Mature leaf & leaflet petioles 30.0%
- Floral buds 13.1%
- Flowers 10.8%
- Large young leaves 9.2%
- Leaves of unknown age 7.7%
- Seeds ---- 3.9% / Mature leaves ---- 3.9% / Petioles of small young leaves 3.9%
- Petioles of large young leaves ---- 3.1% / Moss or lichen ---- 3.1%

Aug. '71 220
- Mature leaf & leaflet petioles 19.1%
- Leaf buds 17.3%
- Floral buds 15.9%
- Ripe fruits 11.4%
- Leaves of unknown age 6.8%
- Medium-sized young leaves 5.0%
- Small young leaves 4.5%
- Mature leaves 4.1%
- Large young leaves ---- 3.6% / Very small young leaves 3.6%

Sept. '71 351
- Leaf buds 19.7%
- Floral buds 16.5%
- Mature leaf & leaflet petioles 14.2%
- Very small young leaves 10.0%
- Medium-sized young leaves --- 9.1% / Leaves of unknown age ---- 9.1%
- Ripe fruits 6.0%
- Small young leaves 4.6%
- Fruits of unknown age ---- 1.7% / Unripe fruits ---- 1.7% / Petioles of small young leaves ---- 1.7% / Mature leaves 1.7%

Oct. '71 212
- Medium-sized young leaves 23.6%
- Large young leaves 21.7%
- Mature leaf & leaflet petioles 12.3%
- Petioles of small young leaves 10.4%
- Leaves of unknown age 9.9%
- Small young leaves 8.0%
- Very small young leaves 4.2%
- Leaf buds 3.3%
- Floral buds 2.4%
- Mature leaves 1.4%

TABLE 39—*Continued*

Month & N	I	II	III	IV	V	VI	VII	VIII	IX	X
Nov. '71 200	Large young leaves 23.5%	Mature leaf & leaflet petioles 17.5%	Floral buds 11.5%	Leaves of unknown age 11.0%	Medium-sized young leaves — 7.0%	Petioles of small young leaves — 7.0%	Small young leaves — 3.5%	Pieces of mature leaves — 3.5%	Petioles of medium-sized young leaves 3.0%	Leaf buds 2.5%
Dec. '71 230	Large young leaves 32.6%	Leaves of unknown age 13.0%	Mature leaf & leaflet petioles 9.1%	Mature leaves 7.4%	Fruits of unknown age 5.7%	Petioles of small young leaves 4.8%	Petioles of medium-sized young leaves 3.9%	Pieces of mature leaves 3.5%	Medium-sized young leaves — 3.0%	Small young leaves — 3.0%; Very small young leaves — 3.0%
Jan. '72 183	Large young leaves 39.9%	Leaves of unknown age 10.4%	Mature leaf & leaflet petioles 9.8%	Flowers 5.5%	Floral buds 4.9%	Pieces of mature leaves 3.8%	Medium-sized young leaves — 3.3%	Apical tips of mature leaves — 3.3%	Petioles of medium-sized young leaves 2.2%	Mature leaves 2.2%; Lichen 2.2%
Feb. '72 297	Leaf buds 34.0%	Leaves of unknown age 22.9%	Mature leaf & leaflet petioles 13.1%	Large young leaves 10.4%	Medium-sized young leaves — 4.7%	Small young leaves — 4.7%	Floral buds 4.0%	Very small young leaves 2.0%	Ripe fruits 1.3%	Seeds — 0.33%; Petioles of large, medium-sized, and small young leaves: each 0.33%; Mature leaves 0.33%; Apical tips of mature leaves 0.33%; Young leaves with galls 0.33%; Small twigs 0.33%
Mar. '72 227	Mature leaf & leaflet petioles 32.6%	Large young leaves 15.4%	Small young leaves 9.3%	Young leaves of unspec. age 8.8%	Leaves of unknown age 8.4%	Medium-sized young leaves 6.6%	Young leaves with galls 5.7%	Petioles of medium-sized young leaves 3.1%	Petioles of small young leaves 2.6%	Mature leaves 2.2%

TABLE 40
Percentage overlap in food habits of *C. b. tephrosceles*, considering specific food items (species and parts eaten)

Month	Dec. '70	Jan. '71	Feb. '71	Mar. '71	Apr. '71	May '71	June '71	July '71	Aug. '71	Sept. '71	Oct. '71	Nov. '71	Dec. '71	Jan. '72	Feb. '72	Mar. '72
Nov. '70	37.0	34.9	33.2	24.9	25.1	25.7	24.0	18.6	18.6	19.9	25.9	21.5	33.9	29.9	24.1	22.2
Dec. '70		27.0	26.2	22.7	17.8	15.6	13.6	16.3	16.3	9.3	19.3	15.6	26.4	30.0	13.7	25.3
Jan. '71			41.5	16.8	29.7	15.3	14.8	14.8	23.9	19.4	13.9	22.5	21.0	25.5	31.0	20.2
Feb. '71				35.8	38.1	21.1	21.8	20.9	18.8	23.8	20.4	18.9	24.8	16.6	32.0	23.3
Mar. '71					24.7	39.8	34.8	23.5	11.3	12.7	19.2	15.8	33.6	27.4	21.3	40.1
Apr. '71						27.5	29.6	18.3	19.2	40.7	20.0	14.7	15.9	13.4	39.3	26.9
May '71							50.0	30.3	12.2	17.0	33.9	32.2	39.8	26.0	17.1	36.6
June '71								27.5	16.5	16.1	26.2	30.1	24.5	18.3	20.8	36.1
July '71									27.1	15.6	21.2	24.3	22.0	25.3	13.4	25.9
Aug. '71										26.3	16.8	21.5	20.5	21.7	17.3	16.0
Sept. '71											22.8	20.3	20.0	15.8	34.1	15.2
Oct. '71												46.1	36.8	27.9	18.7	23.5
Nov. '71													39.3	38.3	15.2	20.6
Dec. '71														48.0	25.9	27.8
Jan. '72															20.2	24.5
Feb. '72																20.8

N = 136

range = 9.3-50.0%

$\overline{X}$ = 24.3%

$$\% \text{ overlap} = \Sigma = 1 - \sum_{ij} \frac{|x_{ij} - y_{ij}|}{2}$$

x_{ij} = frequency of food item x in cell ij

y_{ij} = frequency of food item y in cell ij

TABLE 41

Absolute differences in rainfall in mm (data from Kanyawara Forestry Station)

Month (mm rainfall)	Dec. '70 (58)	Jan. '71 (31)	Feb. '71 (17)	Mar. '71 (75)	Apr. '71 (142)	May '71 (114)	June '71 (65)	July '71 (68)	Aug. '71 (155)	Sept. '71 (138)	Oct. '71 (268)	Nov. '71 (269)	Dec. '71 (12)	Jan. '72 (117)	Feb. '72 (41)	Mar. '72 (71)
Nov. '70 (83)	25	52	66	8	59	31	18	15	72	55	185	186	71	34	42	12
Dec. '70 (58)		27	41	17	84	56	7	10	97	80	210	211	46	59	17	13
Jan. '71 (31)			14	44	111	83	34	37	124	107	237	238	19	86	10	40
Feb. '71 (17)				58	125	97	48	51	138	121	251	252	5	100	24	54
Mar. '71 (75)					67	39	10	7	80	63	193	194	63	42	34	4
Apr. '71 (142)						28	77	74	13	4	126	127	130	25	101	71
May '71 (114)							49	46	41	24	154	155	102	3	73	43
June '71 (65)								3	90	73	103	204	53	52	24	6
July '71 (68)									87	70	200	201	56	49	27	3
Aug. '71 (155)										17	113	114	143	38	114	84
Sept. '71 (138)											130	131	126	21	97	67
Oct. '71 (268)												1	256	151	227	197
Nov. '71 (269)													257	152	228	198
Dec. '71 (12)														105	29	59
Jan. '72 (117)															76	46
Feb. '72 (41)																30

N = 136

range = 1-257 mm

TABLE 42

Selection ratio of *C. b tephrosceles* food habits, April 1971 through March 1972, for certain common tree species

Species	X̄ spead + X̄ depth = crown size for average tree	Density (based on strip enumeration) No./ha.	Crown size × density = cover index in study area	% of total feeding observations for 12-month period	1,000 × % feeding observations ÷ cover index = selection ratio	Visibility class (*average between leafed and leafless condition)
Celtis africana	26.4	0.6	15.84	15.36	969.69	II*
Markhamia	19.2	58.0	1113.6	15.11	13.57	II
Celtis durandii	22.8	34.3	782.04	10.36	13.24	I*
Newtonia	30.6	0.4	12.24	9.85	804.73	I
Aningeria	29.7	0.2	5.94	5.20	875.42	II*
Strombosia	30.8	14.7	452.76	4.59	10.13	IV
Bosqueia	18.6	2.8	52.08	4.06	77.95	II
Teclea	19.1	21.0	401.1	3.78	9.42	IV
Minusops	31.7	2.1	66.57	2.50	37.55	II
Funtumia	13.9	4.5	62.55	1.96	31.33	III
Lovoa	21.3	0.9	19.17	1.35	70.42	I
Parinari	41.3	3.2	132.16	1.09	8.24	I
Diospyros	19.3	65.7	1268.01	0.14	0.11	III

TABLE 43
Nutritional analysis of some *C. b. tephrosceles* foods, in percentages of dry weight

Species and item	Crude protein*	Ether extract (fat)	Dry weight of crude fiber	Ash	Lignin	Phosphorus	Calcium
Celtis durandii young leaves	34.94	0.65	12.61	7.81	5.52	0.59	0.85
Celtis durandii mature leaves	17.58	1.99	14.88	17.30	7.04	0.13	2.69
Markhamia young leaves minus basal 5 cm of petiole	19.55	2.05	23.35	6.42	9.99	0.39	0.60
Markhamia basal 5 cm of young leaf petioles	11.48	0.61		8.46	11.85	0.29	0.89
Markhamia mature leaves minus basal cm of petiole	17.99	0.52	35.46	5.75	15.70	0.23	1.20
Markhamia basal 5 cm of mature leaf petioles	10.53	0.81	38.59	9.86	8.79	0.23	1.48

* The crude protein determinations are based on nitrogen analysis using the Kjeldahl method following the U. S. Association of Official Agricultural Chemists' procedure. The factor used for converting nitrogen to crude protein was 6.25 and assumes 16.0% nitrogen in crude protein. For example, if 2.0% nitrogen was found in the sample, the crude protein was determined as 6.25 x 2.0 = 12.5%.

TABLE 44
Foraging upside down by *C. b. tephrosceles* (48 cases)

	Adult male	Adult female	Subadult or old juvenile	Juvenile	Young juvenile	Clinging infant
% of cases observed	0	4.2%	27.0%	64.6%	4.2%	0
% of cases expected	15%	35%	5%	20%	15%	10%

Expected values are based on average age-sex composition of CW group.

TABLE 45
Plant foods eaten by other *Colobus badius* subspecies

Plant	Part eaten
C. b. temminckii (Senegal)	
Ceiba pentandra (L.) Gaertn.	leaf petioles and floral buds
Combretum cf passargei Aubr.	apical leaf tips
Daniellia oliveri (Rolfe) Hutch. and Dalx.	floral buds
Elaeis guineensis Jacq.	leaflet petiole
Erythrophleum guineense G. Don.	probably pods
Hannoa undulata Planch.	flowers and floral buds
Khaya senegalensis (Desr.) A. Juss.	undetermined part
Lophira lanceolata Van Tiegh	floral buds
Parkia biglobosa Benth.	possibly floral buds
Parinari excelsa Sabine	leaf buds and young leaves
Terminalia macroptera Guill.and Perr.	apical tip and entire leaves
C. b. badius (Ivory Coast)	
Anopyxis klaineana (Pierre) Engl.	fruits
Bombax brevicuspe Sprague	fruits
Bussea occidentalis Hutch.	young leaves
Ceiba pentandra	floral buds
Combretum aphanopetalum Engl. and Diels	seeds
Discoglypremna caloneura (Pax)	fruits
Ficus sp. (strangler fig)	young leaves
Homalium molle (Stapf.)	flowers
Klainedoxa gabonensis Pierre ex. Engl.	fruits (pericarp)
Leptoderris cyclocarpa Dunn	pods
Parinari excelsa	fruits (pericarp)
Parkia bicolor A. Chev.	seeds
Ricinodendron heudelotti (Baill.) Pierre	inner part of fruits and leaf petioles

278

TABLE 45 — *Continued*

Plant	Part eaten
C. b. preussi (Cameroun)	
Antrocaryon klaineanum Pierre	probably apical tip of young leaves
Canarium schweinfurthii Engl.	apical tip mature leaves
Coelocaryon preussii Warb.	possibly young fruits
Irvingia gabonensis (Aubry Lecomte ex. O'Rorke) Baill.	young fruits
Lophira procera A. Chev.	fruits
Scyphocephalium sp.	apical tip of leaves
Staudtia stipitata Warb.	fruits
Vitex sp.	floral buds

TABLE 47

Summary sheet of distribution of time in space for systematic samples, CW group of *C. b. tephrosceles*, Kanyawara, November 1971 through March 1972 (lacking 3 incomplete days of December 1971 systematic samples)

Totals: 97 quadrats
23 days (5 full, 18 incomplete) of the systematic sample
16,797.0 minutes tallied (279.95 hours)
Total observation time in this systematic sample was 12,259 minutes (204 hours, 30 minutes)

	No. mins.	No. days	No. months
Range	2.5 – 1,390.5	1 – 10	1 – 4
Mean	173.2	2.3	1.7

Total minutes (Total days)
Total months

	-20	-19	-18	-17	-16	-15	-14	-13	-12	-11	-10	-9	-8	-7	-6
+8											35.5(1) 1				
+7											130(1) 1	95(1) 1			
+6												22.5(1) 1			
+5									44.5(1) 1						
+4								10(1) 1	388.5(3) 2	49.5(2) 2					
+3									206(3) 2	356.5(4) 1					
+2										241(3) 3					
+1							76(1) 1	45(2) 2	303.5(3) 3	663.5(6) 6	61.5(3) 3				
-1						155(1) 1		362(3) 3	395(5) 5	170(1) 1	145.5(3) 3	304.5(3) 3	132(2) 2		
-2						85(1) 1	181.5(2) 2	663.5(5) 3	264(4) 4	73(1) 1	172.5(2) 2	51.5(2) 2	65(1) 1		
-3						107(2) 2	196(5) 4	179(4) 3	206.5(3) 3	249.5(7) 5	128(3) 3	122.5(2) 2	37(1) 1		
-4						22.5(1) 1	417(7) 4	209(7) 4	279(3) 2	504(6) 4	80(1) 1	77(2) 2			
-5							83(2) 2	1390.5(10) 4	576.5(10) 5	525.5(6) 5	300.5(3) 3	170(2) 2	186(2) 2		
-6							28.5(1) 1	270(3) 3	174(4) 3	202(4) 2	74.5(2) 1		243(2) 1	263(2) 1	
-7							9(1) 1	21.5(1) 1	30(2) 2	41.5(2) 3	10.5(2) 2	28(1) 1	58(1) 1		
-8							99.5(2) 1	30(1) 1	342.5(3) 2	193.5(1) 1	18(1) 1	136(1) 1	95(1) 1	194(1) 1	58(1) 1
-9	102(1) 1	206.5(2) 1	251(2) 1	263(2) 1	102(2) 1	2.5(1) 1			53(1) 1	12(1) 1	34(1) 1	37(1) 1	51.5(1) 1	102(1) 1	5(1) 1
-10	130(1) 1	25.5(1) 1	182(1) 1	175(1) 1	54.5(2) 1	17.5(3) 1						43(1) 1			
-11												95(1) 1			

Note. Table 46 appears on the following pages.

TABLE 46

Summary sheet of time in space for systematic samples, CW group of *C. b. tephrosceles,* Kanyawara, November 1970 through October 1971

Cells show: Total minutes (Total days) / Total months

	-25	-24	-23	-22	-21	-20	-19	-18	-17	-16	-15	-14	-13
+4													
+3													
+2													
+1													166(2) / 2
-1													262(3) / 3
-2											40(1) / 1	165(1) / 1	1042.5(7) / 5
-3											42(1) / 1	411(3) / 3	609.5(4) / 4
-4												94(2) / 2	56(2) / 2
-5												8(1) / 1	199(5) / 5
-6												213(3) / 2	391.5(7) / 4
-7							40(1) / 1					373(4) / 2	507(4) / 3
-8						63(1) / 1	278(3) / 2	174(2) / 1	295(1) / 1	7(1) / 1		150(3) / 2	396(4) / 3
-9						202(1) / 1	236(2) / 2	30(2) / 2	10(1) / 1	52(2) / 2	26.5(2) / 2	631(3) / 2	439(5) / 4
-10						29(1) / 1	107.5(3) / 2	97.5(2) / 2		125.5(3) / 2	112(6) / 4	398(4) / 4	524(8) / 6
-11					6(1) / 1	60(1) / 1	115.5(2) / 2	13(1) / 1	168(3) / 3	108.5(3) / 3	473.5(8) / 5	207.5(5) / 5	430.5(6) / 5
-12					13(1) / 1		20(1) / 1	50(1) / 1	116(2) / 1	584(6) / 3	258.5(6) / 4	177(5) / 4	331.5(6) / 5
-13			4(1) / 1	107(1) / 1	20(2) / 1		78(1) / 1	43(1) / 1	5(1) / 1	381.5(5) / 3	300(4) / 2	323.5(4) / 3	448(5) / 4
-14			44(1) / 1	12.5(1) / 1		15(1) / 1	94(1) / 1	29(2) / 1	7(1) / 1	78.5(2) / 3	393.5(5) / 3	388.5(5) / 2	302.5(3) / 3
-15			74(1) / 1	173(2) / 2	33(2) / 2	59(2) / 2		26.5(1) / 1	55(3) / 3	16(1) / 1	197.5(4) / 3	521(6) / 4	182(3) / 3
-16				66(2) / 2	121(3) / 3	82(1) / 1		132.5(3) / 3	183(3) / 3	147(4) / 1	149(4) / 3	158(5) / 4	64(1) / 1
-17	84.5(1) / 1	8(1) / 1	35(2) / 2	197.5(3) / 2	90(2) / 2		134(1) / 1	36.5(2) / 2	17(1) / 1	243.5(4) / 3	319(4) / 3	833.5(8) / 5	266.5(3) / 2
-18	2.5(1) / 1	62(1) / 1	20.5(2) / 2	256(2) / 2	18(1) / 1		19(1) / 1	20(1) / 1	143(2) / 2	101(2) / 2	307.5(5) / 4	194(3) / 3	51(1) / 1
-19		65(2) / 2	78(2) / 2	95(2) / 2	122(1) / 1		18(1) / 1	15(1) / 1	48(2) / 2	116(2) / 1			
-20		33(1) / 1		101(1) / 1		152(1) / 1	22(2) / 2	118.5(2) / 2	98(2) / 1				
-21						34(1) / 1	76(3) / 2	200(4) / 2	45(1) / 1				
-22						266.5(1) / 1	489.5(4) / 2	882.5(5) / 2	576.5(4) / 2				
-23							26(1) / 1	187(1) / 1					

Legend box: Total minutes (Total days) / Total months

TABLE 46—*Continued*

-12	-11	-10	-9	-8	-7	-6	-5	-4	-3	-2	-1	+1	
		2.5(1)	125.5(1)	27.5(1)	107(1)	10(1)			25(1)	20(1)	12(1)	73(1)	+4
		1	1	1	1	1			1	1	1	1	
	80.5(2)	57(2)	43(2)	791.5(4)	172(2)	32(1)			19.5(1)	307.5(1)	2.5(1)	73(1)	+3
	2	2	2	2	1	1			1	1	1	1	
18(2)	440(5)	224(4)	699.5(6)	1060.5(5)	353(2)	10(1)	10(1)	27.5(1)	28(1)	138(1)			+2
2	3	3	3	3	2	1	1	1	1	1			
316(5)	422(5)	172(3)	450.5(5)	690(4)	89.5(1)	195(2)	454.5(3)	12.5(1)					+1
4	5	3	3	3	1	1	1	1					
336.5(6)	386(7)	562(3)	78(2)	334.5(3)	27.5(1)	114.5(2)	265.5(2)	48(1)					-1
5	6	3	2	2	1	1	1	1					
941(10)	246.5(9)	683(9)	612(4)	659.5(4)	72(1)	94(2)							-2
6	7	6	3	3	1	1							
232.5(5)	540(5)	292.5(7)	467.5(5)	697(3)		2.5(1)	95(1)		207.5(1)				-3
4	5	5	4	3		1	1		1				
478.5(4)	334.5(4)	202(5)	2.5(1)				167.5(1)	37.5(1)	150(1)				-4
4	3	4	1				1	1	1				
135(3)	51(1)	197(3)	114.5(1)	239(2)	432(2)				14.5(1)				-5
3	1	3	1	1	1				1				
405(4)	167(2)	351.5(3)	70(2)	219.5(2)	138.5(1)	30(2)	39(1)		26.5(1)				-6
3	2	3	2	2	1	2	1		1				
108.5(3)	38(2)	105(1)	198.5(2)	429.5(4)	244(2)	492.5(2)	156(1)						-7
3	2	1	2	2	2	2	1						
113(1)	68(1)	187(2)	59(1)	43(1)	13(1)	8(1)	72(1)	72(1)					-8
1	1	2	1	1	1	1	1	1					
338(2)		344.5(3)	59.5(2)	2.5(1)	5(1)								-9
2		2	2	1	1								
178.5(3)			73(1)										-10
3			1										
408.5(5)	297(3)	28(1)											-11
3	2	1											
174(6)	169(2)												-12
4	2												
199(2)													-13
1													
													-14
317.5(1)													-15
1													

Totals: 260 quadrats (50 m x 50 m)
57 systematic sample days (48 full days, 9 incomplete days)
49,618.5 mins. tallied (826.98 hrs.)
Total observation time in this systematic sample
was 39,071 mins. (651 hrs. 10 mins.)

	No. mins.	No. days	No. months
Range	2.5 - 1,060.5	1 - 10	1 - 7
Mean	190.8	2.5	2.0

(-16, -17, -18, -19, -20, -21, -22, -23)

TABLE 48

Summary sheet of distribution of time in space, ST group of *C. b. tephrosceles,* Kanyawara

Box label: Total minutes (Total days) / Total months

	-25	-24	-23	-22	-21	-20	-19	-18	-17	-16	-15	-14
+7												
+6												
+5												
+4												
+3												
+2												
+1												
-1												
-2											43(1) / 1	233(3) / 1
-3												123.5(1) / 1
-4												108.5(1) / 1
-5												
-6												27(1) / 1
-7												112.5(2) / 1
-8												2.5(1) / 1
-9						69.5(2) / 1	12(2) / 1	42.5(1) / 1				92(3) / 2
-10					5(1) / 1	112(4)	95(2)					26.5(2) / 1
-11						15.5(2) / 1	82.5(2) / 1	95(1) / 1	10(1) / 1	25(2) / 1	17(1) / 1	21(1) / 1
-12					2(1) / 1		87.5(2) / 1			33(1) / 1		78.5(1) / 1
-13				12(1) / 1	75.5(2) / 1	25.5(2) / 1	60(1) / 1	102.5(1) / 1		203(1) / 1	4(1) / 1	1(1) / 1
-14				73(3) / 2	22(1) / 1		2.5(1) / 1	3.75(2) / 2		173(1) / 1	2(1) / 1	
-15				166(3) / 1	19.5(1) / 1			26.25(2) / 1	36(1) / 1	23(1) / 1		
-16		68(1) / 1	41.5(2) / 1	41.5(2) / 1				28.5(1) / 1	44(1) / 1	67(2) / 1		
-17		68(1) / 1	2.5(1) / 1	47.5(2) / 1			80(1) / 1	2.5(1) / 1	60(1) / 1	43(1) / 1		
-18			130(1) / 1	104(2) / 1			84(1) / 1	84(1) / 1	55.5(1) / 1	43(1) / 1		
-19			2(1) / 1	168.5(2) / 1					5.5(1) / 1			

TABLE 48 — *Continued*

-13	-12	-11	-10	-9	-8	-7	-6	-5	-4	-3	
			11.5(1) 1								+7
			303.5(2) 1	15(1) 1							+6
		335(1) 1	415.5(3) 1	187(2)			65(1)				+5
			210(3) 1	170(1) 1			65(1) 1	65(1) 1			+4
			106.5(4) 2	175.5(3) 1	17(2) 1						+3
	12(1) 1		30(2) 1	36.5(2) 1	46(2) 1						+2
	16(2) 1	32(1) 1	5(1) 1	136.5(2) 1	146(5) 1	96.5(4) 1	74(2) 1				+1
45.5(4) 1	86.5(6) 1	59(2) 1	74(2)		48(2) 1	55(2) 1	52(2) 1	87(2) 1	84.5(1) 1		-1
457(8) 1	295(10) 1	12(1) 1	82(3)	62(3) 1			32(1) 1	32(1) 1	40(1) 1		-2
131.5(3) 1	204(6) 1	104.5(4) 1	60.5(2) 1				37(1) 1	76.5(2)			-3
121(3) 1	107(3) 1	281.5(4) 1	11.5(2) 1				52(1) 1	67(2) 1			-4
742.5(4) 1	520(4) 1	85(2)	78(2) 1					38(1)	12(1) 1		-5
476.5(4) 1	312.5(3) 1		5(1) 1				34.5(1) 1	203(2) 2	70(2) 1		-6
186(3) 1	112.5(1) 1	11(1) 1	16(2) 1					109(1) 1	0.5(1) 1	19(1) 1	-7
	1(1) 1	11(1) 1	11(1) 1	102(1) 1	88.5(2) 1	80(1) 1	82(1) 1	25(1) 1		16(1) 1	-8
19.5(3) 1	4(2) 1		5.5(1) 1	91(2) 1	15(1) 1	70(1) 1	57(1) 1				-9
261(2) 1											-10
236(2) 1											-11
		45(1) 1	5(1) 1								-12
											-13
											-14
											-15
											-16
											-17
											-18
											-19

Totals:	160 quadrats (50 m x 50 m) 54 days (none full) 13,766.5 mins. tallied	
	No. mins.	No. days
Range	0.5 - 742.5	1 - 10
Mean	86.0	1.85

TABLE 49

Summary sheet of distribution of time in space, BN group of *C. b. tephrosceles*, Kanyawara

Totals box:

Total minutes (Total days)
Total months

Totals: 66 quadrats (50 m x 50 m)
27 days (none full)
2,256.0 mins. tallied

	No. mins.	No. days
Range	0.5 - 205	1 - 4
Mean	34.2	1.34

	-22	-21	-20	-19	-18	-17	-16	-15	-14	-13	-12	-11	-10	-9
-1											49(1)/1			
-2											111(1)/1	47(1)/1	15(1)/1	15(1)/1
-3											112(1)/1	116(1)/1		31(1)/1
-4										14(1)/1		23.5(1)/1		
-5										16(2)/1	77(2)/1	5(1)/1		
-6													9.5(1)/1	2.5(1)/1
-7										65(1)/1	65(1)/1			2(1)/1; 7(1)/1
-8						14.5(1)/1	0.5(1)/1	3(1)/1		3(1)/1	6.5(1)/1			
-9				53(2)/1	203(3)/1	16.5(2)/1	14(1)/1	4(2)/1	24(2)/1	4(1)/1	6.5(1)/1			
-10		5(1)/1		2.5(1)/1					90(2)/1	7(2)/1				
-11				7.5(2)/1	2.5(1)/1	21.5(1)/1	20(1)/1	205(4)/1	98(3)/1	9(2)/1	14(1)/1			
-12														
-13										21(2)/1	21(2)/1			
-14									10(1)/1	52(1)/1				
-15	2(1)/1					3(2)/1								
-16	2.5(1)/1	51.5(1)/1				2.5(1)/1				20.5(3)/1	57.5(2)/1			
-17							31.5(1)/1							
-18							1(1)/1	2(1)/1	2(1)/1	9(1)/1				
-19														
-20														
-21			83(1)/1	39(1)/1	5(1)/1	103(1)/1								
-22			83(1)/1	37(1)/1										

TABLE 50
Indices of diversity (H) for monthly ranging patterns,
CW group of *C. b. tephrosceles*

Month	H
November 1970	2.742
December	3.187
January 1971	3.223
February	3.562
March	3.038
April	3.779
May	3.266
June	3.127
July	3.283
August	3.758
September	2.857
October	3.697
November	3.593
December	3.003
January 1972	3.364
February	2.841
March	2.766

TABLE 51
Percentage overlap in quadrat utilization, CW group of *C. b. tephrosceles*

	Dec. '70	Jan. '71	Feb. '71	Mar. '71	Apr. '71	May '71	June '71	July '71	Aug. '71	Sept. '71	Oct. '71	Nov. '71	Dec. '71	Jan. '72	Feb. '72	Mar. '72
Nov. '70	3.7	0	20.2	33.0	3.3	19.6	0	0	10.2	16.1	3.0	12.8	10.0	16.9	21.2	9.6
Dec. '70		0	9.1	13.5	0	2.0	0	0	4.4	1.2	3.4	2.7	10.1	7.0	1.5	3.2
Jan. '71			11.9	0	17.1	6.6	0.5	9.7	7.4	8.2	4.1	3.4	14.3	2.4	0	2.4
Feb. '71				17.4	0	10.5	0	0	8.0	5.7	4.1	13.8	25.4	19.2	4.3	12.2
Mar. '71					0	24.7	0	0	14.4	5.2	5.4	14.3	16.9	25.1	22.9	24.0
Apr. '71						8.5	25.7	25.2	15.6	7.2	12.7	2.5	0	0	0	0
May '71							1.1	6.4	16.8	26.6	2.8	14.7	13.6	24.5	28.8	21.3
June '71								30.7	29.3	0	0.5	0	0	0	0	0
July '71									6.3	7.0	14.7	9.0	0	0	0	0
Aug. '71										10.8	3.4	19.0	3.5	13.3	19.2	9.7
Sept. '71											9.0	21.4	0	22.6	16.1	10.9
Oct. '71												18.1	2.3	5.7	0.4	1.5
Nov. '71													6.8	27.9	21.7	23.1
Dec. '71														17.7	4.9	8.0
Jan. '72															27.8	34.6
Feb. '72																28.1

TABLE 52
Mean daily distance traveled by CW group of *C. b. tephrosceles*

November 1970	5 days	398.5 m
December	5 days	557.0 m
January 1971	5 days	548.0 m
February	5 days	629.0 m
March	5 days	468.5 m
April	5 days	958.5 m
May	4 days	516.9 m
June	4 days	690.6 m
August	5 days	830.5 m
September	3 days	909.1 m
October	2 days	772.5 m
November	2 days	768.7 m
December	3 days	661.6 m

TABLE 53
Time budget for *C. b. tephrosceles*

	Monthly mean percentages						
	Feeding	Resting	Locomoting	Self-cleaning and grooming	Clinging	Playing	Other
Aug. 1970	48.79	27.98	8.50	6.80	3.92	3.98	0
Sept. 1970	42.40	35.28	12.24	5.72	2.16	2.12	0.08
Oct. 1970	40.64	35.48	9.48	6.78	3.06	4.16	0.40
Nov. 1970	43.84	28.72	12.42	6.94	3.04	4.84	0.20
Dec. 1970	47.88	32.68	10.96	4.16	1.16	2.76	0.40
Jan. 1971	51.28	31.60	7.23	3.28	4.28	2.16	0
Feb. 1971	46.56	36.32	10.48	2.40	2.32	1.36	0.40
Mar. 1971	41.88	43.44	5.96	3.16	2.16	3.20	0.20
Apr. 1971	52.52	28.40	9.12	5.04	2.96	1.96	0
May 1971	43.20	36.08	7.80	4.84	5.04	3.12	0
June 1971	41.76	35.92	7.56	6.56	6.60	1.20	0.40
Aug. 1971	42.32	37.96	6.52	7.80	3.50	0.94	0.96
12-month mean	45.26	34.16	9.03	5.29	3.35	2.65	0.25
S.D.	±4.023	±4.519	±2.144	±1.753	±1.466	±1.246	±0.282
G	3.78	6.42	5.50	6.78	6.88	6.60	
p = (df = 11)	>0.975	>0.5	>0.9	>0.5	>0.5	>0.5	

TABLE 54

Diurnal variation in activity of *C. b. tephrosceles*: ranges, mean, and standard deviation of percentage of monthly observations, August 1970 through August 1971

Time of day	Locomoting			Self-cleaning and grooming			Clinging			Playing		
0700	0–30	10.1	±10.3	0–16	4.1	±6.2	0–10	2.3	±3.4	0–13	1.2	±3.9
0730	0–35	15.3	±10.9	0–10	2.9	±4.4	0–10	2.9	±3.4	0–10	1.6	±3.2
0800	0–21	7.6	±7.7	0–17	5.0	±6.7	0–8	3.4	±2.8	0–17	2.3	±5.5
0830	0–25	12.7	±8.4	0–10	2.5	±4.0	0–12	3.3	±3.4	0–14	2.8	±5.1
0900	0–34	11.9	±10.5	0–20	5.4	±8.1	0–14	5.5	±4.3	0–10	3.3	±4.4
0930	0–20	8.9	±7.0	0–20	2.6	±5.9	0–10	2.7	±3.4	0–13	2.8	±5.0
1000	0–22	5.9	±7.0	0–20	6.3	±6.4	0–11	3.9	±4.6	0–17	4.3	±6.0
1030	0–20	7.3	±6.6	0–15	6.4	±5.8	0–12	4.1	±4.4	0–12	2.8	±4.5
1100	0–19	8.8	±5.7	0–10	3.8	±3.7	0–10	4.1	±3.0	0–13	1.5	±3.9
1130	0–25	9.3	±8.7	0–19	4.9	±6.3	0–5	1.8	±2.3	0–10	2.0	±3.9
1200	0–24	7.2	±8.4	0–19	4.6	±7.2	0–10	3.6	±3.3	0–14	1.8	±4.1
1230	0–22	8.3	±6.8	0–42	9.8	±13.0	0–10	3.1	±3.7	0–10	2.7	±4.1

	range		±	range		±	range		±	range		±
1300	0–14	5.2	± 4.6	0–18	6.3	±6.5	0–10	3.7	±3.3	0–10	3.5	±4.2
1330	0–20	6.8	± 6.7	0–17	7.8	±7.2	0–10	4.2	±3.6	0–20	2.8	±6.2
1400	0–20	6.5	± 5.7	0–15	4.4	±6.1	0–12	2.0	±4.3	0–17	2.3	±5.5
1430	0–15	6.0	± 5.5	0–15	4.8	±5.0	0–8	2.3	±3.0	0–15	3.0	±5.0
1500	0–19	4.8	± 5.7	0–16	4.0	±4.9	0–10	2.7	±3.3	0–25	2.9	±7.2
1530	0–28	10.5	± 9.4	0–19	8.9	±6.5	0–15	3.9	±5.1	0–20	4.8	±6.8
1600	0–20	7.2	± 6.8	0–30	10.7	±9.2	0–16.5	4.8	±4.4	0–16.5	3.5	±5.7
1630	0–29	10.1	± 9.6	0–55	10.5	±15.5	0–15	4.8	±5.2	0–18	5.2	±7.1
1700	0–25	10.3	±10.1	0–19	5.0	±6.1	0–10	3.7	±3.6	0–13	1.9	±4.5
1730	5–29	13.7	± 6.7	0–12	3.0	±4.2	0–8	3.0	±2.9	0–5	0.7	±1.6
1800	4–25	12.5	± 8.2	0–10	3.4	±4.4	0–10	2.3	±3.2	0–10	2.0	±3.3
1830	4–20	11.8	± 5.6	0–10	3.3	±4.3	0–5	2.7	±2.4	0–20	2.8	±5.8
1900	0–30	7.8	± 8.0	0–10	1.6	±3.7	0–15	3.1	±4.4	0–14	1.9	±4.6

TABLE 55

Summary of census data for anthropoid primates in Kibale Forest

Area, sample size, and species	% of censuses containing species	X̄ no. of groups per census	Range in no. groups per census	±S.D.	C.V. = $\frac{S.D.}{\bar{X}}$ × 100	X̄ no. of groups per km of transect	X̄ no. of groups in 100 ha. (1 km²) based on estimated area of census
Compartment 30 (N=44)							
Colobus badius	95.5%	3.80	0-8	1.75	46.1%	0.95	9.45
Colobus guereza	61.4%	0.84	0-3	0.83	98.8%	0.21	2.09
Cercopithecus ascanius	100.0%	2.77	1-8	1.52	54.9%	0.69	6.89
Cercopithecus mitis	93.2%	1.80	0-4	0.95	52.8%	0.45	4.48
Cercopithecus lhoesti	29.5%	0.32	0-2	0.52	162.5%	0.080	0.80
Cercocebus albigena	31.8%	0.39	0-2	0.53	135.9%	0.097	0.97
Pan troglodytes	11.4%	0.14	0-2	0.41	292.9%	0.035	0.35
Compartments 13, 12, and 17 (N = 11)							
Colobus badius	100.0%	3.09	1-6	1.64	53.1%	0.51	2.53
Colobus guereza	100.0%	4.90	3-7	1.14	23.3%	0.80	4.02
Cercopithecus ascanius	90.9%	2.73	0-6	2.10	76.9%	0.45	2.24

Cercopithecus mitis	81.8%	1.27	0-4	1.10	86.6%	0.21	1.04
Cercocebus albigena	18.2%	0.18	0-1	0.40	222.2%	0.030	0.15
Papio anubis	9.1%	0.09	0-1	0.30	333.3%	0.015	0.074
Tura River Bridge Area (N = 5)							
Colobus badius	80%	1.40	0-2			0.31	
Colobus guereza	40%	0.40	0-1			0.089	
Cercopithecus ascanius	100%	3.80	2-5			0.84	
Cercopithecus lhoesti	40%	0.40	0-1			0.089	
Cercocebus albigena	80%	0.80	0-1			0.18	
Papio anubis	20%	0.20	0-1			0.044	
Pan troglodytes	40%	0.60	0-2			0.13	

TABLE 56
Biomass estimates for anthropoid primates in compartment 30, Kibale Forest

	X̄ group size	X̄ group composition and (weight in kg.)			X̄ group biomass (kg.)	X̄ no. groups per 100 ha.	Total biomass per 100 ha.
		Adult male	Adult female	Immatures			
Colobus badius	50	15% (10.5)	35% (7)	50% (3.5)	295.8	5.95	1,760.0
Colobus guereza	9	11.1% (10.5)	33.3% (7)	55.5% (3.5) *	49.0	1.32	64.7
Cercopithecus ascanius	15	6.7% (4)	33% (2.9)	60.3% (1.5)	37.3	4.34	161.9
Cercopithecus mitis	15	6.7% (6)	33% (3)	60.3% (1.5)	45.6	2.82	128.6
Cercopithecus lhoesti	10	10% (6)	33.3% (3)	60.6% (1.5)	25.1	0.50	12.6
Cercocebus albigena	15	20% (10.5)	46.7% (7)	33.3% (3.5)	98.5	0.61	60.1
Pan troglodytes	4	50% (45)		50% (22)	134.0	0.22	29.5
							2,217.4 kg

* This average weight of 3.5 kg. is based on the following from J. F. Oates referring to a "typical" guereza group: 1 large subadult male 7.0 kg., 1 subadult male and 1 subadult female each 4 kg., 1 juvenile 2 kg., and 1 infant 0.5 kg.

TABLE 57
Tendency of primate species to form polyspecific and monospecific associations, Kibale Forest

Area, species, and (no. of associations)	Frequency in poly-specific associations: observed and (expected)	Frequency in mono-specific associations: observed and (expected)	Chi-square	Probability
Compartment 30 (44 censuses)				
Colobus badius (144)	93 (72)	51 (72)	12.25	p < 0.001
Colobus guereza (35)	13 (17.5)	22 (17.5)	2.3	p > 0.10
Cercopithecus ascanius (107)	97 (53.5)	10 (53.5)	70.7	p < 0.001
Cercopithecus mitis (75)	71 (37.5)	4 (37.5)	59.9	p < 0.001
Cercopithecus lhoesti (13)	4 (6.5)	9 (6.5)	1.9	p > 0.20
Cercocebus albigena (14)	11 (7)	3 (7)	4.6	0.05 > p > 0.02*
Pan troglodytes (6)	1	5		
Compartments 13, 12 and 17 (11 censuses)				
Colobus badius (25)	9 (12.5)	16 (12.5)	1.96	p > 0.10
Colobus guereza (48)	14 (24)	34 (24)	8.3	0.01 > p > 0.001
Cercopithecus mitis (9)	7	2	**	
Cercopithecus ascanius (20)	12 (10)	8 (10)	0.80	p > 0.30
Cercocebus albigena (2)	2	0		
Papio anubis (1)	0	1		

* G = 3.66 (0.10>p>0.05) with Yates's correction
** G = 1.84 (0.50>p>0.10) with Yates's correction

TABLE 58

Diurnal distribution of *Cercopithecus* associations with CW group of *C. b. badius* (during complete sample days and only when times were determined with certainty)

30-minute time period	Cercopithecus ascanius		Cercopithecus mitis	
	Formation	Disbandment	Formation	Disbandment
0700	11.5%		17.6%	
0730	8.2%	6.1%	14.7%	12.5%
0800	9.8%	3.0%	2.9%	6.3%
0830	8.2%	12.1%	5.9%	
0900	13.1%	12.1%	2.9%	12.5%
0930	1.6%	9.1%		12.5%
1000	3.3%	9.1%	8.8%	
1030	3.3%	3.0%	5.9%	6.3%
1100	1.6%	3.0%		12.5%
1130	3.3%	3.0%	2.9%	12.5%
1200	3.3%	3.0%	2.9%	
1230	1.6%		8.8%	
1300	1.6%			
1330	1.6%			6.3%
1400	3.3%	6.1%	2.9%	
1430	3.3%		5.9%	
1500	4.9%	3.0%	2.9%	6.3%
1530			2.9%	
1600	1.6%		5.9%	
1630	1.6%			
1700	3.3%	3.0%		
1730	1.6%	3.0%	2.9%	
1800	3.3%	9.1%		12.5%
1830	4.9%	6.1%	2.9%	
1900		6.1%		
Total	61	33	34	16

Notes

CHAPTER 3

1. Copulation bouts consisted of one to 6 sexual mounts between the same male and female and all occurring within a period of less than 9 minutes. Bouts between the same partners were separated from one another by 10 minutes and usually much longer.

2. The total number of recipients listed here equals 81 because in one case males LB and SAM were the simultaneous recipients of harassment.

3. In view of the relatively small sample size (3–5 days per month on average) the possibility should not be ruled out that some few but very important social interactions occurred between SAM and male CW outside of my monthly samples, which greatly affected their relationship and effected SAM's rise in the hierarchy. A serious fight might have been such a decisive interaction.

4. These estimates and others relating to births and infant ontogeny are based on data compiled up to the time of writing, 7 July 1973. It should be noted that BT disappeared after 10 June 1973, which terminated the duration of her second inter-birth interval at 8 months. PGCW disappeared in October 1972 without giving birth since she was first recognized on 30 September 1971. SK was not recognizable until 31 August 1972, when she was in possession of the infant who was presumably the same as the one born between 10 and 17 April 1971. My inability to distinguish this infant weakens the estimate for SK.

5. The last estimate of 1–4 months was imprecise because I was absent from the CW group for nearly 3 months when it underwent the color change. A more accurate estimate of the age at which this infant changed color is probably between 3 and 3.5 months, which brings the overall mean closer to 3.75 months.

6. Expected frequencies in this matrix were computed as the product of the probability by chance of a particular monkey's being the neighbor and the number of times that monkeys were within 2.5 m of the particular sample animal concerned. Because the group was comprised of 19 monkeys, the probability by chance alone of one particular monkey being the neighbor of another particular monkey was $1/18$. For example, adult male LB had 15 neighbors within 2.5 m of him during all of his samples, so that the expected frequency with which male CW would be within 2.5 m of LB is $1/18$ x 15 or 0.83. In cases where individuals of a particular age-sex class could not be distinguished, the expected value was computed on the basis of the number of animals in that class. Thus, in the same example of LB, the expected frequency with which 4 indistinguishable adult females would be within 2.5 m of him was $4/18$ x 15 or 3.33. Another point needs clarification. Comparison of the 2 cells between any 2 distinguishable individuals in table 22 reveals that the cells differ, depending on which individual was the sampled monkey. For example, when male CW was the sample animal, male LB was observed within 2.5 m of him on 2 occasions, whereas when LB was the sample animal, CW was within 2.5 m of him on 3 occasions. These apparent discrepancies are caused by the fact that in some cases both recognizable members of a pair were not always sampled in the same spatial arrangements; e.g., in a given sample period, LB may be scored as being within 2.5 m of CW when CW is the sample animal, but before LB becomes the sample animal the spatial arrangement changes so that CW is no longer within 2.5 m of LB.

CHAPTER 4

1. This notation follows Marler 1970.
2. See footnote 1.
3. In Marler's (1970) classification, rapid quavers would be included in his chook category, but

I prefer not to use his term because I feel it is a poor onomatopoeic rendition and because he also includes calls in this category which I classify as barks.

4. These data were collected on the following dates: 4-8 November, 1-5 December 1970; 3-7 January, 2-5 February, 1-5 March, 6-10 April, and 12-15 May 1971.

5. This operational definition of peak was selected because it is equivalent to defining a peak as the mean plus one standard deviation where the coefficient of variation (standard deviation/mean x 100) equals 50%. I consider this a rigid criterion that is not likely to confuse spurious fluctuations in the frequency distribution with "real" peaks.

6. Measurements were made from daily range maps on which the direction, distance, and time of move were systematically plotted.

CHAPTER 5

1. *Newtonia* leaves of unknown age require special comment. In general it was difficult to determine if the badius were eating only *Newtonia* leaves. While feeding in this species, they were high up in the tree, making observation difficult, and, in addition, the actual feeding motion was rapid. They generally fed on the terminal leaves, and I suspect they were also eating that part of the stem which would produce floral bud spikes.

2. The phenological condition of 85 trees representing 11 species was evaluated at the end of each month and just prior to the systematic sample of the CW group at the beginning of each month. These species and their corresponding sample sizes are: *Celtis durandii* (10), *Celtis africana* (5), *Markhamia* (10), *Lovoa* (5), *Aningeria* (5), *Teclea* (5), *Diospyros* (10), *Parinari* (10), *Strombosia* (10), *Symphonia* (5), and *Funtumia* (10). Each tree was individually marked with a metal plate. All phytophases were evaluated for each tree on the basis of 5 classes of relative abundance (the scores 0-4, representing zero to many in figs. 14-27).

3. The Shannon-Wiener information theory was used to compute indices of diversity, employing the following formula:

$$H = - \sum_{i=1}^{n} p_i \log_e p_i$$

where p_i equals the proportion of observation for each species or item (from Wilson and Bossert 1971).

4. It might be argued that the product of these two means gives a better indication of crown size than does their sum. However, it makes little difference for this analysis because comparison of the sum of these two means with their product for each of the 13 species gives $r_s = 0.98$, which is highly significant ($p < 0.01$); i.e., there is a positive correlation between the sums and products.

5. Dr. D. A. Christensen analyzed these specimens for me while he was with the Department of Animal Science and Production, Makerere University of Kampala, and I am most grateful for his cooperation.

6. *Newtonia* is a tree typical of mature rain forest. Its apparent role as a staple food item for the badius at Gombe and Kanyawara may, in part, account for the apparent dependency of badius on mature rain forest in these areas. Populations of badius are much lower or absent from young forest (see section on censuses).

7. Tree density estimates for these 2 quadrats were determined by a slightly different method than for the 3 preceding examples. Only half of each 0.25 ha. quadrat was enumerated for trees at least 10 m tall. Consequently, the total area enumerated for the 2 quadrats combined was only 0.25 ha. instead of 0.50 ha.

8. Only the top 4 food species were considered for each month, because, on a monthly average, they made up 58.7% of the monthly diet (range 43.4-78.7%).

9. When one of the top 4 food species is a species for which data are not available to compute a cover index, then the 5th or 6th ranking species is used. This was the case in 9 of the 17 months but was not considered to bias the results, because even in these 9 months the 4 species used for computing the cover index made up, on a monthly average, 47.1% of the diet.

10. Daylight is approximately restricted to this time period, with official sunrise fluctuating during the year from 0640 to 0710 hours and sunset from 1847 to 1917 hours, as determined from the Astronomical Ephemeris of 1971 for a location at the equator and 30° E. (J. F. Oates, pers. comm.).

11. The distribution of the activity samples for *C. b. tephrosceles* is as follows, with the time of day followed in parentheses by the number of observations, and then the month followed in parentheses by the number of observations:

0700 hrs. (250), 0730 hrs. (251), 0800 hrs. (306), 0830 hrs. (315), 0900 hrs. (297), 0930 hrs. (286), 1000 hrs. (291), 1030 hrs. (326), 1100 hrs. (329), 1130 hrs. (276), 1200 hrs. (289), 1230 hrs. (288), 1300 hrs. (276), 1330 hrs. (268), 1400 hrs. (272), 1430 hrs. (261), 1500 hrs. (252), 1530 hrs. (257), 1600 hrs. (261), 1630 hrs. (252), 1700 hrs. (258), 1730 hrs. (304), 1800 hrs. (274), 1830 hrs. (276), 1900 hrs. (268). Total 6,983.

August 1970 (500), September (975), October (545), November (715), December (532), January 1971 (561), February (505), March (502), April (506), May (514), June (501), August (627). Total 6,983.

12. Two or more species were considered to be in association when the different species were spatially intermingled, i.e., spatially placed between conspecifics.

13. This test requires that the expected values be five or greater. Furthermore, statistical independence must be satisfied, which means that each primate association considered in the analysis must be different from the others. In this case these criteria mean that the test can only be validly applied to data from one census and only for species of which at least 10 groups occurred in that census. In none of the censuses did any of the species appear in 10 groups. Strictly speaking, then, this test cannot be applied to the census data. However, some insight can be gained by applying the test to the pooled data from all censuses of a given area and comparing the relative values for each species and for the different areas.

CHAPTER 6

1. It is suggested that these habitats are marginal and thereby contain low food densities for red colobus because of the less diverse plant community, lower density of food species (especially in the savanna woodland of Senegal) and the small area of the forest patches.

2. These five species are: *Markhamia platycalyx*, *Celtis durandii*, *Newtonia buchanani*, *Celtis africana*, and *Millettia dura*.

References

Aldrich-Blake, F. P. G. 1968. A fertile hybrid between two *Cercopithecus* spp. in the Budongo Forest, Uganda. *Folia primat.* 9:15-21.

— — —. 1970. Problems of social structure in forest monkeys. In *Social behaviour in birds and mammals*, ed. J. H. Crook, pp. 79-99. London and New York: Academic Press.

Altmann, S. A. 1962. A field study of the sociobiology of rhesus monkeys, *Macaca mulatta*. *Ann. N.Y. Acad. Sci.* 102:338-435.

Altmann, S. A., and Altmann, J. 1970. *Baboon ecology.* Chicago and London: University of Chicago Press.

Barnicot, N. A., and Hewett-Emmett, D. 1972. Red cell and serum proteins of *Cercocebus, Presbytis, Colobus* and certain other species. *Folia primat.* 17:442-57.

Bauchop, T., and Martucci, R. 1968. Ruminant-like digestion of the langur monkey. *Sci.* 161:698-700.

Bernstein, I. S. 1967. Intertaxa interactions in a Malayan primate community. *Folia primat.* 7:198-207.

— — —. 1968. The lutong of Kuala Selangor. *Behaviour* 32:1-16.

Boelkins, R. C., and Wilson, A. P. 1972. Intergroup social dynamics of the Cayo Santiago rhesus (*Macaca mulatta*) with special reference to changes in group membership by males. *Primates* 13:125-40.

Booth, A. H. 1956. The distribution of primates in the Gold Coast. *J. W. Afr. Sci. Ass.* 2:122-33.

— — —. 1957. Observations on the natural history of the olive colobus monkey, *Procolobus verus* (van Beneden). *Proc. Zool. Soc. London* 129:421-30.

Brown, L. 1970. *African birds of prey.* London: Collins.

Carpenter, C. R. 1934. A field study of the behavior and social relations of howling monkeys. *Comp. Psychol. Monogr.* 10, 2:1-168.

— — —. 1935. Behavior of red spider monkeys in Panama. *J. Mammal.* 16:171-80.

— — —. 1940. A field study in Siam of the behavior and social relations of the gibbon, *Hylobates lar. Comp. Psychol. Monogr.* 16, 5:1-212.

Chalmers, N. R. 1968. Group composition, ecology, and daily activities of free living mangabeys in Uganda. *Folia primat.* 8:247-62.

Clutton-Brock, T. H. 1972. Feeding and ranging behaviour in the red colobus monkey. Unpublished Ph.D. thesis, Cambridge University, England.

Crook, J. H. 1970. The socio-ecology of primates. In *Social behaviour in birds and mammals*, ed. J. H. Crook, pp. 103–66. London and New York: Academic Press.

Crook, J. H., and Gartlan, J. S. 1966. The evolution of primate societies. *Nature* (London) 210:1200–1203.

Dandelot, P. 1968. *Preliminary identification manual for African mammals.* Washington, D. C.: Smithsonian Institution.

DeVore, I. 1962. The social behavior and organization of baboon troops. Unpublished Ph.D. thesis, University of Chicago.

———. 1963. A comparison of the ecology and behavior of monkeys and apes. In *Classification and human evolution*, ed. S. L. Washburn, pp. 301–19. Chicago: Aldine.

De Vos, A., and Omar, A. 1971. Territories and movements of Sykes monkeys (*Cercopithecus mitis kolbi* Neuman) in Kenya. *Folia primat.* 16:196–205.

Dobzhansky, T. 1970. *Genetics of the evolutionary process.* New York and London: Columbia University Press.

Dorst, J., and Dandelot, P. 1970. *A field guide to the larger mammals of Africa.* London: Collins.

Drawert, F.; Kuhn, H.-J.; and Rapp, A. 1962. Gaschromatographische Bestimmung der niederflüchtigen Fettsäuren im Magen von Schlankaffen (*Colobinae*). Sonderdruck, *Hoppe-Seyler's Zeitschrift für physiologische Chemie* 329:84–89.

Eisenberg, J. F.; Muckenhirn, N. A.; and Rudran, R. 1972. The relation between ecology and social structure in primates. *Science* 176:863–74.

Epstein, S. S. 1970. A family likeness. *Environment* 12:16–25.

Fager, E. W. 1957. Determination and analysis of recurrent groups. *Ecology* 38:586–95.

Friedmann, H., and Williams, J. G. 1970. *Additions to the known avifauna of the Bugoma, Kibale, and Impenetrable Forests, West Uganda.* Contributions in Science No. 198, Los Angeles County Museum.

Furuya, Y. 1961–62. The social life of the silvered leaf monkeys (*Trachypithecus cristatus*). *Primates* 3:41–60.

Gartlan, J. S., and Brain, C. K. 1968. Ecology and social variability in *Cercopithecus aethiops* and *C. mitis*. In *Primates: Studies in adaptation and variability*, ed. P. C. Jay, pp. 253–92. New York and London: Holt, Rinehart, and Winston.

Gartlan, J. S., and Struhsaker, T. T. 1972. Polyspecific associations and niche separation of rain-forest anthropoids in Cameroon, West Africa. *J. Zool.* (London) 168:221–66.

Gautier, J. P., and Gautier-Hion, A. 1969. Les associations polyspecifiques chez les Cercopithecidae du Gabon. *La Terre et la Vie* 116:164–201.

Gautier-Hion, A. 1971. L'écologie du talapoin du Gabon. *La Terre et la Vie* 25:427–90.

Goodall, J. 1965. Chimpanzees of the Gombe Stream Reserve. In *Primate behavior: Field studies of monkeys and apes*, ed. I. DeVore, pp. 425–73. New York: Holt, Rinehart, and Winston.

Greig-Smith, P. 1964. *Quantitive plant ecology.* 2d ed. New York: Plenum Press.

Haddow, A. J. 1952. Field and laboratory studies on an African monkey, *Cercopithecus ascanius schmidti* Matchie. *Proc. Zool. Soc. London* 122:297–394.

Hall, K. R. L., and DeVore, I. 1965. Baboon social behavior. In *Primate behavior*, ed. I. DeVore, pp. 53–110. New York: Holt, Rinehart, and Winston.

Harvard Computational Laboratory. 1955. *Tables of the cumulative binomial probability distribution*. Cambridge: Harvard University Press.

Hausfater, G. 1972. Intergroup behavior of free-ranging rhesus monkeys (*Macaca mulatta*). *Folia primat.* 18:78–107.

Hill, W. C. Osman. 1952. The external and visceral anatomy of the olive colobus monkey (*Procolobus verus*). *Proc. Zool. Soc. London* 127–86.

———. 1966. *Primates: Comparative anatomy and taxonomy*. Vol. 6. *Cercopithecoidea*. Edinburgh: Edinburgh University Press.

Hill, W. C. Osman, and Booth, A. H. 1957. Voice and larnyx in African and Asian Colobidae. *J. Bombay Nat. Hist. Soc.* 54:309–21.

Hladik, A., and Hladik, C. M. 1969. Rapports trophiques entre végétation et primates dans la forêt de Barro Colorado (Panama). *La Terre et la Vie* 1:25–117.

Hladik, C. M., and Hladik, A. 1972. Disponibilités alimentaires et domaines vitaux des primates à Ceylan. *La Terre et la Vie* 26:149–215.

Holmes, R. T., and Pitelka, F. A. 1968. Food overlap among coexisting sandpipers on northern Alaskan tundra. *System. Zool.* 17:305–18.

Jay, P. 1962. Aspects of maternal behavior among langurs. *Ann. N.Y. Acad. Sci.* 102:468–76.

———. 1965. The common langur of North India. In *Primate Behavior*, ed. I. DeVore, pp. 197–249. New York: Holt, Rinehart, and Winston.

Jolly, C. J. 1966. Introduction to the Cercopithecoidea with notes on their use as laboratory animals. *Symp. Zool. Soc. London* 17:427–57.

Jones, C. 1972. Natural diets of wild primates. In *Pathology of simian primates*, ed. R. N. T-W-Fiennes, vol. 1, pp. 58–77. Basel: Karger.

Jones, C., and Sabater Pi, J. 1968. Comparative ecology of *Cercocebus albigena* (Gray) and *Cercocebus torguatus* (Kerr) in Rio Muni, West Africa. *Folia primat.* 9:99–113.

Kaufmann, J. H. 1962. Ecology and social behavior of the Coati, *Nasua narica*, on Barro Colorado Island, Panama. *Univ. Calif. Pub. Zool.* 60, 3:95–222.

Kawabe, M., and Mano, T. 1972. Ecology and behavior of the wild proboscis monkey, *Nasalis larvatus* (Wurmb), in Sabah, Malaysia. *Primates* 13, 2:213–28.

Keay, R. W. J. 1959. *Vegetation map of Africa south of the Tropic of Cancer*. London: Oxford University Press.

Kingdon, J. 1971. *East African mammals*. Vol. 1. London and New York: Academic Press.

Kingston, B. 1967. *Working plan for Kibale and Itwara Central Forest Reserves*. Entebbe, Uganda: Forest Department.

Kingston, T. J. 1971. Notes on the black and white colobus monkey in Kenya. *E. Afr. Wildl. J.* 9:172–75.

Klein, L. L., and Klein, D. J. In press. Neotropical primates: Aspects of habitat usage, population density, and regional distribution in La Macarena, Colombia.

Koelmeyer, K. O. 1959. The periodicity of leaf change and flowering in the principal forest communities of Ceylon. *The Ceylon Forester* 4, nos. 2, 4.

Kuhn, H. -J. 1964. Zur Kenntnis von Bau and Funktion des Magens der Schlankaffen (Colobinae). *Folia primat.* 2:193-221.

— — —. 1967. Zur Systematik der Cercopithecidae. In *Neue Ergebnisse der Primatologie*, ed. D. Starck, R. Schneider, and H.-J. Kuhn, pp. 25-46. Stuttgart: Fischer.

— — —. 1972. On the perineal organ of male *Procolobus badius. J. Human Evolution* 1:371-78.

Lack, D. 1968. *Ecological adaptations for breeding in birds.* London: Methuen.

Leskes, A., and Acheson, N. H. 1971. Social organization of a free-ranging troop of black and white colobus monkeys (*Colobus abyssinicus*). *Proc. 3rd Int. Congr. Primat.*, Zurich 1970, 3:22-31.

Malbrant, R., and Maclatchy, A. 1949. Faune de l'équateur Africain Français. In *Encyclopédie biologique*, vol. 2, *Mammifères*. Paris: Paul Le Chevalier.

Marler, P. 1969. *Colobus guereza:* Territoriality and group composition. *Science* 163:93-95.

— — —. 1970. Vocalizations of East African monkeys. I. Red Colobus. *Folia primat.* 13:81-91.

— — —. 1973. A comparison of vocalizations of red-tailed monkeys and blue monkeys (*Cercopithecus ascanius* and *C. mitis*) in Uganda. *Z. Tierpsychol.* 33:223-47.

Mason, W. A. 1968. Use of space by Callicebus groups. In *Primates: Studies in adaptation and variability,* ed. P. C. Jay, pp. 200-216. New York and London: Holt, Rinehart, and Winston.

Mohnot, S. M. 1971. Some aspects of social changes and infant killing in the Hanuman langur (*Presbytis entellus*) (primates: *Cercopithecidae*) in Western India. *Mammalia* 35:175-98.

Mosteller, F.; Rourke, R. E. K.; and Thomas, G. B., Jr. 1961. *Probability: A first course.* Reading, Mass., and London: Addison-Wesley.

Napier, J. R., and Napier, P. H. 1967. *A handbook of living primates.* London and New York: Academic Press.

Neville, M. K. 1972. Social relations within troops of red howler monkeys (*Alouatta seniculus*). *Folia primat.* 18:47-77.

Nishida, T. 1972. A note on the ecology of the red colobus monkeys (*Colobus badius tephrosceles*) living in the Mahali Mountains. *Primates* 13:57-64.

Oates, J. F. 1974. The ecology and behaviour of the black-and-white colobus monkey (*Colobus guereza* Rüppell) in East Africa. Unpublished Ph.D. thesis, University of London, England.

Odum, H. T. 1970. Summary: An emerging view of the ecological system at El Verde. In *A tropical rain forest: A study of irradiation and ecology at El Verde, Puerto Rico,* ed. H. T. Odum. Oak Ridge, Tenn.: Division of

Technical Information, U.S. Atomic Energy Commission.

Orians, G. H., and Pfeiffer, E. W. 1970. Ecological effects of the war in Vietnam. *Science* 168:544-54.

Oxnard, C. E. 1969. A note on the ruminant-like digestion of langurs. *Laboratory Primate Newsletter* 8:24-25.

Pocock, R. I. 1935. External characters of a female specimen of a red colobus monkey. *Proc. Zool. Soc. London,* 939-40.

Poirier, F. E. 1968a. Analysis of a Nilgiri langur (*Presbytis johnii*) home range change. *Primates* 9:29-43.

— — —. 1968b. The Nilgiri langur (*Presbytis johnii*) mother-infant dyad. *Primates* 9:45-68.

— — —. 1968c. Nilgiri langur (*Presbytis johnii*) territorial behavior. *Primates* 9:351-64.

— — —. 1969. The Nilgiri langur troop: Its composition, structure, function, and change. *Folia primat.* 10:20-47.

— — —. 1970. The Nilgiri langur (*Presbytis johnii*) of South India. In *Primate behavior: Development in field and laboratory research,* ed. L. A. Rosenblum, vol. 1, pp. 251-383. New York: Academic Press.

Rahm, U. H. 1970. Ecology, zoogeography, and systematics of some African forest monkeys. In *Old World monkeys,* ed. J. R. Napier and P. H. Napier, pp. 589-626. London and New York: Academic Press.

Ripley, S. 1967. Intertroop encounters among Ceylon gray langurs (*Presbytis entellus*). In *Social communication among primates,* ed. S. A. Altmann and J. A. Altmann, pp. 237-53. Chicago: University of Chicago Press.

— — —. 1970. Leaves and leaf monkeys: The social organisation of foraging in gray langurs, *Presbytis entellus thersites.* In *Old World monkeys,* ed. J. R. Napier and P. H. Napier, pp. 481-512. London and New York: Academic Press.

Rowell, T. E. 1966. Forest living baboons in Uganda. *J. Zool.* (London) 149:344-64.

Rudran, R. 1973a. The reproductive cycles of two subspecies of purple-faced langurs (*Presbytis senex*) with relation to environmental factors. *Folia primat.* 19:41-60.

— — —. 1973b. Adult male replacement in one-male troops of purple-faced langurs (*Presbytis senex senex*) and its effect on population structure. *Folia primat.* 19:166-92.

Schenkel, R., and Schenkel-Hulliger, L. 1967. On the sociology of free-ranging *Colobus* (*Colobus guereza caudatus* Thomas 1855). In *Progress in primatology,* ed. D. Starck, R. Schneider, and H. J. Kuhn, pp. 185-94. Stuttgart: Fischer Verlag.

Siegel, S. 1956. *Nonparametric statistics.* New York: McGraw-Hill.

Sokal, R. R., and Rohlf, F. J. 1969. *Biometry: The principles and practice of statistics in biological research.* San Francisco: W. H. Freeman.

Southwick, C. H.; Beg, M. A.; and Siddiqi, M. R. 1965. Rhesus monkeys in north India. In *Primate behavior,* ed. I. DeVore, pp. 111-59. New York: Holt, Rinehart, and Winston.

Southwick, C. H., and Cadigan, F. C., Jr. 1972. Population studies of Malaysian primates. *Primates* 13:1-18.

Struhsaker, T. T. 1967a. Social structure among vervet monkeys (*Cerco-pithecus aethiops*). *Behaviour* 29:83–121.

— — —. 1967b. Ecology of vervet monkeys (*Cercopithecus aethiops*) in the Masai-Amboseli Game Reserve, Kenya. *Ecology* 48:891–904.

— — —. 1969. Correlates of ecology and social organization among African Cercopithecines. *Folia primat.* 11:80–118.

— — —. 1970. Phylogenetic implications of some vocalizations of *Cerco-pithecus* monkeys. In *Old World monkeys,* ed. J. R. Napier and P. H. Napier, pp. 365–444. New York: Academic Press.

— — —. 1971. Social behaviour of mother and infant vervet monkeys (*Cerco-pithecus aethiops*). *Anim. Behav.* 19:233–50.

— — —. 1972. Rain-forest conservation in Africa. *Primates* 13:103–9.

— — —. 1973. A recensus of vervet monkeys in the Masai-Amboseli Game Reserve, Kenya. *Ecology* 54.

— — —. 1974. Correlates of ranging behavior in a group of red colobus monkeys (*Colobus badius tephrosceles*). *Amer. Zool.* 14:177–84.

Struhsaker, T. T., and Oates, J. F. 1975. Comparison of the behavior and ecology of red colobus and black-and-white colobus monkeys in Uganda: A summary. In *Socio-ecology and psychology of primates,* ed. R. H. Tuttle. The Hague: Mouton.

Sugiyama, Y. 1966. An artificial social change in a hanuman langur troop (*Presbytis entellus*). *Primates* 7:41–72.

— — —. 1967. Social organization of hanuman langurs. In *Social communi-cations among primates,* ed. S. A. Altmann and J. A. Altmann, pp 221–36. Chicago: University of Chicago Press.

Sugiyama, Y.; Yoshiba, K.; and Parthasarathy, M. D. 1965. Home range, mating season, male group, and intertroop relations in hanuman langurs (*Presbytis entellus*). *Primates* 6:73–106.

Tanaka, J. 1965. Social structure of Nilgiri langurs. *Primates* 6:107–22.

Thorington, R. W., Jr. 1967. Feeding and activity of *Cebus* and *Saimiri* in a Colombian forest. In *Neue Ergebnisse der Primatologie,* ed. D. Starck, R. Schneider, and H.-J. Kuhn, pp. 180–84. Stuttgart: Fischer Verlag.

Ullrich, W. 1961. Zur Biologie und Soziologie der Colobusaffen (*Colobus guereza caudatus* Thomas 1885). *Der Zool. Garten* 25:305–68.

Verheyen, W. N. 1962. Contribution à la craniologie comparée des primates. *Musée Roy. Afrique Centrale-Tervuren, Belgique Ser. 8 Sci. Zool.* 105:1–256.

Vogel, C. 1971. Behavioral differences of *Presbytis entellus* in two different habitats. *Proc. 3rd Int. Congr. Primat.,* Zurich 1970, 3:41–47. Basel: Karger.

Washburn, S. L., and DeVore, I. 1961. Social behavior of baboons and early man. In *Social life of early man,* ed. S. L. Washburn, pp. 91–105. New York: Viking Fund Publications in Anthropology, no. 31.

Watson, R. M.; Graham, A. D.; and Parker, I. S. C. 1969. A census of the large mammals of Loliondo controlled area, Northern Tanzania. *E. Afr. Wildl. J.* 7:43–59.

Wilson, E. O., and Bossert, W. H. 1971. *A primer of population biology.* Stamford, Conn.: Sinauer Associates.

Wing, L. D., and Buss, I. O. 1970. *Elephants and forests.* Wildlife Monographs, no. 19. Washington, D.C.: The Wildlife Society.

Yoshiba, K. 1967. An ecological study of hanuman langurs, *Presbytis entellus. Primates* 8:127-54.

— — —. 1968. Local and intertroop variability in ecology and social behavior of common Indian langurs. In *Primates: Studies in adaptation and variability,* ed. P. C. Jay, pp. 217-42. London and New York: Holt, Rinehart, and Winston.

Index

(Except where other species are names, all entries refer to *Colobus badius*)